Thermal and Electro-Thermal System Simulation

Thermal and Electro-Thermal System Simulation

Special Issue Editors

Márta Rencz
Lorenzo Codecasa

MDPI • Basel • Beijing • Wuhan • Barcelona • Belgrade

MDPI

Special Issue Editors

Márta Rencz
Budapest University of Technology and Economics
Hungary

Lorenzo Codecasa
Politecnico di Milano
Italy

Editorial Office
MDPI
St. Alban-Anlage 66
4052 Basel, Switzerland

This is a reprint of articles from the Special Issue published online in the open access journal *Energies* (ISSN 1996-1073) in 2019 (available at: https://www.mdpi.com/journal/energies/special_issues/ thermal_electro_thermal_system)

For citation purposes, cite each article independently as indicated on the article page online and as indicated below:

LastName, A.A.; LastName, B.B.; LastName, C.C. Article Title. *Journal Name* **Year**, *Article Number*, Page Range.

ISBN 978-3-03921-736-6 (Pbk)
ISBN 978-3-03921-737-3 (PDF)

Contents

About the Special Issue Editors vii

Preface to "Thermal and Electro-Thermal System Simulation" ix

Lorenzo Codecasa, Salvatore Race, Vincenzo d'Alessandro, Donata Gualandris,
Arianna Morelli and Claudio Maria Villa
TRAC: A Thermal Resistance Advanced Calculator for Electronic Packages [†]
Reprinted from: *Energies* **2019**, *12*, 1050, doi:10.3390/en12061050 1

Lisa Mitterhuber, Elke Kraker and Stefan Defregger
Structure Function Analysis of Temperature-Dependent Thermal Properties of Nm-Thin Nb_2O_5
Reprinted from: *Energies* **2019**, *12*, 610, doi:10.3390/en12040610 11

Genevieve Martin, Christophe Marty, Robin Bornoff, Andras Poppe, Grigory Onushkin,
Marta Rencz and Joan Yu
Luminaire Digital Design Flow with Multi-Domain Digital Twins of LEDs [†]
Reprinted from: *Energies* **2019**, *12*, 2389, doi:10.3390/en12122389 26

András Poppe, Gábor Farkas, Lajos Gaál, Gusztáv Hantos, János Hegedüs and Márta Rencz
Multi-Domain Modelling of LEDs for Supporting Virtual Prototyping of Luminaires
Reprinted from: *Energies* **2019**, *12*, 1909, doi:10.3390/en12101909 54

Anton Alexeev, Grigory Onushkin, Jean-Paul Linnartz and Genevieve Martin
Multiple Heat Source Thermal Modeling and Transient Analysis of LEDs
Reprinted from: *Energies* **2019**, *12*, 1860, doi:10.3390/en12101860 86

Robin Bornoff
Extraction of Boundary Condition Independent Dynamic Compact Thermal Models of
LEDs—A Delphi4LED Methodology
Reprinted from: *Energies* **2019**, *12*, 1628, doi:10.3390/en12091628 114

Marcin Janicki, Tomasz Torzewicz, Przemysław Ptak, Tomasz Raszkowski,
Agnieszka Samson and Krzysztof Górecki
Parametric Compact Thermal Models of Power LEDs
Reprinted from: *Energies* **2019**, *12*, 1724, doi:10.3390/en12091724 124

Paweł Górecki and Krzysztof Górecki
Modelling a Switching Process of IGBTs with Influence of Temperature Taken into Account [†]
Reprinted from: *Energies* **2019**, *12*, 1894, doi:10.3390/en12101894 134

Krzysztof Górecki and Kalina Detka
Influence of Power Losses in the Inductor Core on Characteristics of Selected
DC–DC Converters
Reprinted from: *Energies* **2019**, *12*, 1991, doi:10.3390/en12101991 146

Gabor Farkas, Zoltan Sarkany and Marta Rencz
Structural Analysis of Power Devices and Assemblies by Thermal Transient Measurements
Reprinted from: *Energies* **2019**, *12*, 2696, doi:10.3390/en12142696 161

Andreas Nylander, Josef Hansson, Majid Kabiri Samani, Christian Chandra Darmawan, Ana Borta Boyon, Laurent Divay, Lilei Ye, Yifeng Fu, Afshin Ziaei and Johan Liu
Reliability Investigation of a Carbon Nanotube Array Thermal Interface Material [†]
Reprinted from: *Energies* **2019**, *12*, 2080, doi:10.3390/en12112080 . . . . **183**

David C. Deisenroth and Michael Ohadi
Thermal Management of High-Power Density Electric Motors for Electrification of Aviation and Beyond
Reprinted from: *Energies* **2019**, *12*, 3594, doi:10.3390/en12193594 . . . . **193**

About the Special Issue Editors

Márta Rencz (Prof.) received the Electrical Engineering degree and the PhD degree from the Budapest University of Technology and Economics. She was PI in numerous international research projects, mostly in the field of investigating, measuring and modeling multi-physical effects in electronics. She has published her theoretical and practical results in more than 300 technical papers. She was a co-founder and CEO of Micred Ltd, which is now part of Mentor, a Siemens Business, where she still holds a research director position. She holds various awards of excellence, among others Harvey Rosten award (2001) and the Allan Krauss thermal management award of ASME (2015). In 2013 she has received the Doctor Honoris Causa degree from the Tallinn University of Technology in Estonia. In 2019 she received the Thermal Hall of Fame, Lifetime achievement award at Semitherm in San Jose, CA, USA.

Lorenzo Codecasa (Prof.) received a Ph.D. degree in Electronic Engineering from Politecnico di Milano in 2001. From 2002 to 2010, he worked as Assistant Professor of Electrical Engineering with the Department of Electronics, Information, and Bioengineering of Politecnico di Milano. Since 2010, he has worked as Associate Professor of Electrical Engineering in the same department. His main research contributions are in the theoretical analysis and in the computational investigation of electric circuits and electromagnetic fields. In his research on heat transfer and thermal management of electronic components, he has introduced original industrial-strength approaches to the extraction of compact thermal models, currently also available in commercial software. For these activities, in 2016, he received the Harvey Rosten Award for Excellence. He has been serving as Associate Editor for the IEEE Transactions of Components, Packaging and Manufacturing Technology. He is served as Program Chair of the conference Thermal Investigation of Integrated Circuits (THERMINIC). In his research areas, he has authored or co-authored over 190 papers in refereed international journals and conference proceedings.

Preface to "Thermal and Electro-Thermal System Simulation"

Microelectronics thermal experts from four continents met in the fall of 2018 at the 24th THERMINIC Workshop in Stockholm to discuss the latest issues in the design, characterization, and simulation of thermal and reliability problems in electronic devices and systems. These subjects are gaining more and more importance with increasing power density in electronics systems. The workshop is largely application-oriented and shows a rare balance of contributions from both academy and industry often in great synergy. The workshop had participants from 23 countries. At THERMINIC 2018, significant results were presented about thermal and electro-thermal simulations. For this reason, it was thought to organize a Special Issue of Energies entitled "Thermal and Electro-Thermal System Simulation", where external contributions were also invited. In this Special Issue, papers have been accepted irrespective of whether they were derived from THERMINIC 2018 contributions or not. At the end of a rigorous revision process, twelve papers have been selected to be published in this issue. Most of the papers here are extended versions of papers presented at THERMINIC 2018, thus, confirming the fact that THERMINIC 2018 was a stage of choice for presenting outstanding contributions on thermal and electro-thermal simulation in electronic systems. In particular, the papers selected for this Special Issue testify to the broad activity that is currently pursued in parametric thermal and electro-thermal modeling, multi-physics simulation of LEDs, and electro-thermal simulation of power electronics applications. We hope that all the selected papers will provide useful information to readers who are interested in these recent important questions of microelectronics thermal issues.

Márta Rencz, Lorenzo Codecasa
Special Issue Editors

energies

MDPI

Article

TRAC: A Thermal Resistance Advanced Calculator for Electronic Packages [†]

Lorenzo Codecasa [1,*]**, Salvatore Race** [2]**, Vincenzo d'Alessandro** [2]**, Donata Gualandris** [3]**, Arianna Morelli** [3] **and Claudio Maria Villa** [3]

[1] Department of Electronics, Information and Bioengineering, Politecnico di Milano, 20133 Milan, Italy
[2] Department of Electrical Engineering and Information Technology, University Federico II, 80125 Naples, Italy; salv.race@gmail.com (S.R.); vindales@unina.it (V.d.)
[3] STMicroelectronics, 20864 Agrate Brianza, Italy; donata.gualandris@st.com (D.G.); arianna.morelli@st.com (A.M.); claudio-maria.villa@st.com (C.M.V.)
* Correspondence: lorenzo.codecasa@polimi.it
† This manuscript is based on the conference paper "Thermal Resistance Advanced Calculator (TRAC)" in the Proceedings of the 24th International Workshop on Thermal Investigations of ICs and Systems.

Received: 13 February 2019; Accepted: 17 March 2019; Published: 19 March 2019

Abstract: This paper presents a novel simulation tool named thermal resistance advanced calculator (TRAC). Such a tool allows the straightforward definition of a parametric detailed thermal model of electronic packages with Manhattan geometry, in which the key geometrical details and thermal properties can vary in a chosen set. Additionally, it can apply a novel model-order reduction-based approach for the automatic and fast extraction of a parametric compact thermal model of such packages. Furthermore, it is suited to automatically determine the joint electron device engineering council (JEDEC) thermal metrics for any choice of parameters in a negligible amount of time. The tool was validated through the analysis of two families of quad flat packages.

Keywords: electronic packages; JEDEC metrics; model-order reduction; thermal simulation

---

1. Introduction

In the last two decades many efforts have been made to improve the way semiconductor vendors deliver thermal data of electronic components to their customers. This has led to the introduction of boundary condition independent (BCI) compact thermal models (CTMs) of the components [1]. However, nowadays customers prefer to request joint electron device engineering council (JEDEC) thermal metrics [2] to vendors instead of BCI CTMs. Consequently, some vendors have expressed the need for an approach suited to achieve these metrics for any product in the package families they sell in a fully automatic way and in a small amount of time. Unfortunately, each product of a package platform differs from the others for the values of selected geometrical dimensions and thermal properties that can vary in an *a-priori* known set.

The current Delphi-like approach to the extraction of BCI CTMs does not seem suitable for meeting these needs due to the following reasons:

- Delphi BCI CTMs can only consider fixed values of geometrical dimensions and thermal properties;
- often they are not as accurate as requested;
- their extraction can be very time-consuming, and most of the effort is spent for coping with boundary conditions (BCs) that are not relevant for the extraction of JEDEC thermal metrics;
- they cannot be used for electronic components with multiple heat sources (HSs) [1].

Some of the authors have recently developed novel approaches [3–9] relying on model-order reduction (MOR) to the extraction of various families of CTMs. For the specific case of parametric CTMs (pCTMs) [4], the proposed MOR-based technique starts from the detailed thermal model (DTM) of an electronic component, some parameters of which (geometrical dimensions and thermal properties) are assumed to vary in a chosen set, either finite or infinite. Then it fully automatically extracts a BCI pCTM, which depends on the assigned set of parameters and ensures a selected level of accuracy. The obtained pCTM does not lose information with respect to the original DTM, since it allows entirely reconstructing the space-time distribution of temperature rise in the modeled electronic component from its few degrees of freedom (DoFs).

The aim of this paper is to extend [10], where a simulation tool for electronic packages, referred to as thermal resistance advanced calculator (TRAC), was presented. TRAC allows a straightforward definition of a steady-state parametric DTM (pDTM) of a package with Manhattan geometry, and is also equipped with the option of using a pCTM extracted from the pDTM. Moreover, it is suited to automatically calculate the JEDEC thermal metrics of any product of the assigned family of packages from the above parametric models in a very low (pDTM) or negligible (pCTM) amount of time. By virtue of such appealing features, TRAC can be particularly helpful for vendors and customers in the semiconductor industry.

The first TRAC release described in [10] was developed to simulate a family of exposed pad (epad) quad flat packages (QFPs), the key parameters of which, namely, the package type and size, the number of leads, the epad size, the die attach material and the die dimensions, were assumed to vary in an assigned set. The improved version proposed here makes use of an advanced variant [9] of the order-reduction algorithm in [4] for deriving the pCTM from the pDTM; additionally, it also allows (i) describing quad flat no-leads packages (QFNs); and (ii) defining a rectangular dissipation region with arbitrary size and position over the die.

The paper is articulated as follows. In Section 2, the thermal metrics and the pDTM of the package families under test are introduced. Section 3 details the extraction of the pCTM. Section 4 discusses the numerical results obtained with both the pDTM and pCTM. Conclusions are then drawn in Section 5.

2. Parametric Detailed Thermal Model

The steady-state thermal behavior of a family of electronic components with Manhattan geometry can be straightforwardly modeled by assigning:

- a sequence of parallelepipeds of chosen size and position, including the one representing the dissipation region, hereinafter referred to as HS;
- the thermal conductivity of all materials;
- the BCs.

For the QFPs and QFNs that can be handled by the latest TRAC version, the size and positions of parallelepipeds, as well as the thermal conductivities, correspond to parameters to be selected in a chosen set. A rectangular HS with arbitrary size, position, and dissipated power can be defined on the top surface of the die.

TRAC is suited to automatically compute the JEDEC metrics ϑ_{JA}, Ψ_{JB}, Ψ_{JCtop}, ϑ_{JB}, ϑ_{JCtop}, $\vartheta_{JCbottom}$ [2] in 4 ambients, which differ in terms of thermal path followed by the heat generated within the HS to emerge from the die; more specifically: the ambient to evaluate $\vartheta_{JCbottom}$ requires a cold plate in intimate contact with the package backside; for the computation of ϑ_{JCtop} the plate is located over the top surface; in the ambient for determining ϑ_{JB}, a cold ring surrounds the package; no cooling systems are exploited in the ambient common to ϑ_{JA}, Ψ_{JB}, Ψ_{JCtop}. In all ambients, the board over which the package is mounted is thermally modeled with a single finely-meshed parallelepiped with a thermal conductivity adjusted to account for the aggregate effect of metal traces and vias, the detailed representation of which would have unnecessarily made the thermal problem much more complex. As shown in Figure 1, the metrics are calculated from the temperatures probed in four positions, namely: (1) the point of the die where the peak ("junction") temperature is reached; (2) the center of the top of the case; (3) the center of the bottom of the case; and (4) at the foot of the package lead half way along the side of the package (QFP) or within 1 mm of the package body (QFN). ϑ_{JA} is computed from (1) and the ambient temperature; Ψ_{JB} from (1) and (4); Ψ_{JCtop} from (1) and (2); ϑ_{JB} from (1) and (4); ϑ_{JCtop} from (1) and (2); $\vartheta_{JCbottom}$ from (1) and (3). As far as the metric ϑ_{JCtop} is concerned, a calibrated layer was interposed between the epad and the high-conductivity cold plate to emulate the epad-plate contact resistance. For all other metrics, the heat emerging from the die flows through the low-conductivity board, and the contact resistance epad-board was not accounted for, since it plays a negligible role.

As far as the BCs are concerned, specific values of heat transfer coefficients are applied to all surfaces of any structure (i.e., package and ambient) under test for each ambient; such values were preliminarily calibrated by comparing the JEDEC metrics simulated with commercial numerical programs with the experimental counterparts (see e.g., [2] for the measurement procedures) for a broad variety of package families.

For each parallelepiped, a mesh step size can be defined for each axis direction; as a result, a Cartesian mesh is automatically extracted. A finite integration technique (FIT) discretization [11] of the heat conduction problem is then generated in the form

$$\mathbf{K}(\mathbf{p})\vartheta(\mathbf{p}) = \mathbf{G}(\mathbf{p})\mathbf{P} \tag{1}$$

where $\vartheta(\mathbf{p})$ is the $N \times 1$ vector with the DoFs of the temperature rise distribution, $\mathbf{K}(\mathbf{p})$ is the N-order stiffness matrix, $\mathbf{G}(\mathbf{p})$ is a $N \times n$ power density matrix, $\mathbf{p}$ is the $\mathbf{p} \times 1$ parameter vector varying in a set $\mathcal{P}$, and $\mathbf{P}$ is the $n \times 1$ vector containing the powers P_i (with $i = 1, \ldots, n$) dissipated by the n independent HSs in the structure. These equations define the pDTM.

3. Parametric Compact Thermal Model

From the achieved pDTM, a pCTM can be extracted in a pre-processing stage. An $N \times q$ matrix $\mathbf{U}$ is determined, which allows approximating the $N \times 1$ temperature rise vector in the form

$$\vartheta(\mathbf{p}) = \mathbf{U}\hat{\vartheta}(\mathbf{p}) \tag{2}$$

for all $\mathbf{p} \in \mathcal{P}$, in which $\hat{\vartheta}(\mathbf{p})$ is a $q \times 1$ vector with $q \ll N$. The pCTM is derived from Equation (1) using Equation (2) and the Galerkin's projection. In this way it results in

$$\hat{\mathbf{K}}(\mathbf{p})\hat{\vartheta}(\mathbf{p}) = \hat{\mathbf{G}}(\mathbf{p})\mathbf{P} \tag{3}$$

in which

$$\hat{\mathbf{K}}(\mathbf{p}) = \mathbf{U}^T\mathbf{K}(\mathbf{p})\mathbf{U} \tag{4}$$

$$\hat{\mathbf{G}}(\mathbf{p}) = \mathbf{U}^T\mathbf{G}(\mathbf{p}) \tag{5}$$

are approximated by the technique described by some of the Authors in [9], which allows applying a *fully generic* transformation to the reference package; this improves the merely Cartesian transformation [4] adopted for the previous TRAC version presented in [10]. This system of equations, defining the pCTM, has formally the same structure of the system of equations defining the pDTM, but benefits from a significantly reduced complexity since $q \ll N$. Such a pCTM is updated at each iteration of the parametric MOR method, as in Algorithm 1. The iterations are stopped when, for novel values of the parameter vector $\mathbf{p}$, the relative residual ζ does not exceed the assigned value Ξ, so that the pCTM does not vary any longer.

Algorithm 1. Parametric model-order reduction (MOR) iteration.

Step	
1	Pick up a value of **p** ε $\mathcal{P}$ if a pCTM has already been extracted **then** solve pCTM Equations (3) for $\hat{\vartheta}(\mathbf{p})$ determine $\vartheta(\mathbf{p})$ from Equation (2) using $\hat{\vartheta}(\mathbf{p})$
2	Determine the relative residual ζ of Equations (1) using $\vartheta(\mathbf{p})$ if $\zeta > \Xi$ **then**
3	Solve pDTM Equations (1) for $\vartheta(\mathbf{p})$
4	Update **U** using $\vartheta(\mathbf{p})$
5	Update the pCTM using **U**

In Algorithm 1, at step 1, the elements of **p** ε $\mathcal{P}$ are chosen equal to the values of the parameters defining the specimen in the family of packages for which the JEDEC thermal metrics must be computed. This strategy minimizes the time needed to evaluate the metrics for this case.

At step 2, the relative residual is determined as

$$\zeta = \frac{\|\mathbf{V}^{-1}(\mathbf{K}(\mathbf{p})\vartheta(\mathbf{p}) - \mathbf{G}(\mathbf{p})\mathbf{P})\|_{\mathbf{V}}}{\|\mathbf{V}^{-1}\mathbf{G}(\mathbf{p})\mathbf{P}\|_{\mathbf{V}}} \tag{6}$$

V being the N-order diagonal matrix with the measures of the N volumes introduced in the FIT discretization.

At step 3, the solution of Equations (1) is ensured by a multigrid iterative solver, with a computational complexity linearly increasing with the dimension N of the problem.

At step 4, the **U** matrix is updated by appending a column orthonormal to the columns of the initial **U** matrix in the $\|\bullet\|_{\mathbf{V}}$ norm, such that the columns of the final **U** matrix span $\vartheta(\mathbf{p})$.

At step 5, the updated **U** matrix is used for reconstructing the pCTM. Exploiting the fact that the last column of **U** is changed, the update of the pCTM does *not* require recomputing the whole model.

4. Numerical Results

The latest TRAC release allows describing and simulating the families of epad QFPs and QFNs, although the tool can be extended to other electronic components with relatively little effort. In particular, the package thickness [low-profile QFP (LQFP) and thin QFP (TQFP) types are considered, the thicknesses of which are 1.4 and 1 mm, respectively] and size, the number of leads, the epad size, the type of glue, the three dimensions of the die, as well as the HS size and position, are defined by twelve (QFP) or fourteen (QFN) parameters. The parameters can be easily selected through a user-friendly graphical interface, which also checks whether the whole parameter set corresponds to a package manufactured by the vendor. Some examples are as follows: the horizontal die dimensions can vary in a continue range between a minimum and a maximum value, the latter fixed by project rules to ensure the attachment between die and epad; the QFP die thickness can be chosen among 100, 280, 375 (value selected for the numerical results shown later), and 580 μm; the glue types used to attach die and pad, which differ in terms of thickness and thermal conductivity (the type with conductivity amounting to 6 W/mK was chosen).

Figure 1 shows the 3-D schematic representation of two specimens of the LQFP and QFN families.

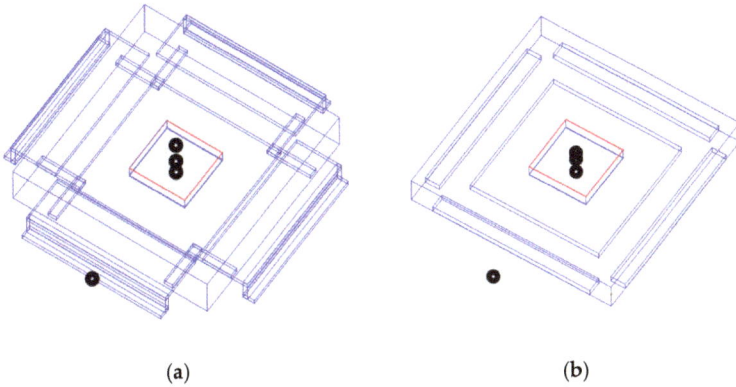

(a)   (b)

Figure 1. Specimens of the LQFP (**a**) and QFN (**b**) families sharing the same size of package (6×6 mm^2), epad (4.5×4.5 mm^2), and die (2×2 mm^2). Black circles represent the temperature probes needed to determine the thermal metrics.

The inherent symmetry of the packages under test allowed meshing and simulating only a quarter of each structure, thus reducing the computational burden; the missing portions were virtually restored by applying adiabatic BCs (i.e., zero heat flux) over the planes of symmetry.

A preliminary convergence analysis of the spatial mesh discretization of the constructed pDTMs was performed for chosen packages; more specifically, the calculated thermal metrics were monitored by increasing the DoFs until a negligible mesh sensitivity was observed. Then the discretization leading to only <0.1% inaccuracy was selected not to face the computational effort required by extremely fine meshes, this choice being also justified by the higher uncertainty of the experimental data. Figure 2 illustrates the convergence analysis performed for an LQFP with a 10×10 mm^2 package, a 4.5×4.5 mm^2 epad, and a 2×2 mm^2 die; as can be seen, the mesh leading to $\vartheta_{JA} = 28.87$ K/W (1.487×10^6 DoFs for a quarter of the structure) was chosen to avoid the huge number of DoFs ($>50 \times 10^6$) required to obtain a negligibly more accurate ϑ_{JA} value.

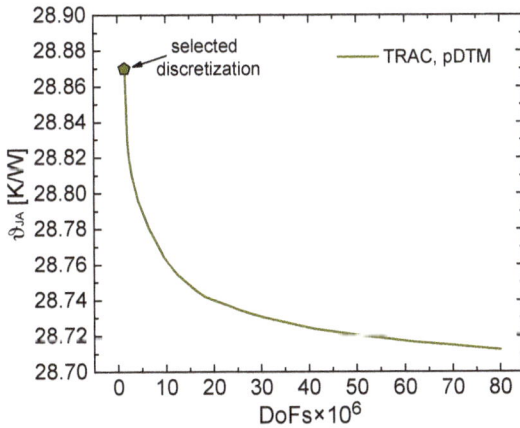

Figure 2. JEDEC thermal metric ϑ_{JA} as a function of the number of DoFs, as evaluated through the pDTM for a quarter of the 10×10 mm^2 LQFP with a 4.5×4.5 mm^2 pad and a 2×2 mm^2 die. The selected discretization is indicated.

The accuracy of the JEDEC thermal metrics computed by TRAC using the pDTM is witnessed by a comparison with a finite volume (FV) commercial software. This is shown in Figure 3, which reports the metrics ϑ_{JA}, ϑ_{JB}, and ϑ_{JCtop} (the others were not represented not to overcrowd the graphs) for a 10×10 mm^2 LQFP and a 14×14 mm^2 TQFP (Figure 3a), as well as for a 6×6 mm^2 QFN (Figure 3b). The slight discrepancy between TRAC and the FV software must be attributed: (i) to the different mesh styles of the compared tools; and (ii) to the fineness degree adopted in both of them, which was not extremely high to prevent unnecessarily long CPU times. As can be seen, the thermal metrics decrease with increasing the die size (i.e., the HS size) due to the lower dissipated power density and the enhanced lateral heat spreading.

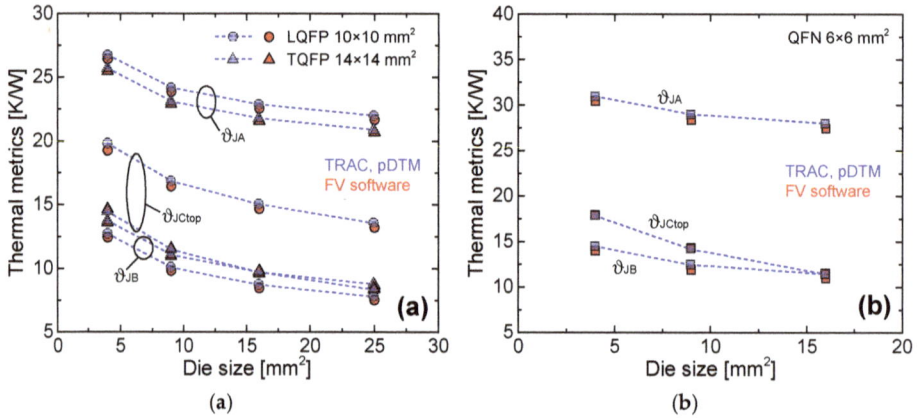

Figure 3. Some JEDEC thermal metrics against die size, as determined from the pDTM (blue) and a FV commercial software (red): (**a**) 10×10 mm^2 LQFP (circles) and 14×14 mm^2 TQFP (triangles), both with a 6×6 mm^2 epad; (**b**) 6×6 mm^2 QFN (squares) with a 4.7×4.7 mm^2 epad.

Figure 4 reports the metrics ϑ_{JA} and ϑ_{JCtop} as a function of die size for 10×10 mm^2 LQFPs and TQFPs commonly sharing a 6×6 mm^2 epad. It is inferred that the impact of the package thickness is significant only for ϑ_{JCtop} (for TQFPs, a 25–35% reduction of this metric is observed with respect to LQFPs), since in this case the heat generated in the die flows toward the cold plate placed on the top crossing the whole package; a marginal influence is instead found for all other metrics (including ϑ_{JA}), where the heat propagates mostly downward.

Lastly, using another available glue type with a reduced thermal conductivity (2 W/mK) for the die-epad attach was found to increase the downward-heat metrics by 15–25%, while ϑ_{JCtop} remains almost unaffected.

Differently from the previous tool version [10], where the power dissipation region was forced to coincide with the whole die, in the latest TRAC release it is also possible to select an HS with arbitrary size and position within the die, thus allowing the representation of more realistic conditions. Figure 5 shows the metric ϑ_{JA} as a function of HS size, the HS being centered in the die (which offers the possibility of meshing and simulating one quarter of the structure), for a 10×10 mm^2 LQFP with a 4.5×4.5 mm^2 epad and a 2×2 mm^2 die; the assigned dissipated power amounts to 1 W regardless of the HS size. As expected, ϑ_{JA} markedly increases with reducing the HS size, which implies a growth in power density. This analysis allows concluding that a correct representation of the HS geometry (which depends on the specific application) is of utmost importance for an accurate evaluation of the thermal metrics of electronic packages.

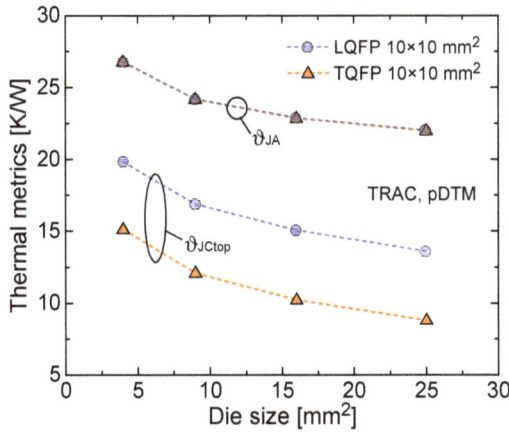

Figure 4. JEDEC thermal metrics ϑ_{JA} and ϑ_{JCtop} vs. die size evaluated with the pDTM; results obtained for 10×10 mm^2 LQFPs (blue circles) are compared with the TQFP counterparts (orange triangles); in both cases, a 6×6 mm^2 epad is considered.

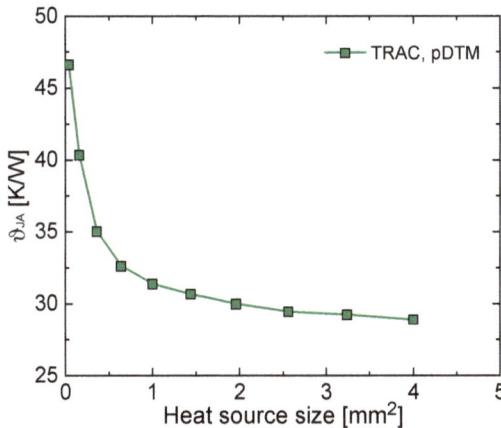

Figure 5. JEDEC thermal metric ϑ_{JA} vs. HS size for a square HS centered in the die, as evaluated through the pDTM for a 10×10 mm^2 LQFP with a 4.5×4.5 mm^2 epad and a 2×2 mm^2 die.

Lastly, a pCTM of any of the derived pDTMs, ensuring accuracy better than 0.5% in the reconstruction of the thermal field, was extracted in less than 20 min on an iMac with a 3.5 GHz Intel Core i7 according to Algorithm 1; the resulting pCTMs enjoy less than 70 DoFs (to be compared with a few millions for the pDTMs). It is worth noting that the steady-state thermal simulation corresponding to a given set of parameters requires 10–20 s for the pDTM (depending on the package), 30–60 s for the FV commercial software, and less than 0.2 s using the pCTM in TRAC. A comparison of the results from the pDTM and the pCTM is shown in Figure 6 for various QFPs.

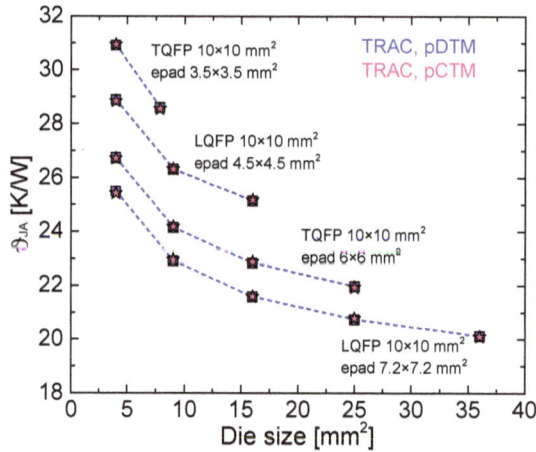

Figure 6. Comparison of the JEDEC thermal metric ϑ_{JA} provided by both a pDTM and a pCTM for various QFPs.

5. Conclusions

In this paper, a tool denoted as TRAC has been presented. TRAC determines the steady-state thermal behavior of electronic packages with Manhattan geometry, the key geometrical and material parameters of which vary in an assigned range. The latest TRAC release allows the straightforward definition of pDTMs, as well as the automatic and fast (about 20 min) extraction of pCTMs in a pre-processing stage, for two families of packages. In addition, it automatically provides the JEDEC thermal metrics corresponding to selected packages, thereby favoring an easy evaluation of the influence of the key parameters (e.g., die size, package thickness, die-epad attach material). Compared to a FV commercial software, a simulation performed by TRAC using a pDTM takes about $1/3$ of the CPU time, while the adoption of a pCTM leads to almost instantaneous results without any loss of accuracy. Moreover, TRAC is equipped with a user-friendly graphical interface that simplifies the choice of parameters and thus the construction of the geometry and mesh of the corresponding package. Owing to its features, TRAC can be used by semiconductor vendors with a two-fold aim: (1) to let customers effortlessly evaluate the thermal metrics of the packages they want to purchase through, for example, a web application; and (2) to support the design of more complex electronic systems.

TRAC variants suited to simulate other package families, handle multiple heat sources, and carry out dynamic thermal analyses are currently under development.

Author Contributions: Methodology, L.C.; Software, L.C. and S.R.; Validation, D.G., A.M., and C.M.V.; Investigation, L.C., S.R., and V.d.; Writing—Original Draft Preparation, V.d. and S.R.; Writing—Review & Editing, V.d. and S.R.; Supervision, C.M.V.

Funding: This research received no external funding.

Conflicts of Interest: The authors declare no conflict of interest.

Nomenclature

ϑ_{JA} (K/W)	junction to ambient thermal resistance
Ψ_{JB} (K/W)	thermal characterization parameter to report the difference between junction temperature and the temperature of the board measured at the top surface of the board
Ψ_{JCtop} (K/W)	thermal characterization parameter to report the difference between junction temperature and the temperature at the top center of the outside surface of the component package
ϑ_{JB} (K/W)	junction to board thermal resistance

ϑ_{JCtop} (K/W)	junction to case top thermal resistance
$\vartheta_{JCbottom}$ (K/W)	junction to case bottom thermal resistance
TRAC	thermal resistance advanced calculator
BC	boundary condition
BCI	BC independent
CTM	compact thermal model
pCTM	parametric CTM
DTM	detailed thermal model
pDTM	parametric DTM
DoF	degree of freedom
epad	exposed pad
FIT	finite integration technique
FV	finite volume
HS	heat source
JEDEC	joint electron device engineering council
QFP	quad flat package
LQFP	low-profile (thick) QFP
TQFP	thin QFP
QFN	quad flat no-leads package
MOR	model-order reduction

References

1. Lasance, C. Ten years of boundary-condition-independent compact thermal modeling of electronic parts: A review. *Heat Transfer Eng.* **2008**, *29*, 149–169. [CrossRef]
2. JESD51-12. *Guidelines for Reporting and Using Electronic Package Thermal Information*; JEDEC: Arlington, VA, USA, 2005.
3. Codecasa, L.; d'Alessandro, V.; Magnani, A.; Rinaldi, N.; Zampardi, P.J. FAst Novel Thermal Analysis Simulation Tool for Integrated Circuits (FANTASTIC). In Proceedings of the International Workshop on THERMal INvestigation of ICs and systems (THERMINIC), London, UK, 24–26 September 2014.
4. Codecasa, L.; d'Alessandro, V.; Magnani, A.; Rinaldi, N. Parametric compact thermal models by moment matching for variable geometry. In Proceedings of the International Workshop on THERMal INvestigation of ICs and systems (THERMINIC), London, UK, 24–26 September 2014.
5. Janssen, J.H.J.; Codecasa, L. Why matrix reduction is better than objective function based optimization in compact thermal model creation. In Proceedings of the International Workshop on THERMal Investigation of ICs and systems (THERMINIC), Paris, France, 30 September–2 October 2015.
6. Codecasa, L.; d'Alessandro, V.; Magnani, A.; Rinaldi, N. Matrix reduction tool for creating boundary condition independent dynamic compact thermal models. In Proceedings of the International Workshop on THERMal INvestigation of ICs and systems (THERMINIC), Paris, France, 30 September–2 October 2015.
7. Codecasa, L.; d'Alessandro, V.; Magnani, A.; Rinaldi, N. Structure preserving approach to parametric dynamic compact thermal models of nonlinear heat conduction. In Proceedings of the International Workshop on THERMal INvestigation of ICs and systems (THERMINIC), Paris, France, 30 September–2 October 2015.
8. Rogié, B.; Codecasa, L.; Monier-Vinard, E.; Bissuel, V.; Laraqi, N.; Daniel, O.; D'Amore, D.; Magnani, A.; d'Alessandro, V.; Rinaldi, N. Delphi-like dynamical compact thermal models using model order reduction. In Proceedings of the International Workshop on THERMal INvestigation of ICs and systems (THERMINIC), Amsterdam, The Netherlands, 27–29 September 2017. (best paper award).
9. Codecasa, L.; d'Alessandro, V.; Magnani, A.; Rinaldi, N. Novel approach for the extraction of nonlinear compact thermal models. In Proceedings of the International Workshop on THERMal INvestigation of ICs and systems (THERMINIC), Amsterdam, The Netherlands, 27–29 September 2017.

10. Codecasa, L.; Race, S.; d'Alessandro, V.; Gualandris, D.; Morelli, A.; Villa, C.M. Thermal Resistance Advanced Calculator (TRAC). In Proceedings of the International Workshop on THERMal INvestigation of ICs and Systems (THERMINIC), Stockholm, Sweden, 26–28 September 2018.

11. Codecasa, L.; Specogna, R.; Trevisan, F. Base functions and discrete constitutive relations for staggered polyhedral grids. *Comput. Meth. Appl. Mech. Eng.* **2009**, *198*, 1117–1123. [CrossRef]

energies

MDPI

Article

Structure Function Analysis of Temperature-Dependent Thermal Properties of Nm-Thin Nb_2O_5

Lisa Mitterhuber *, Elke Kraker and Stefan Defregger

Materials Center Leoben Forschung GmbH, Roseggerstrasse 12, 8700 Leoben, Austria;
Elke.Kraker@mcl.at (E.K.); stefan.defregger@mcl.at (S.D.)
* Correspondence: lisa.mitterhuber@mcl.at

Received: 29 December 2018; Accepted: 14 February 2019; Published: 15 February 2019

Abstract: A 166-nm-thick amorphous Niobium pentoxide layer (Nb_2O_5) on a silicon substrate was investigated by using time domain thermoreflectance at ambient temperatures from 25 °C to 500 °C. In the time domain thermoreflectance measurements, thermal transients with a time resolution in (sub-)nanoseconds can be obtained by a pump-probe laser technique. The analysis of the thermal transient was carried out via the established analytical approach, but also by a numerical approach. The analytical approach showed a thermal diffusivity and thermal conductivity from 0.43 mm^2/s to 0.74 mm^2/s and from 1.0 W/mK to 2.3 W/mK, respectively to temperature. The used numerical approach was the structure function approach to map the measured heat path in terms of a $R_{th}C_{th}$-network. The structure function showed a decrease of R_{th} with increasing temperature according to the increasing thermal conductivity of Nb_2O_5. The combination of both approaches contributes to an in-depth thermal analysis of Nb_2O_5 film.

Keywords: thermal conductivity; niobium pentoxide; structure function; time domain thermoreflectance; thin film

---

1. Introduction

Energy efficiency and saving in microelectronic devices go along with thermal management, as their failure rate increases exponentially with the operating temperature. The miniaturization and increase of device packing density trigger the importance of the heat dissipation, as well as the thermal management in microelectronics. Thus, it is necessary to develop heat dissipation strategies, which requires knowledge of devices' thermophysical properties. The devices themselves are composed of multiple layers of submicrometer thin films. The thermal properties of thin films can deviate from their bulk values and thermal boundary resistance becomes more dominant for the heat dissipation [1].

In this study, nm-thin Niobium pentoxide (Nb_2O_5) layer was characterized by thermal properties and their temperature dependencies were presented in the temperature range of 25 °C to 500 °C. Nb_2O_5 films can be found in optical filter, electrochromic device, sensors, capacitors, and microelectronic devices. Therefore, lots of investigations were done in terms of Nb_2O_5's optical and structural properties [2,3]. However, as far as we know, investigations about their thermal properties and temperature dependency can rarely be found in the literature, although these are important parameters for a device's efficiency and reliability [4].

The Nb_2O_5 films were thermally investigated with the time domain thermal reflectance (TDTR) method [5]. The TDTR records the response of a thin film in high speed (down to picoseconds (ps)) after a heating laser pulse. In this work, in addition to the common analytical evaluation of TDTR measurements, a numerical analysis is presented. This numerical analysis maps the sample's heat path in a one-dimensional way as the structure function, introduced by Székely et al. [6]. The structure function is a numerical approach, containing values of the materials' thermal resistance (R_{th}) and thermal capacitance (C_{th}) [7,8]. Here, a modified calculation of the structure function was applied to the TDTR measurements. This structure function overcomes certain up-front assumptions, which are necessary when applying analytical solution. Furthermore, the TDTR measurements were simulated via finite volume simulations to understand the structure function and get absolute R_{th} values [9].

2. Materials and Methods

An amorphous 166-nm-thin Nb_2O_5 film on Si was investigated at temperatures from 25 °C to 500 °C, obtaining the Nb_2O_5 temperature dependent thermal properties. The amorphous nature of the Nb_2O_5 was confirmed by Raman measurements (see Appendix A—Figure A1). The experimental thermal investigations were done by two TDTR systems (PicoTR and NanoTR from Netzsch Group in representative of PicoTherm [10]) with different time resolutions and, therefore, different heat penetration depths. These TDTR measurements were additionally simulated to assess temperatures of the thermal transient. Both, the TDTR measurement and simulation were transformed into the structure function for thermal analysis.

2.1. Time Domain Thermoreflectance

Both TDTR systems, PicoTR and NanoTR (see Table 1), provide a non-contact method in the field of thin film thermal properties metrology. The TDTR systems are based on a pump–probe technique with fiber laser beams. A pulse of the pump beam heats up the sample with a power of 25 mW on the Pt-layer (Front Heating). The probe beam, which has a lesser power of 0.8 mW, has to be focused on the heated area of the pump beam. The probe beam detects the temperature change caused by the pump beam. The temperature change is monitored by the reflectivity change with a differential photodetector (Front Detection). The photodetector's signal is used for further signal processing. So there are no absolute temperature values for the thermal transient, but the amplitude is directly proportional to temperature (thermo-reflectance principle).

Table 1. Technical data of the time domain thermal reflectance (TDTR) measurement setups.

	λ_{Pump} (nm)	λ_{Probe} (nm)	P_{Pump} (mW)	$\varnothing_{Pump}$ (µm)	$\varnothing_{Probe}$ (µm)	t_p (s)	T_p (s)
PicoTR	1550	775	25	45	25	5×10^{-13}	5×10^{-8}
NanoTR	1550	775	25	100	50	1×10^{-9}	2×10^{-5}

λ_{Pump}: wavelength of the pump laser; λ_{Probe}: wavelength of the probe laser; P_{Pump}: heating power of pump laser; $\varnothing_{Pump}$: diameter of the pump beam; $\varnothing_{Probe}$: diameter of the probe beam; t_p: pulse width of the pump beam; T_p: repetition time of pump laser (max. measurement time).

To measure the temperature dependency of the thermal properties, there is an oven integrated into the TDTR systems. The measurements with the NanoTR were performed from room temperature to 300 °C and the PicoTR from room temperature up to 500 °C. All the measurements were done under nitrogen atmosphere to avoid oxidation.

Nb_2O_5 is transparent for both laser wavelengths, λ_{Probe} and λ_{Pump}, so an optical transducer on the top surface of the sample was needed to get an optical reflectance of the probe laser. Therefore, a Pt-layer (100 nm thickness) was sputtered above the Nb_2O_5 [11].

The PicoTR and NanoTR measurement systems vary by their time resolution. The PicoTR records the response of pulse heating in picoseconds (ps) and NanoTR in nanoseconds (ns). Both systems also have a different repetition time of pulse heating (T_p), which restricts the maximal recording time of the thermal transient. These facts enable an investigation of the sample in different length scales.

2.2. Finite Volume Simulation

To validate the application of the proposed methodology, thermal simulations were carried out, visualizing the heat path of the Nb_2O_5 film.

A fully 3D geometrical model of the Nb_2O_5 sample was built in FloTHERM® (12.1., Mentor Graphics, Wilsonville, OR, USA) (see schematic in Figure 1). The initial thermal properties of the materials (e.g., thickness (d), cross-sectional area (A), density (ρ), specific heat (c_p)) were added to the simulation model according to literature values and experimental results. The thermal conductivity (λ) and the thermal interface resistance (TIR) were adjusted to the measurements. The experimental parameters, as time resolution (t_p), length of measurement period (T_p), laser power (P_{Pump}), and illuminated area ($\emptyset_{Pump}$) were reconstructed in the simulation according to PicoTR and NanoTR. The input of thermal power was placed from the surface 10 nm deep into the transducer, taking into account the optical penetration depth of the pump laser in Pt [12]. The temperature change was recorded with monitor points in the middle of the simplified uniformly heated area of $\emptyset_{Pump}$.

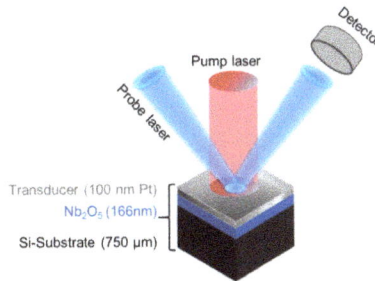

Figure 1. Schematic of the investigated sample and the representation of the TDTR measurement.

2.3. Theoretical Background

An established analysis for thermal transients, typically used for thermal transients of microelectronic devices, is the structure function. The thermal transient is transformed into a one-dimensional heat path representation. The heat path is represented in the structure function as a ladder network of R_{th}s and C_{th}s. These network elements contain information about the involved materials. The R_{th} is a function of $d/(\lambda \cdot A)$ and the C_{th} is $c_p \cdot d \cdot A \cdot \rho$ [8].

In this work, the transformation of the thermal transient into its structure function was applied on the thermal transient of a TDTR measurement. The transformation is valid when the heat source as well as the probing of the thermal transient happens on the same place of the sample. For the Nb_2O_5 sample, this happened at the Pt-transducer. Furthermore, the validity of a diffusive and a one-dimensional heat transport has to be assured, using the structure function for validation. These demands were met, as the diameter of the laser spot is larger than the thermal diffusion length of the excited layer. [13] The error caused by the assumption of the one-dimensional heat transport was estimated by the comparison of the fully 3D simulation model with the 1D simulation model. The relative error was 3% for the PicoTR measurement and 8% for the NanoTR measurement [14].

The thermal path investigation of microelectronic devices is realized by recording the switching-off response of the device [15,16]. So the power excitation is a step-function. However, in the TDTR measurement, the thermal transient originates a laser pulse excitation. [10] Therefore, the calculation of the structure function has to be modified, which is described in the following sections.

2.3.1. Structure Function for Dirac-Delta Function Excitation (NanoTR)

To apply the same calculation of the structure function as it is used for a step-function thermal transient, the calculations of the structure function have to be modified. The thermal excitation of the pump laser has to be regarded as a Dirac delta function (δ) [17].

The thermal transient $T(t)$ is the product of power P and thermal impedance Z_{th}. Z_{th} in the Laplace domain as a function of the complex frequency (s), is written in Equation (1). $Z_{th}(s)$ can be transformed into an analogous $R_{th}C_{th}$-network [18], a Foster-type network, where $\tau_i = R_{th,i} \cdot C_{th,i}$ [19].

$$Z_{th}(s) = \sum_i \frac{R_{th,i}}{(1 + \tau_i s)}, \tag{1}$$

The power excitation in the form of a Heaviside step-function (H) is in the time domain $P(t) = P_0 \cdot H(t)$ and in the Laplace domain $P(s) = P_0 \cdot 1/s$. P_0 is the absolute power value. The temperature change in terms of the $R_{th}C_{th}$-network is:

$$T_H(s) = P_0 \sum_i \frac{R_{th,i}}{s(1 + \tau_i s)}, \tag{2}$$

$$T_H(t) = P_0 \sum_i R_{th,i} \left(1 - e^{-\frac{t}{\tau_i}}\right) \tag{3}$$

The power excitation by δ can be written as $P(t) = P_0 \delta(t)$ or as $P(s) = P_0$, so the temperature change is Equation (4) and its inverse Laplace transformation is Equation (5).

$$T_\delta(s) = P_0 \sum_i \frac{R_{th,i}}{(1 + \tau_i s)} \tag{4}$$

$$T_\delta(t) = P_0 \sum_i R_{th,i} \frac{e^{-\frac{t}{\tau_i}}}{\tau_i} \tag{5}$$

Equation (3) is the time derivative of Equation (5); the integration of the temperature change of a $\delta(t)$-excitation over time results in a temperature change of a $H(t)$-excitation. Therefore, the time integration of the thermal transients of a TDTR measurement allows for calculating the time constant spectrum in the same way as it is done for the step-function thermal transient (NID-method [20]). The time constant spectrum is a function of $R_{th,i}$ over τ_i, and hence, the values for the Foster network after discretization. The Foster network [21] has to be transformed into a Cauer network to get a physical meaning of R_{th} and C_{th}. The cumulative Cauer network's C_{th} as a function of its cumulative R_{th} is the structure function [22].

2.3.2. Structure Function for Cycled Pulsed Excitation (PicoTR)

The above-mentioned structure function calculation is only valid if the repetitive laser pulse excitations do not affect each other. Therefore, the thermal transient has to reach a thermal equilibrium during T_p. This demand is not always fulfilled with regards to a TDTR measurement—especially a PicoTR measurement with a T_p of 50 ns.

Therefore, the power excitation has to be assumed as a single Dirac-delta function, but as a pulsed cycled excitation event (Figure 2). The thermal transient of the TDTR has to be regarded as a pulsed thermal transient [23,24]. There, the repeated pulse train of the pump beam (T_p) is assumed to appear in a square shape, with a certain peak width (t_p). Now, $Z_{th}(t)$ is not only a function of time but also a function of the duty, the ratio between t_p and T_p. $Z_{th}(t)$ of an intercycle is described in Equation (6).

$$Z_{th}\left(t, \frac{t_p}{T_p}\right) = \sum_j Z_{th}(jT_p) - Z_{th}(jT_p - t_p) \tag{6}$$

With the use of Equation (3), the equation above can be rewritten as,

$$Z_{th,i}\left(t, \frac{t_p}{T_p}\right) = \sum_j R_{th,i}\left(1 - e^{\frac{jT_p}{\tau_i}}\right) - R_{th,i}\left(1 - e^{\frac{j(T_p - t_p)}{\tau_i}}\right) = R_{th,i}\left(e^{\frac{T_p}{\tau_i}} - 1\right) \sum_j e^{\frac{jT_p}{\tau_i}} \tag{7}$$

and with the identity $\sum_j a^{-j} = \frac{1}{a-1}$, Equation (7) can be transformed to Equation (8).

$$Z_{th,i}\left(t,\frac{t_p}{T_p}\right) = R_{th,i}\,\frac{e^{\frac{t_p}{\tau_i}}-1}{e^{\frac{T_p}{\tau_i}}-1} = R_{th,i}\,\frac{e^{\frac{t}{\tau_i}}-1}{e^{\frac{t}{(T_p\backslash t_p)\tau_i}}-1} \overset{\sum_i}{\Rightarrow}$$

$$Z_{th}\left(t,\frac{t_p}{T_p}\right) = \sum_i R_{th,i}\,\frac{e^{\frac{t}{\tau_i}}-1}{e^{\frac{t}{(T_p\backslash t_p)\tau_i}}-1}$$

(8)

To get the time constant spectrum out of this equation, the logarithm of the time variables ($z = \log(t)$; $\zeta = \log(\tau)$) is used:

$$Z_{th} = \sum_i R_{th,i}\,\frac{e^{e^{z-\zeta}}-1}{e^{(T_p\backslash t_p)e^{z-\zeta}}-1} \Rightarrow R(z) \otimes \frac{e^{e^z}-1}{e^{(T_p\backslash t_p)e^z}-1}$$

(9)

Equation (9) takes into account the influence of repetitive pump laser pulses.

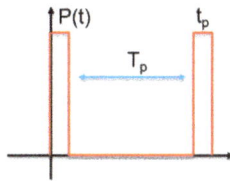

Figure 2. Schematic diagram of a pulsed cycled excitation.

3. Results

3.1. Analytical Solution of Fourier's Law to Evaluate Thermal Diffusivity

The Nb_2O_5 film was measured with both PicoTR and NanoTR at different ambient temperatures. The measured and normalized thermal transients are shown in Figure 3. An established way to obtain the thermal properties of the Nb_2O_5 film is fitting Fourier's heat equation to these thermal transients [5]. As there is Pt as the transducer [25] and Si as the substrate, the heat equation of multiple layers was used for further analysis. The temperature-dependent properties of Nb_2O_5 [26], Pt [27–29] and Si [29–31] were based on values from the literature. The values for α of Nb_2O_5 resulted from the fit are presented in Figure 4. From α, the λ can be calculated:

$$\alpha = \frac{\lambda}{c_p \cdot \rho}$$

(10)

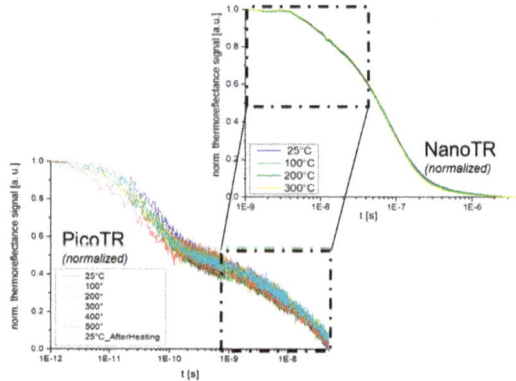

Figure 3. Normalized thermal transients of the Pt-Nb$_2$O$_5$-Si sample at different ambient temperatures from PicoTR (25 °C–500 °C) and NanoTR (25 °C–300 °C).

The c_p of Nb$_2$O$_5$ as a function of temperature was taken from Reference [32]. This increasing behaviour of c_p with temperature was confirmed by Reference [33] due to the excitation of more high-energy phonon modes. The Nb$_2$O$_5$ density was measured via XRR (Rigaku SmartLab 5-circle diffractometer). XRR curves were recorded over a range of 6 degrees and evaluated by GenX software [34], resulting in a value of 4750 kg/m^3.

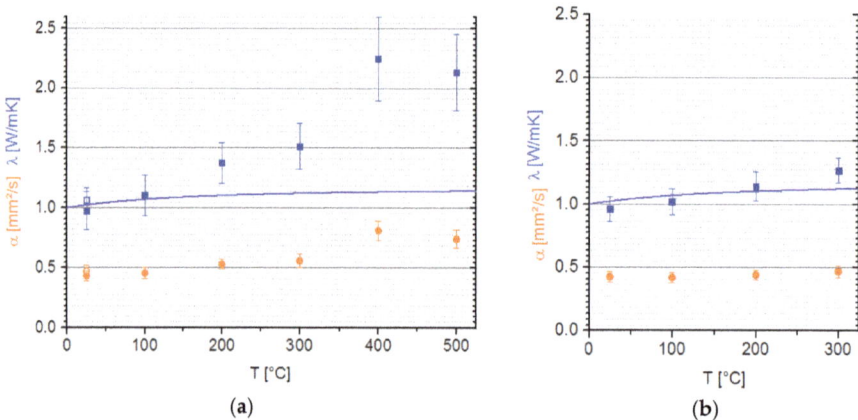

Figure 4. The thermal properties of Nb$_2$O$_5$ presented as a function of temperature from PicoTR (a) and NanoTR (b). The y-axis stands for λ marked as squares and α as circles. The open marker in (a) is the measurement after the heating procedure. α and λ of the PicoTR and NanoTR measurements show the same trend. The solid line shows the minimal thermal conductivity model by Cahill et al. [35]—Equation (11).

The sample was measured by the PicoTR (Figure 4a). Figure 4a shows that α increased with the ambient temperature from 0.43 mm^2/s to 0.74 mm^2/s. The thermal interface resistance (TIR) between Pt and Nb$_2$O$_5$ of 18 Km^2/GW was obtained by the evaluation of the thermal transient. The TIR showed no significant temperature dependency in the measured temperature range. The value of the TIR was in a comparable range to the results of References [36,37].The thermal conductivity was calculated with Equation (10) and had a value of (0.98 ± 0.16) W/mK at 25 °C. The thermal conductivity values at 25 °C and 100 °C agreed with the measurements of Reference [33]. The error of the PicoTR measurements was dominated by the signal to noise ratio of the thermal transient and the uncertainty of the Nb$_2$O$_5$

density. The thermal conductivity showed the characteristic temperature dependence of an amorphous solid: an increase of the thermal conductivity with increasing temperature [38,39]. An increase of thermal conductivity by more than a factor of two was observed from 25 °C to 500 °C.

For amorphous solids, Cahill et al. [35] developed a minimum thermal conductivity model, which was compared to the measurement results.

$$\lambda_{min} = \left(\frac{\pi}{6}\right)^{\frac{1}{3}} k_B n^{\frac{2}{3}} \sum_i v_i \left(\frac{T}{\theta_{D,i}}\right)^2 \int_0^{\frac{\theta_{D,i}}{T}} \frac{x^3 e^x}{(e^x - 1)^2} \, dx, \tag{11}$$

where k_B is the Boltzmann constant and n is the number density of atoms ($n = 7.53 \times 10^{22}$ cm^{-3}). Equation (11) considers the three sound modes (two transverse and one longitudinal) with the speed of sound v_i. The transverse speed of sound had a value of 3202 m/s and was calculated with the shear modulus of 49 GPa [40]; the longitudinal speed of sound was 5311 m/s according to Young's modulus of 134 GPa [41]. $\theta_{D,i}$, the cutoff frequency for each mode expressed as a temperature unit, was calculated via $\theta_{D,i} = v_i (\hbar/k_B)(6\pi^2 n)^{\frac{1}{3}}$. The calculated and measured thermal conductivities of Nb$_2$O$_5$ (Figure 4) are comparable at 25 °C and 100 °C. However, at 200 °C, the measured thermal conductivity started to increase, whereas the calculated thermal conductivity stayed nearly constant ($\Delta\lambda = 0.04$ W/mK between 200 °C and 500 °C). This discrepancy is still under investigation and not clearly understood yet.

After the PicoTR measurement, the sample was again measured at 25 °C (Figure 4a—open marker), to see if the sample was affected by the temperature treatment during measurement. α and therefore λ, were nearly equal before and after the heating: 0.43 mm^2/s to 0.47 mm^2/s and 1.0 W/mK to 1.1 W/mK. The variation of these values were within the measurement error range. So further investigation was carried out at the NanoTR on the same sample. The measurements with the NanoTR were done up to an ambient temperature of 300 °C. The NanoTR error arose mainly from the uncertainty of the Nb$_2$O$_5$ thickness and density. The results of α and λ of Nb$_2$O$_5$ fitted to the ones of PicoTR, according to the measurement error range. Differences between NanoTR and PicoTR may appear due to the different time resolution and time range.

3.2. Structure Function to Analyse the Heat Path in Nb$_2$O$_5$

3.2.1. Structure Function for Delta Function Excitation of Nanosecond Thermal Transients

The structure function calculation of Section 2.3.1. was applied on the normalized thermal transients of a NanoTR measurement. The trend of the thermal properties respectively to temperature can be seen here in terms of total R_{th} (Figure 5) like in Section 3.1. The resulting time constant spectrum of the NanoTR measurements showed two peaks, at 100 ns and 900 ns (Figure 5 inlet). The structure function comprised two step-like features, accordingly. It could be assumed that two different materials, dominating the thermal path, contribute to these features. However, the thermal transients of the NanoTR provided only relative temperature values. Therefore, a scaling factor (x) had to be applied on the structure function to get quantitative values ($x^{-1} \cdot R_{th}$ and $x \cdot C_{th}$).

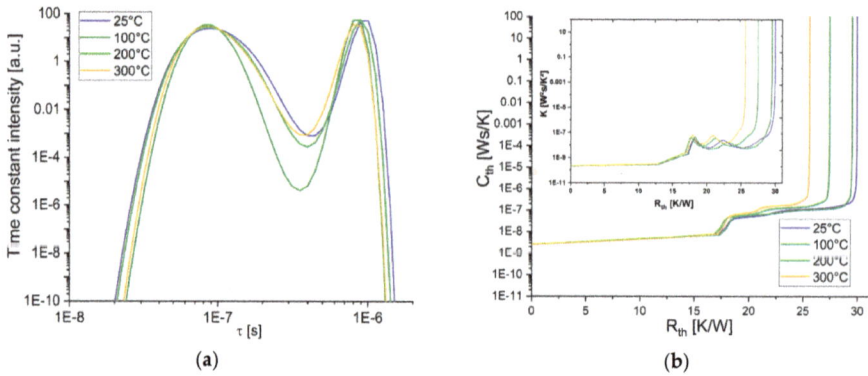

Figure 5. (a) The logarithmic time constant spectrum of NanoTR measurements, showing two dominant time constant peaks. (b) The structure function and its differential representation (inlet), where $K = dC_{th}/dR_{th}$, shows the lowest R_{th} at 300 °C and the highest R_{th} at 25 °C.

To verify this assumption and determine the scaling factor x, the NanoTR measurement was reconstructed by numerical simulation. The simulation showed that the time resolution of NanoTR was not able to catch the heat path of the transducer, the Pt-layer. The characteristic time of a 100 nm Pt-layer lies under 1 ns according to Equation (1). The simulation revealed that the first step-like feature in the structure function depended on both TIR, between Pt and Nb_2O_5, between Nb_2O_5 and Si, and on the Nb_2O_5 film itself. These parameters were relevant for the R_{th} value of the step-feature. The λ of Nb_2O_5 also influenced the ratio between R_{th} and C_{th}, not only the position on the x-axis but also the height of the step-feature. The rear part of the structure function appeared to be sensitive to the thermal properties of the Si substrate (Figure 6).

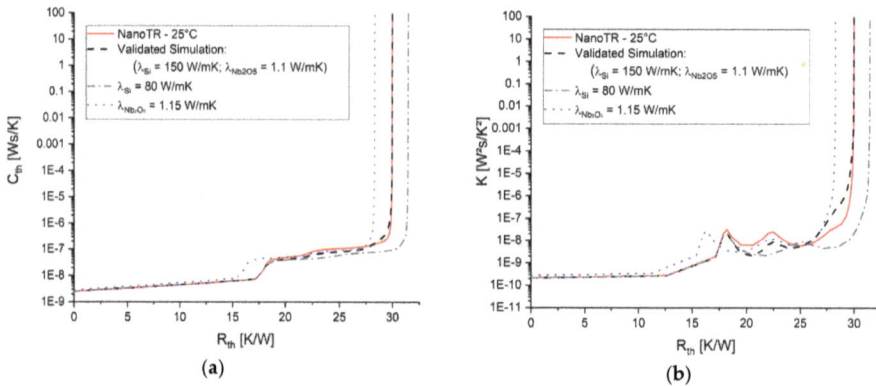

Figure 6. Structure function (a) and its differential representation (b) of the NanoTR measurement at 25 °C with its validated simulation and variations of the simulation to see the influence of the input parameters.

Both the measured at 25 °C and simulated thermal transients were normalized, from which the structure functions were calculated. Then, the structure function of the simulation was adjusted to the measured one. The resulting parameters of the best structure functions' match (validated simulation) were: $TIR_{Pt-Nb_2O_5} = 13$ Km^2/GW; $\lambda_{Nb_2O_5} = 1.0$ W/mK; $TIR_{Pt-Nb_2O_5} = 15$ Km^2/GW; $\lambda_{Si} = 150$ W/mK. The used parameters for the validated simulation were in accordance with the analytical solutions (compare to Figure 4). After validation, the scaling factor x and the absolute values of R_{th} and C_{th} can be determined using the non-normalized simulated thermal transient. The scaling factor x in the case of NanoTR measurements had a value of 4.53×10^{-9}. With the assumption that the thermoreflectance coefficient stayed the same for all the NanoTR measurements (Figure 5)—their R_{th} values are 30 K/W to 26 K/W from 25 °C and 300 °C respectively.

3.2.2. Structure Function for Pulsed Cycled Excitation of Picosecond Thermal Transients

The thermal transients of the PicoTR measurements showed a signal to noise ratio of 16 dB (Figure 3). Before calculating the structure functions with the thermal transients of the PicoTR, they had to be smoothed via a Savitzky-Golay filter [42]. Here, a moving window of 351 points and a polynomial degree of 4 was applied to get rid of the noise. The calculation of Section 2.3.1. was carried out for the PicoTR thermal transients, resulting in a structure function without any features. This did not represent reality, since the sample consisted of multiple layers, multiple peaks in the time constant spectrum, and features in the structure function were expected. Therefore, the calculation of the structure function was carried out accordingly to Section 2.3.2. There, the influence of the repetitive excitation is taken into account, due to the high pump laser repetition rate of 20MHz at the PicoTR [43,44]. This effect is visualized in Figure 7.

The resulting time constant spectrum of the PicoTR measurements (Figure 8) showed three peaks. The first at a τ of (100–200) ps is comparable to the characteristic diffusion time of the Pt-layer. The second and third peak can be found at 1 ns and 15 ns. The three peaks of time constant spectrum can be seen as three humps in the structure function at 5 K/W, 6 K/W, and 7 K/W (Figure 8 inlet). The x-axis of the structure function reflected the results from the analytical solution. The structure function of the 500 °C measurement showed the lowest and the one of 25 °C measurement the highest R_{th} (Figure 8).

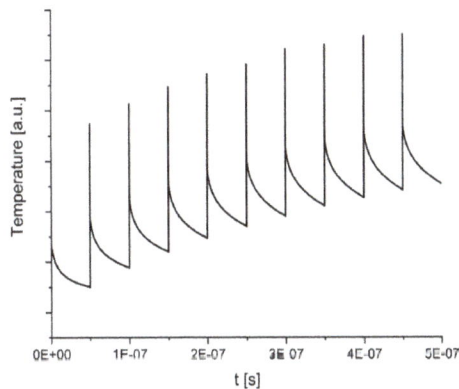

Figure 7. The sample heats up during measurement according to the repetitive pulses of 20 MHz.

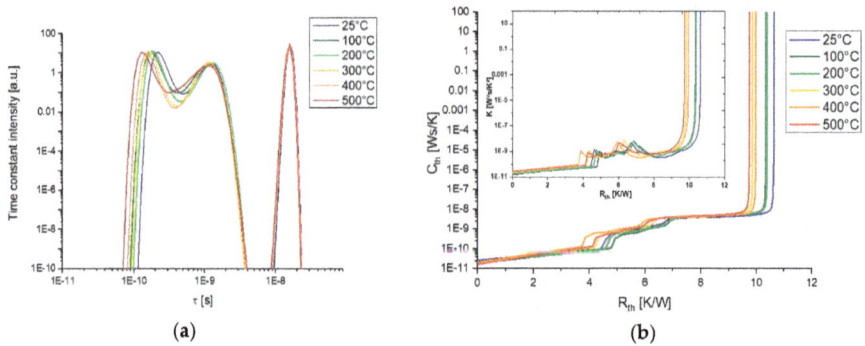

(a) (b)

Figure 8. (a) The logarithmic time constant spectrum of NanoTR measurements, showing three dominant time constants. (b) The structure functions and its differential representation (inlet) of the PicoTR measurements resulted in different R_{th}.

To analyze the structure functions of the PicoTR measurements, thermal simulations had to be carried out too. The simulation thermal transient was recorded after 10 ps of excitation. In the simulation, the thermal transient was recorded in the Pt-layer below the optical penetration depth as recommended by Reference [45]. Then the structure function was calculated from both the normalized PicoTR measurement and simulation. The validated simulation parameters were: $\lambda_{Pt} = 55$ W/mK; $TIR_{Pt-Nb_2O_5} = 10$ Km^2/GW; $\lambda_{Nb_2O_5} = 0.95$ W/mK; $TIR_{Pt-Nb_2O_5} = 15$ Km^2/GW; $\lambda_{Si} = 150$ W/mK. The validation of the measurement revealed that the first step-like feature is sensitive to the Pt (5 K/W), whose thermal conductivity determines its plateau lengths (Figure 9). Whereas the Nb$_2$O$_5$ is responsible for the second step-feature (7 K/W). Both TIR are formative for the slope of both step-features. The Si substrate of the sample had no influence on the structure function. This is in accordance with the Nb$_2$O$_5$ heat diffusion time of 78 ns. The determining parameter for the total R_{th} of these structure functions was the thermal conductivity of Nb$_2$O$_5$.

The validated non-normalized thermal transient of the simulation revealed the absolute temperature and hence, the R_{th} values. The scaling factor here was 0.016. The total R_{th} decreased from 10.6 K/W to 9.7 K/W with increasing temperature (Figure 8). The analytical solution of the measurement at 400 °C showed the highest thermal conductivity. However, the structure function analysis showed that the total R_{th} is lower than that of the 500 °C measurement. The structure function revealed that the TIR had a lower value, or a higher thermal conductivity of Pt or Nb$_2$O$_5$ at 400 °C than the rest. The first step appeared at lower R_{th} (<4 K/W).

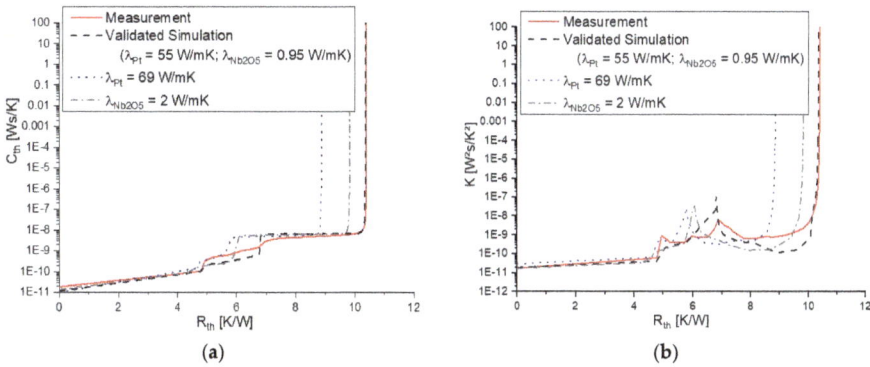

Figure 9. Structure function (**a**) and differential structure function (**b**) of the PicoTR measurement of 100 °C with its validated simulation and variations of the simulation, to see the influence of the materials.

4. Discussion

The temperature-dependent thermal conductivity of 166-nm-thin Nb_2O_5 was investigated via two TDTR systems. The combination of both PicoTR and NanoTR enabled us to investigate the complete heat path of Nb_2O_5 films, from ps to µs. The numerical analysis and structure function visualized the Nb_2O_5 temperature-dependent heat path [46]. The structure function of the PicoTR measurements showed the heat path of the Pt-layer in the ps time regime and the Nb_2O_5 film in the ns time regime. The NanoTR structure function displayed heat path of the Nb_2O_5 and its substrate. The temperature dependency of the structure functions is in accordance with the analytical solutions of the thermal conductivity. The indirect proportionality between the thermal conductivity of the analytical solutions and R_{th} of the structure is a function of temperature and can be seen in Figure 10. In order to assess the Nb_2O_5 performance and its stability, further investigations have to be carried out upon thermal fatigue cycling.

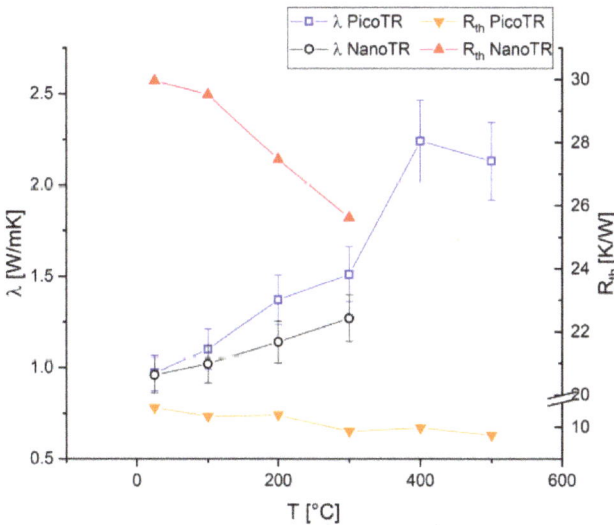

Figure 10. Comparison of the measured thermal conductivity (left-hand axis) and calculated thermal resistance (right-hand axis) as a function of temperature.

5. Conclusions

The temperature-dependent thermal properties of a 166-nm-thin Nb_2O_5 film were investigated via an analytical approach and via structure function by using the signals of two different TDTR systems (ns and ps time resolution).

The analytical approach, based on the multi-layer heat equation, revealed that the thermal diffusivity and thermal conductivity increased with temperature from 0.43 mm^2/s to 0.74 mm^2/s and from 1.0 W/mK to 2.3 W/mK.

The TDTR measurement transformed into a structure function and reflected the difference in the heat path as K_{th} and C_{th}. The measurements at 25 °C to 500 °C showed a decrease of R_{th} in the structure function. The PicoTR and NanoTR showed different length scales in the heat path, ranging from nm to μm. PicoTR showed the heat path of the transducer to the Nb_2O_5 and NanoTR of Nb_2O_5 to the Si substrate. The structure function offered an identification tool via heat path analysis to localize the material's change within the thin film multilayer stack. This approach is also applicable for other material systems in the nanometer range.

Author Contributions: Conceptualization, S.D.; methodology, L.M.; simulations, L.M.; validation, L.M., formal analysis, L.M.; investigation, L.M.; experiments, L.M.; writing—original draft preparation, L.M.; writing—review and editing, E.K.; visualization, L.M.; supervision, E.K.; project administration, S.D.; funding acquisition, S.D.

Funding: The authors gratefully acknowledge the financial support under the scope of the COMET program within the K2 Center "Integrated Computational Material, Process and Product Engineering (IC-MPPE)" (Project No 859480) This program is supported by the Austrian Federal Ministries for Transport, Innovation and Technology (BMVIT) and for Digital and Economic Affairs (BMDW), represented by the Austrian research funding association (FFG), and the federal states of Styria, Upper Austria and Tyrol.

Acknowledgments: Juraj Todt (Erich Schmid Institute of Materials Science) is gratefully acknowledged for the XRR measurement. Vignaswaran K. Veerapandiyanis gratefully acknowledged for the Raman measurement. René Hammer and Prof. Heinz Krenn is gratefully acknowledged for the fruitful discussions.

Conflicts of Interest: The authors declare no conflict of interest.

Appendix A

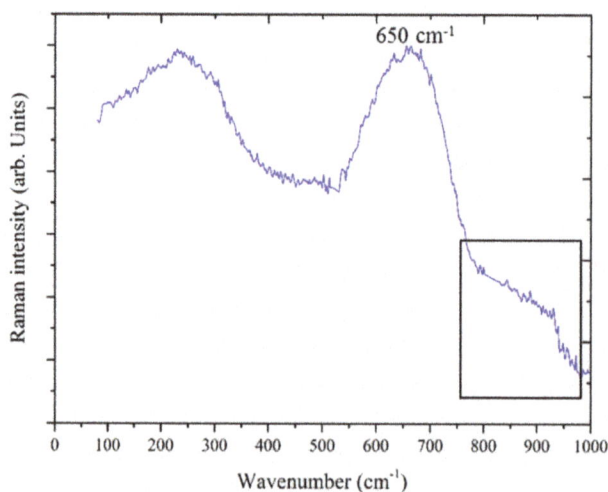

Figure A1. Raman spectra of Nb_2O_5 film.

The Raman spectra showed broad band at 650 cm^{-1} and 900 cm^{-1}, which indicates the amorphous nature of the Nb_2O_5 film.

References

1. Jones, R.E.; Duda, J.C.; Zhou, X.W.; Kimmer, C.J.; Hopkins, P.E. Investigation of size and electronic effects on Kapitza conductance with non-equilibrium molecular dynamics. *Appl. Phys. Lett.* **2013**, *102*, 183119. [CrossRef]
2. Coşkun, Ö.D.; Demirel, S.; Atak, G. The effects of heat treatment on optical, structural, electrochromic and bonding properties of Nb2O5 thin films. *J. Alloys Compd.* **2015**, *648*, 994–1004. [CrossRef]
3. Lorenz, R.; O'Sullivan, M.; Fian, A.; Sprenger, D.; Lang, B.; Mitterer, C. A comparative study on NbO$_x$ films reactively sputtered from sintered and cold gas sprayed targets. *Appl. Surf. Sci.* **2018**, *436*, 1157–1162. [CrossRef]
4. Magnien, J.; Mitterhuber, L.; Rosc, J.; Schrank, F.; Hörth, S.; Goullon, L.; Hutter, M.; Defregger, S.; Kraker, E. Reliability and failure analysis of solder joints in flip chip LEDs via thermal impedance characterisation. *Microelectron. Reliab.* **2017**, *76–77*, 601–605. [CrossRef]
5. Taketoshi, N.; Baba, T.; Ono, A. Development of a thermal diffusivity measurement system for metal thin films using a picosecond thermoreflectance technique. *Meas. Sci. Technol.* **2001**, *12*, 2064–2073. [CrossRef]
6. Székely, V.; Van Bien, T. Fine Structure of Heat Flow Path in Semiconductor Devices: A Measurement and Identification Method. *Solid. State. Electron.* **1988**, *31*, 1363–1368. [CrossRef]
7. Rencz, M.; Poppe, A.; Ress, S. A Procedure to Correct the Error in the Structure Function Based Thermal Measuring Methods. In Proceedings of the Twentieth Annual IEEE Semiconductor Thermal Measurement and Management Symposium, San Jose, CA, USA, 11 March 2004.
8. Rencz, M.; Székely, V.; Morelli, A.; Villa, C. Determining Partial Thermal Resistances with Transient Measurements, and Using the Method to Detect Die Attach Discontinuities. In Proceedings of the Eighteenth Annual IEEE Semiconductor Thermal Measurement and Management Symposium, San Jose, CA, USA, 12–14 March 2002; pp. 15–20.
9. Mitterhuber, L.; Defregger, S.; Hammer, R.; Magnien, J.; Schrank, F.; Hörth, S.; Hutter, M.; Kraker, E. Validation methodology to analyze the temperature-dependent heat path of a 4-chip LED module using a finite volume simulation. *Microelectron. Reliab.* **2017**, *79*, 462–472. [CrossRef]
10. Netzsch (Ed.) *Thermoreflectance by Pulsed Light Heating Thermoreflectance—NanoTR/PicoTR*. Available online: https://www.netzsch-thermal-analysis.com/media/thermal.../Thermoreflectance.pdf (accessed on 14 February 2019).
11. Wilson, R.B.; Apgar, B.A.; Martin, L.W.; Cahill, D.G. Thermoreflectance of metal transducers for optical pump-probe studies of thermal properties. *Opt. Express* **2012**, *20*, 28829–28838. [CrossRef]
12. Werner, W.S.M.; Glantschnig, K.; Ambrosch-Draxl, C. Optical constants and inelastic electron-scattering data for 17 elemental metals. *J. Phys. Chem. Ref. Data* **2009**, *38*, 1013–1092. [CrossRef]
13. Hopkins, P.E.; Serrano, J.R.; Phinney, L.M.; Kearney, S.P.; Grasser, T.W.; Harris, C.T. Criteria for Cross-Plane Dominated Thermal Transport in Multilayer Thin Film Systems During Modulated Laser Heating. *J. Heat Transf.* **2010**, *132*, 081302. [CrossRef]
14. Chen, H.; Lu, Y.; Gao, Y.; Zhang, H.; Chen, Z. The performance of compact thermal models for LED package. *Thermochim. Acta* **2009**, *488*, 33–38. [CrossRef]
15. Székely, V.; Szalai, A. Measurement of the time-constant spectrum: Systematic errors, correction. *Microelectron. J.* **2012**, *43*, 904–907. [CrossRef]
16. Székely, V.; Rencz, M.M. Increasing the Accuracy of Thermal Transient Measurements. *IEEE Trans. Compon. Packag. Technol.* **2002**, *25*, 539–546. [CrossRef]
17. Ezzahri, Y.; Shakouri, A. Application of network identification by deconvolution method to the thermal analysis of the pump-probe transient thermoreflectance signal. *Rev. Sci. Instrum.* **2009**, *80*, 074903. [CrossRef] [PubMed]
18. Russo, S. Measurement and Simulation of Electrothermal Effects in Solid-State Devices for RF Applications. Ph.D. Thesis, Universita Degli Studi di Napoli Ferderico II, Napoli, Italy, 2010.
19. Glavanovics, M.; Zitta, H. Thermal Destruction Testing: An Indirect Approach to a Simple Dynamic Thermal Model of Smart Power Switches. In Proceedings of the 27th European Solid-State Circuits Conference, Villach, Austria, 18–20 September 2001; Volume 2.
20. Székely, V. A new evaluation method of thermal transient measurement results. *Microelectron. J.* **1997**, *28*, 277–292. [CrossRef]

21.   Jakopovid, Z.; Bencic, Z.; Koncar, R.; Jakopovid, R.K.Z.; Bencic, Z. Identification of thermal equivalent-circuit parameters for semiconductors. In Proceedings of the 1990 IEEE Workshop on Computers in Power Electronics, Lewisburg, PA, USA, 5–7 August 1990; pp. 251–260.

22.   Masana, F.N. A straightforward analytical method for extraction of semiconductor device transient thermal parameters. *Microelectron. Reliab.* **2007**, *47*, 2122–2128. [CrossRef]

23.   Stout, R.P. *How to Generate Square Wave, Constant Duty Cycle, Transient Response Curves*; Semiconductor Component Industries, LLC: Phoenix, AZ, USA, 2006.

24.   Székely, V.; Rencz, M. Thermal Dynamics and the Time Constant Domain. *IEEE Trans. Compon. Packag. Technol.* **2000**, *23*, 587–594. [CrossRef]

25.   Norris, P.M.; Caffrey, A.P.; Stevens, R.J.; Klopf, J.M.; McLeskey, J.T.; Smith, A.N. Femtosecond pump-probe nondestructive examination of materials (invited). *Rev. Sci. Instrum.* **2003**, *74*, 400–406. [CrossRef]

26.   Douglass, D.L. The thermal expansion of niobium pentoxide and its effect on the spalling of niobium oxidation films. *J. Less-Common Met.* **1963**, *5*, 151–157. [CrossRef]

27.   Laubitz, M.J.; van der Meer, M.P. The thermal conductivity of platinum between 300 and 1000 K. *Can. J. Phys.* **1966**, *44*, 3173–3183. [CrossRef]

28.   Nakamura, F.; Taketoshi, N.; Yagi, T.; Baba, T. Observation of thermal transfer across a Pt thin film at a low temperature using a femtosecond light pulse thermoreflectance method. *Meas. Sci. Technol.* **2011**, *22*, 024013. [CrossRef]

29.   Engineering ToolBox. Coefficients of Linear Thermal Expansion. 2003. Available online: https://www. engineeringtoolbox.com/linear-expansion-coefficients-d_95.html (accessed on 21 November 2018).

30.   Ho, C.Y.; Powell, R.W.; Liley, P.E. Thermal conductivity of the Elements. *J. Phys. Chem. Ref. Data* **1972**, *1*, 279–421. [CrossRef]

31.   Endo, R.K.; Fujihara, Y.; Susa, M. Calculation of the density and heat capacity of silicon by molecular dynamics simulation. *High Temp. High Press.* **2003**, *35–36*, 505–511. [CrossRef]

32.   Chase, M.W. *NIST-JANAF Thermochemical Tables 2 Volume-Set (Journal of Physical and Chemical Reference Data Monographs)*; American Institute of Physics: College Park, MD, USA, 1998.

33.   Cheng, Z.; Weidenbach, A.; Feng, T.; Tellekamp, M.B.; Howard, S.; Wahila, M.J.; Zivasatienraj, B.; Foley, B.; Pantelides, S.T.; Piper, L.F.; et al. Diffuson-driven Ultralow Thermal Conductivity in Amorphous Nb_2O_5 Thin Films. *arXiv*, 2018; arXiv:1807.05483.

34.   Björck, M.; Andersson, G. GenX: An extensible X-ray reflectivity refinement program utilizing differential evolution. *J. Appl. Crystallogr.* **2007**, *40*, 1174–1178. [CrossRef]

35.   Cahill, D.G.; Watson, S.K.; Pohl, R.O. Lower limit to the thermal conductivity of disordered crystals. *Phys. Rev. B* **1992**, *46*, 6131–6140. [CrossRef]

36.   Bai, S.; Tang, Z.; Huang, Z.; Yu, J. Thermal Characterization of Si_3N_4 Thin Films Using Transient Thermoreflectance Technique. *IEEE Trans. Ind. Electron.* **2009**, *56*, 3238–3243.

37.   Chien, H.C.; Yao, D.J.; Hsu, C.T. Measurement and evaluation of the interfacial thermal resistance between a metal and a dielectric. *Appl. Phys. Lett.* **2008**, *93*, 12–15. [CrossRef]

38.   Nath, P.; Chopra, K.L. Thermal conductivity of amorphous vs. crystalline Ge and GeTe films. *Jpn. J. Appl. Phys.* **1974**, *13*, 781–784. [CrossRef]

39.   Cahill, D.G. Thermal conductivity measurement from 30 to 750 K: The 3omega method. *Rev. Sci. Instrum.* **1990**, *61*, 802–808. [CrossRef]

40.   Gaillac, R.; Pullumbi, P.; Coudert, F.-X. ELATE: An open-source online application for analysis and visualization of elastic tensors. *J. Phys. Condens. Matter* **2016**, *28*, 275201. [CrossRef]

41.   Shcherbina, O.B.; Palatnikov, M.N.; Efremov, V.V. Mechanical properties of Nb_2O_5 and Ta_2O_5 prepared by different procedures. *Inorg. Mater.* **2012**, *48*, 433–438. [CrossRef]

42.   Savitzky, A.; Golay, M.J.E. Smoothing and Differentiation of Data by Simplified Least Squares Procedures. *Anal. Chem.* **1964**, *36*, 1627–1639. [CrossRef]

43.   Ezzahri, Y.; Pernot, G.; Joulain, K.; Shakouri, A. Capturing the Cumulative Effect in the Pump Probe Transient Thermoreflectance Technique using Network Identification by Deconvolution Method. *Mater. Res. Soc. Symp. Proc.* **2011**, *1347*, 26–33. [CrossRef]

44.   Dilhaire, S.; Rampnoux, J.-M.; Grauby, S.; Pernot, G.; Calbris, G. Nanoscale Thermal Transport Studied With Heterodyne Picosecond Thermoreflectance. In Proceedings of the ASME 2009 Second International Conference on Micro/Nanoscale Heat and Mass Transfer, Shanghai, China, 18–21 December 2009; pp. 451–456.

Energies **2019**, *12*, 610

45. Dilhaire, S.; Pernot, G.; Calbris, G.; Rampnoux, J.M.; Grauby, S. Heterodyne picosecond thermoreflectance applied to nanoscale thermal metrology. *J. Appl. Phys.* **2011**, *110*, 114314. [CrossRef]
46. Yagi, T.; Tamano, K.; Sato, Y.; Taketoshi, N.; Baba, T.; Shigesato, Y. Analysis on thermal properties of tin doped indium oxide films by picosecond thermoreflectance measurement. *J. Vac. Sci. Technol. A Vac. Surf. Film* **2005**, *23*, 1180–1186. [CrossRef]

![energies logo] *energies*

MDPI

Article

Luminaire Digital Design Flow with Multi-Domain Digital Twins of LEDs [†]

Genevieve Martin [1,*], Christophe Marty [2], Robin Bornoff [3], Andras Poppe [4,5], Grigory Onushkin [1], Marta Rencz [4,5] and Joan Yu [1]

[1]   Signify, Hightech Campus 45, 5056AE Eindhoven, The Netherlands; grigory.onushkin@signify.com (G.O.); joan.yu@signify.com (J.Y.)
[2]   Ingelux, 69120 Vaulx en Velin, France; c.marty@ingelux.com
[3]   Mentor, a Siemens Business, 81 Bridge Rd, Molesey, East Molesey KT8 9HH, UK; robin_bornoff@mentor.com
[4]   Department of Electron Devices, Budapest University of Technology and Economics; Magyar tudósok körútja 2, bldg. Q, 1117 Budapest, Hungary; poppe@eet.bme.hu or andras_poppe@mentor.com (A.P.); rencz@eet.bme.hu or marta_rencz@mentor.com (M.R.)
[5]   Mentor, a Siemens Business, Mechanical Analysis Division, Gábor Dénes utca 2, 1117 Budapest, Hungary
*   Correspondence: genevieve.martin@signify.com
[†]   This paper is an extended version of our paper published in the 2018 24rd International Workshop on Thermal Investigations of ICs and Systems (THERMINIC), Stockholm, Sweden, 26–28 September 2018.

Received: 9 May 2019; Accepted: 18 June 2019; Published: 21 June 2019

Abstract: At present, when designing a Light Emitting Diode (LED) luminaire, different strategies of development are followed depending on the size of the company. Since on LED datasheets there is only limited information provided, companies designing LED luminaires spend a lot of effort gathering the required input of LED details to be able to design reliable products. Small and medium size enterprises (SMEs) do not have the bandwidth to gather such input and solely rely on empirical approaches leading to approximated luminaire designs, while larger companies use advanced hardware and software tools to characterize parts, design versions, and finally optimize all design steps. In both cases, considerable time and money is spent on prototyping, sampling, and laboratory testing. Digitalization of the complete product development (also known as Industry 4.0 approach) at all integration levels of the solid state lighting (SSL) supply chain would provide the remedy for these pains. The Delphi4LED European project aimed at developing multi-domain compact models of LED (for a consistent, combined description of electronic, thermal, and optical properties of LEDs) as digital twins of the physical products to support virtual prototyping during the design of luminaires. This paper provides an overview of the Delphi4LED approach aimed at supporting new, completely digital workflows both for SMEs and larger companies (Majors) along with some comparison with the traditional luminaire design. Two demonstration experiments are described: One to show the achievable benefits of the approach and another one to demonstrate the ease of use and ability to be accommodated in a larger scale product design for assessing design choices like e.g., number and type of LEDs versus electrical/thermal conditions and constraints, in a tool agnostic manner.

Keywords: LED digital twin; design flow; multi-domain compact model; tool agnostic; multi-LED

1. Introduction

The past two decades have seen growing adoption of light emitting diodes (LEDs) thanks to their improved efficacy and decreasing cost. According to the Strategies in Light conference in 2019, the trend is to move towards LED components commoditization, and the lighting industry is entering another era with more demanding customers, market acceleration, and integration that extends beyond lighting (e.g., sensors). Industry 4.0, the fourth industrial revolution that exploits automation and data exchange, further facilitates acceleration and customization. To stay competitive, the lighting industry should embrace the digitalized early phase of development, which is currently in its infancy, to create space for new innovation beyond lighting (e.g. LiFi). A primary challenge to this approach is establishing ways for system integrators to have sufficiently detailed information on components without requiring LED suppliers to disclose proprietary information. The Delphi4LED H2020 ECSEL R&D project [1,2] of the EU proposes means to accelerate the transition to this new paradigm by creating the very first multi-domain LED "digital twin" from measurements and characterisations, and by suggesting new workflows using such digital twins [3]. One of the project's main objectives was to develop testing and modelling methodologies aimed at multi-domain characterisation of LED-based products at different levels of integration along the solid state lighting (SSL) supply chain, starting from modelling LED chips up to creating complete system level models of LED luminaires.

1.1. Digital Twins, Virtual Prototyping, and Digitalized Design Flow

Schluse provides a definition of digital twins [3]. In "Industry 4.0" context, a "digital twin" is a virtual representation of a real world subject (person, software system, ...) or a real world object (machine, component, part of the environment, ...). A digital twin contains models of its "data" (geometry, structure, ...), its functionality (data processing, behaviour, ...) and its communication interfaces. Figure 1 represents the digital twin components.

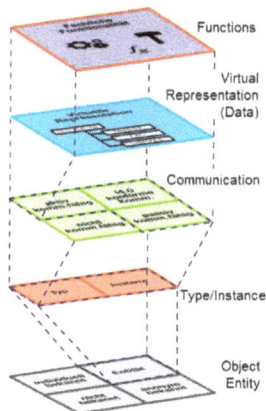

Figure 1. The "Industry 4.0 component" [4].

Research on all levels is required to allow for building full-blown digital twins of the product development cycle, the resulting products or the production environments.

When applied to LEDs and solid-state lighting product design, the LED digital twin is the multi-domain model of the packaged LED chips, including the physical structure of the package. In the context of the work presented in this paper virtual prototyping means testing of different versions of complete luminaire designs with the help of the luminaires' digital twins in different application scenarios by means of computer simulation. In a wider design flow context product optimization is also performed with the help of such digital twins. Development workflows relying completely on digital twins are called fully digitalized development workflows. Figure 2 represents the different dimensions involved in the creation of the LED digital twin (after [3]).

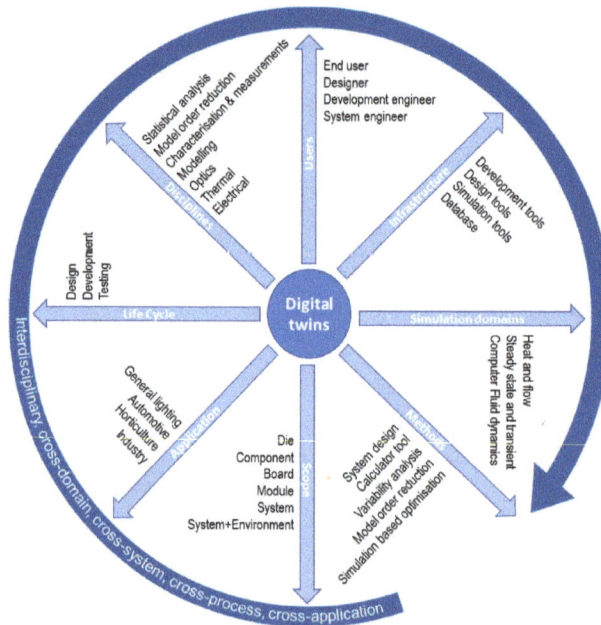

Figure 2. Different dimensions considered for the creation of the light emitting diode (LED) digital twin [3].

In other words, in the context of the lighting industry, major properties (such as the total emitted luminous flux) of a foreseen new luminaire can by identified under different combinations of application conditions through computer simulation and the effects of different design choices can be assessed, therefore design parameters can be optimized.

1.2. Objectives

The main purpose and novelty of the Delphi4LED project was the development of a comprehensive methodology to create a multi-domain LED digital twins (capturing LED behaviour regarding the electrical, thermal, and optical properties consistently in its entire operating domain) and provide a systematic, bottom-up hierarchical modelling approach that matches the computerized design environments, design, and simulation tools forming a complete design and simulation workflow that brings Industry 4.0 to the design of solid-state lighting products. To the best of the authors' knowledge, such a comprehensive approach is still missing. In view of this, the present paper proposes a generic methodology applicable to situations ranging from LED packages up to complete luminaires by creating a "digital twin" that properly represents its real, physical counterpart for simulation based experiments and optimization.

In this paper, we demonstrate the new "Industry 4.0"-like approach developed within the Delphi4LED project through the examples of LED-based luminaires. This paper extends earlier work [5], to validate with more LEDs, making it tool-agnostic and addressing the impact on the company digital footprint. In this study, the digital flow, its approach, and impacts are further presented, analysed, and discussed. The paper highlights the three key pillars for ensuring the sustainability of the approach in the lighting industry: Modelling of the LED packages to create their "digital twins" for the proposed design flow and the actual implementation of these models; the workflow together with involved personas at the different stages of workflow and the impact of the LED digital twins; and the proposed new product design workflows on the company digital footprint.

2. Materials and Methods

To protect their intellectual property, LED suppliers do not share sensitive and proprietary product details such as material properties or the architecture of their LED components. To develop LED-based products that are both reliable and cost-effective, detailed modelling is required early in the design process to predict performance so that appropriate design choices can be made as early as possible during the product development process. In the early product design phase, to properly represent the behaviour of an LED, a full LED model has to be built. Currently, this is done using reverse engineering to identify the LED build-up, dimensions, and material properties prior to modelling. This step adds cost, time, and uncertainty to the product design phase. The vision of the Delphi4LED consortium was to reduce time to market while increasing the accuracy of predictions by "digitizing" the early design phase: Replacing characterisation of physical product samples by their digital twins aimed for computer simulation. One key to a future with fully digitalized LED product development is the availability of compact, multi-domain (thermal-optical-electrical) models of LED components. The goal of the proposed approach is to create a multi-domain digital twin of LED packages directly from measurements [2]. (The term "LED package" is used by different professional organizations in the lighting world to refer to an LED chip fully encapsulated in a mechanical enclosure, the "package", complete with electrical connections and optical parts to form a fully functional LED device that can be used further on in product integration, as the basic component.)

Figure 3 illustrates the hierarchical, bottom-up modelling approach from LED chip level to LED module level. The multi-domain operation is captured on LED chip level. Such a chip level model combined with the compact thermal model of the mechanical structure of the package forms the package level LED model, also called LED digital twin. At the module level, we create the compact thermal model of the module substrate as a multi heat-source model, where the heat-sources are the pads of the LED packages through which the power dissipated by the LEDs of the populated module enter the substrate. The luminaire's digital twin (virtual prototype) includes the luminaire's thermal model, considering also the LED driver and luminaire optics and their losses. As it will be shown later in this paper, the level of detail in the luminaire digital twin depends on the actual application and on the available CAD tools with which the proposed workflow is implemented.

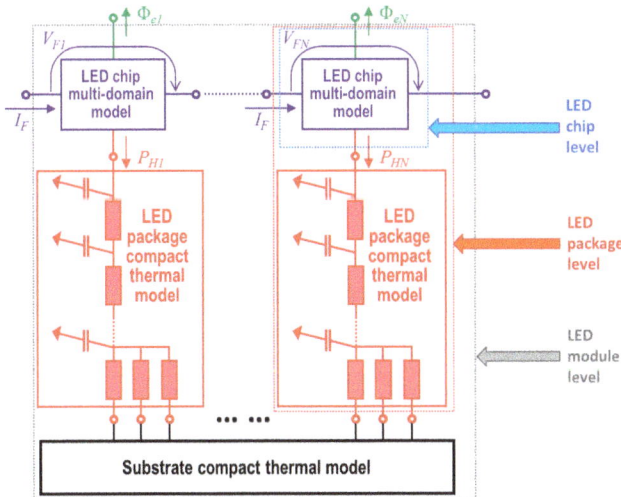

Figure 3. Bottom-up, hierarchical approach of modelling from LED chip to LED module level, with multi-domain operation captured on the chip.

Figure 4 illustrates through what steps the physical LED devices are linked to the module or luminaire level of product design and simulation tools in a fully Industry 4.0-like solid-state lighting eco-system. As illustrated here, the interface between the physical devices and their digital twins is represented by future electronic LED datasheets. The first step towards achieving widespread use of such datasheets is the LED suppliers' and LED users' consensus on LED test data reporting, both in terms of the type and number of tests data included and in terms of machine readable file formats such as XML. Work regarding recommendations on LED test data reporting is in progress at the International Commission on Illumination (CIE) [6].

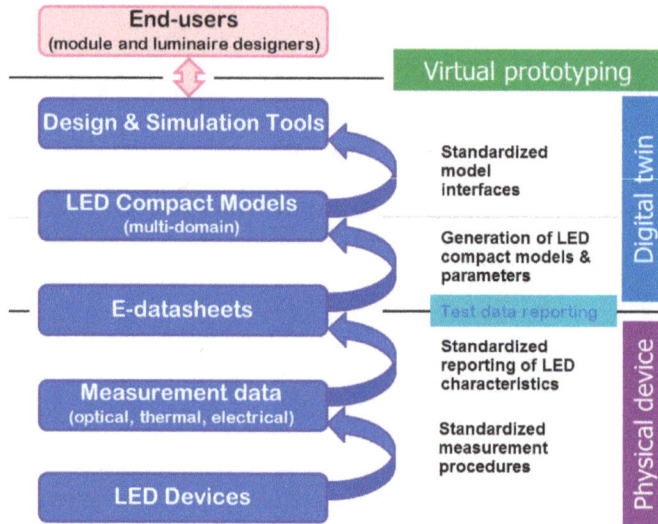

Figure 4. The Delphi4LED approach to digitalization of LED application design by creating digital twins of physical samples of packaged LEDs. The interface between the physical LEDs and their digital twins is represented by the electronic LED datasheets. The interface between LED vendors and end-users is represented by multi-domain LED compact models: the LED digital twins.

Figure 5 provides the major steps of creating the multi-domain compact models of LED packages. The LED package (multi-domain) compact model is comprised of; (1) a boundary condition-independent compact thermal model of the physical LED package and (2) a multi-domain, so called Spice-like, model that describes the LED operation at the chip level [7]. The former can be created with the help of the measured real thermal impedance, $Z_{th}(t)$ of the LED package in question while the parameters of a chip level LED model can be identified from the so called isothermal current-voltage-flux characteristics of the LED device (see later).

Temperature is the most important operating parameter that defines the operation of LEDs. Thermal modelling of LED packages, modules, and luminaires is a cornerstone at all integration levels from LED chips to LED luminaires. We treat the multi-domain operation basically at the LED chip level/LED package level, as shown in Figure 3 and in Figure 5.

At the chip level, a physics-based model is created by using characteristics derived from measurements while at the package level a dynamic thermal compact model is created in two steps. First a detailed thermal model of the LED package structure is set up and "calibrated" with the help of structure functions derived from the measured thermal impedance, $Z_{th}(t)$ of the package and then, in a subsequent step, thermal capacitance and thermal resistance values of a thermal RC network model of the package are optimized to best represent the thermal properties of the LED package under any boundary conditions at the package footprints and the LED's dome (lens) [8].

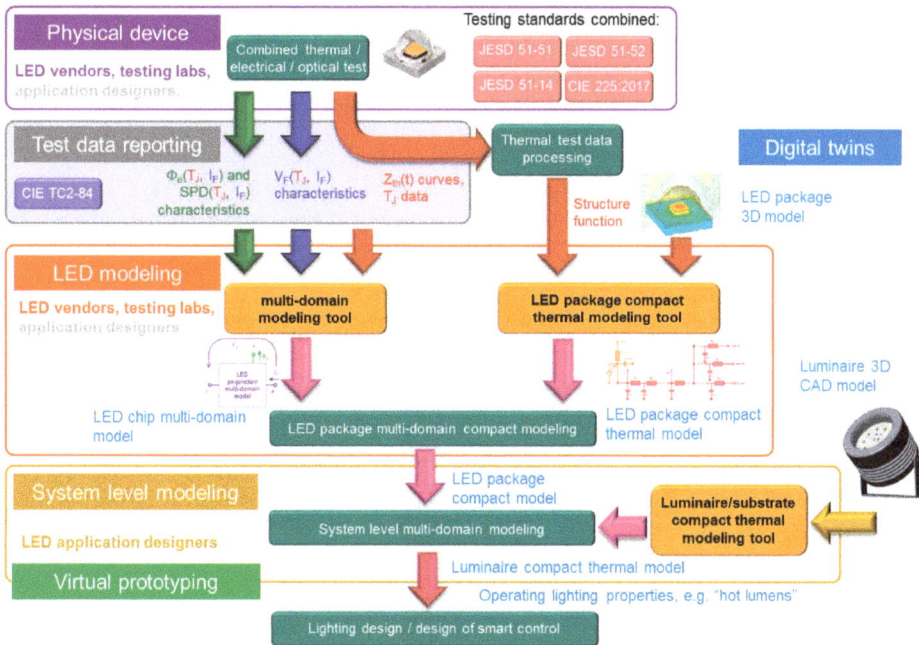

Figure 5. Delphi4LED workflow for generation and application of digital twins of LED packages, modules, and luminaires.

From LED to luminaire (Figure 5), the Delphi4LED approach consists of the following key steps:

1. Standardized measurements and characterisation of LEDs combining the testing standards JEDEC JESD51-14, JESD 51-51, JESD 51-52, and CIE 225-2017 [9–12]; for a summary on LED testing see [13].

2. Characteristics are reported in a standard way through an electronic datasheet (e-datasheet); work towards recommendations on such LED test data reporting is in progress at the CIE TC2-84 technical committee [6].

3. A LED digital twin, also called multi-domain compact model [7,14–17], is created. This includes a dynamic compact thermal model of the mechanical structure of the LED package [8].

4. The LED digital twin is implemented in a calculation tool or simulation tool together with the light engine and luminaire twins [5] through a standardized interface.

Step one through three are executed typically by LED vendors or testing labs, as indicated also in Figure 5. Step four is performed together with the luminaire design by luminaire makers. As will be discussed later, we categorize these luminaire manufacturers into two groups: Small and medium size companies (SMEs) and large international companies (Majors).

In the following subsections we detail these four major steps in greater detail, from a product design perspective, supported also by a case study.

2.1. Measurements and Characterisation of LEDs, Test Data Reporting

To describe the multi-domain feature of LED operation, parameters including light output, thermal, and electrical characteristics are measured in a consistent way.

New testing protocols and modelling methodologies were developed to allow easy creation of compact models based on measurement data [16–19].

Measurements take into account the latest LED testing methods, such as JEDEC's thermal testing standards for power semiconductors [9] and power LEDs [10,11] as well as CIE's LED optical testing recommendations for high power LEDs [12]. Applying these recommendations, so called isothermal current-voltage-flux characteristics (IVL characteristics) are measured in combination with the real thermal impedance, $Z_{th}(t)$ of the package. For thermal modelling purposes, the equivalent representation, so called structure functions, of the impedance curves are used [13].

The basis of the Delphi4LED approach of LED application design is that LED vendors provide the necessary data on packaged LED chips (LED packages) in a standard way; e.g., an electronic format that can be further processed by software tools for extracting the right LED digital twins from these. Currently, the TC2-84 technical committee of Division 2 of CIE deals with establishing recommendations on test data reporting of LED packages [6].

2.2. Creation of the LED Digital Twin

Poppe et al. described die level modelling and integration with package modelling, up to a system level compact model of a complete luminaire [20,21]. In the subsequent subsection, following the bottom-up hierarchy shown in Figure 3, we present, how this earlier concept was implemented in the Delphi4LED design flow.

As mentioned earlier, the multi-domain model of an LED package combines the digital twin at chip level and the dynamic thermal compact model of the LED package.

2.2.1. Digital Twin of LED Chip

In the Delphi4LED project two methods for chip level multi-domain modelling and simulation were developed: A quasi black-box model of LEDs was proposed through appropriate extension of the built-in diode model of generic Spice solvers, on the one hand; while on the other hand, a new Spice-like LED model has also been proposed that is more closely based on the physics of today's heterojunction based multiple quantum well structures of high power LEDs [7]. Both models have been implemented as LTspice circuit macros and the model equations were also implemented in form of Visual Basic macros aimed at the implementation in an Excel spreadsheet based luminaire design calculation too [7]. Both implementations share the same set of model parameters that can be extracted straight from the measured LED characteristics as indicated in Figure 5. As a result, basic electrical, thermal, and optical behaviour of LEDs is described for the full working space in the LED application, which enables efficient simulation of LED applications on higher product integration levels.

These chip level multi-domain models have evolved from the previously published ones [14–16]. As hinted by both Figures 4 and 5, in the future, the required measurements and model parameter identification will be expected from LED vendors and the resulting LED characteristics/model parameter sets reported in a standardized format so that they can ultimately be used in a luminaire design tool or directly integrated in simulation tools in a way outlined in [5,7] and [8].

2.2.2. Digital Twin of the LED Package

In the Delphi4LED project, dynamic compact thermal models (DCTMs) of the LED packages are created (see Figure 6b).

The process of creating thermal digital twins of the LED packages starts with creating calibrated detailed 3D models of the LED packages (such as presented in Figure 6a) with the help of the structure functions [22]. These are dependent on, but do not explicitly require proprietary information of the LED: The detailed geometry and material properties.

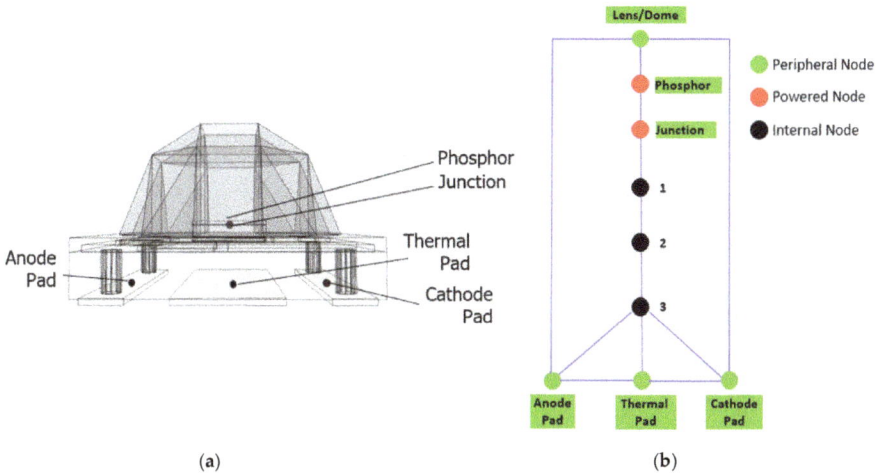

(a)   (b)

Figure 6. (a) Outlines of major shapes in a detailed 3D LED package model; (b) topology of the corresponding LED package dynamic compact thermal model (node J represents the LED junction; every internal node also represents a thermal capacitance which are omitted here) [8].

According to the original DELPHI methodology standardized in the electronics cooling simulation industry [23] and also in our present approach, compact thermal models are identified through an optimization process [8] using calibrated detailed thermal models [22].

These are later used in luminaire level analysis (following the hierarchical approach proposed in [20] and [21]) as the actual digital twins of the LED packages representing their true thermal behavior.

A package DCTM connected to a chip level multi-domain LED model is the ultimate digital twin of a packaged LED, provided e.g. as a Spice netlist (see Figure 3). For the detailed description of the methodology of creating LED package dynamic compact thermal models refer to the paper of R. Bornoff [6] published in this special Journal issue.

2.2.3. Secondary Heat Source: Phosphor

As hinted by Figure 6a, in case of phosphor converted white LEDs, the phosphor plays a significant role as secondary heat source which cannot be neglected [24]. Especially when the phosphor layer is thick and/or has low thermal conductivity, the energy loss and the corresponding temperature rise in the phosphor cannot be lumped with the dissipation and the temperature of the LED's pn-junction. Besides that, further optical losses also need to be accounted for and modelled. On the following we provide a brief summary of the results achieved within the Delphi4LED consortium.

The light trapped in the dome and light conversion losses in the phosphor create secondary heat sources P_{hC} and P_{hD} inside an LED package (Figure 7). The package level light extraction efficiency may be as low as 65% for LEDs with flat lenses. Thus, up to 35% of the light emitted by the pn-junction may be trapped in the package that is converted into heat. These additional, secondary heat sources have a significant impact on the structure functions [24–26] identified from the temperature transient measured at the LED's junction, thus, affect the accuracy of the calibration of the detailed 3D model of the LED package. Therefore, to properly calibrate the thermal parameters of the main heat-flow path, these secondary heat sources have to be accounted for.

The part of the junction temperature transient response associated with the smallest thermal time-constants is known to be dependent on the geometrical features and thermal parameters of the die. As a consequence, any heat generation in the phosphor has negligible effect on the part of the structure function prior to the die attach [24–26]. To account for the thermal effect of the phosphor, the following method is used to calibrate the 3D finite element model and separate the main and the secondary heat sources [24]:

1. Define the precise chip substrate geometry with x-ray measurements.
2. Assign arbitrary thermal properties of the chip substrate and of the dome materials (within an expected final range).
3. Calibrate the P_{hJ} power source by matching the initial junction temperature transient response of the model to the measured one.
4. Define the secondary heat source power as the difference between, P_{hJ} and the measured heat total dissipation, P_h.
5. Perform a calibration of the thermal properties of the rest of the model.
6. Both the heat generated in the phosphor and the lens must be calibrated by fitting the long time-constants part (0.01 s $< t <$ 10 s) of the time-constant spectrum of the transient response, or, in other words, by fitting to the "tails" of the structure function.

This methodology prevents an unphysical calibration of the thermal parameters that might be due to the significant presence of the secondary heat sources within an LED package.

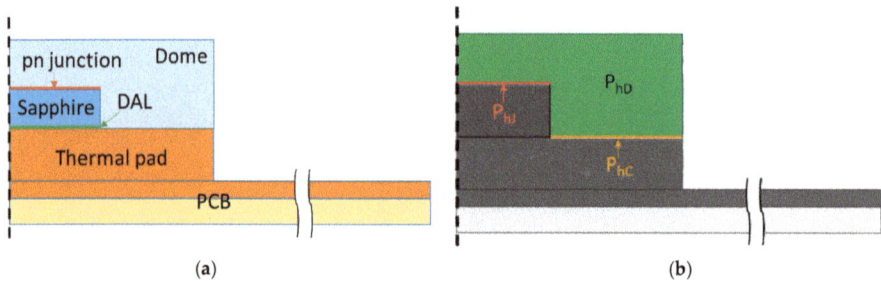

(a)   (b)

Figure 7. (a) Cross section of an LED package (DAL stays for die attach layer); (b) The different heat generating regions (heat sources) in the model.

2.2.4. Variability Assessment of the LED Digital Twin

In an industrial environment, LEDs do not have perfect characteristics; particularly at package level which manufacturing process is less well controlled than the semiconductor part chip level. Their variability might have a significant impact on the product performance. Variability among the different LED samples was also investigated [27–31]. In our actual study 11 royal blue LED samples and 11 samples of their phosphor converted white counterparts (LED having the blue peak at the same wavelength as the peak wavelength of the royal blue) have been measured and modelled. One sample with a die attach failure, as an obvious outlier was later omitted from the statistical analysis of the thermal test results.

Methods to assess the variability of the thermal properties of LED packages have been proposed. One approach uses a rough, step-wise approximation of the structure function (e.g., by means of a Cauer-type RC ladder with about a dozen stages) [28], see also Figure 8a. Based on this, a Monte Carlo simulation technique was developed to study the possible changes of the $Z_{th}(t)$ thermal impedance curves resulting from the local thermal resistance variations along the heat-flow path. The results are histograms about the distributions of the values of the local partial thermal resistances of the different heat-flow path sections (Figure 8b) and the transient temperature responses at the different locations of the heat-flow path, represented by the different nodes of the ladder model (Figure 8c).

(a)

(b)

(c)

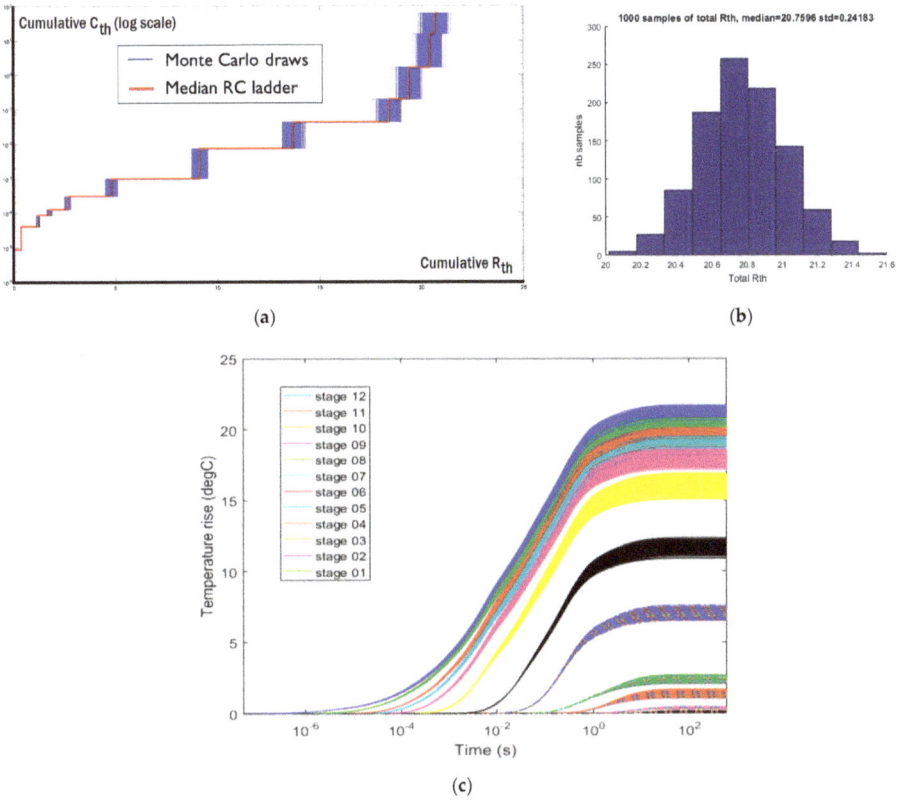

Figure 8. (**a**) Step-wise approximation of the structure function of the "median" sample of a population of 11 royal blue LEDs with Monte Carlo modelled scatter of the individual partial thermal resistances of the heat-flow path; (**b**) The histogram of the simulated value distribution of one of the stages RC ladder model of the structure function; (**c**) Simulated transient responses at the different nodes of the 12-stage RC ladder model of the heat-flow path, for all Monte-Carlo draws of local thermal resistance values of the heat-flow path sections [28].

Another method derives a new representation of the heat-flow path from the structure function which presents the changes of the local thermal resistance [29,30].

Variability of the isothermal IVL characteristics is also studied, see Figure 9. The statistical analysis of the measured isothermal IVL characteristics of the same 11 royal blue and 10 phosphor converted white LEDs that have been studied from thermal point of view is in progress, the first results are provided elsewhere [31].

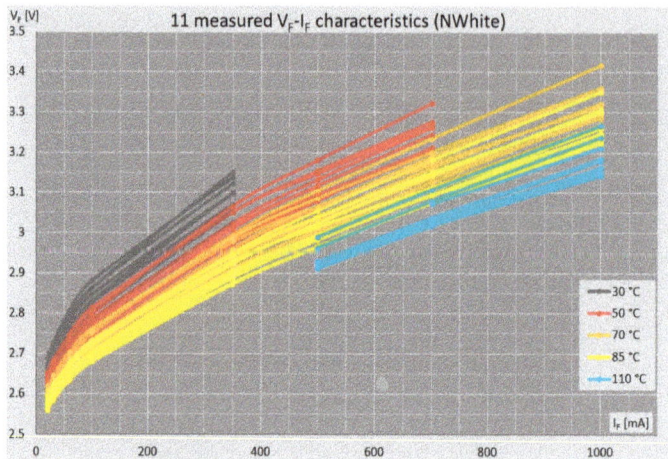

Figure 9. Isothermal forward voltage – forward current characteristics of 11 royal blue LED samples, measured at different junction temperatures between 30 °C and 110 °C.

2.3. Using the LED Digital Twin for Real Design Tasks

The advantage of the compact models is that they no longer carry proprietary information of the LED vendors, thus, LED vendors can share them with end-users without sharing their sensitive IP. A package DCTM connected to a chip level multi-domain LED model is the ultimate digital twin of a packaged LED, provided e.g., as a Spice netlist or implemented in a simple Excel spreadsheet based "luminaire design calculation tool" [5]. In our prior publication [5] we presented a simple application use-case, the design of a "10 W outdoor LED luminaire". In the following we briefly describe a design task where the LED digital twin embedded in a "calculation tool" was used to perform a more complex optimization task of a new light engine aimed as an LED based replacement of "70 W high-pressure sodium (HpS) lamps".

2.4. Modelling and Simulation of LED Modules

On the way towards multi-domain modelling and simulation of a complete LED based application, like a "70 W HpS lamp replacement" LED based solution, the next step level in the bottom-up hierarchy is an LED module realised on a substrate such as a printed circuit board (PCB). The quoted example, the "70 W HpS lamp replacement" LED light engine is quite a complex one, containing the LED driver electronics, the cap fitting into an Edison-type socket on one end, and proper cooling assembly on the other end. The middle part of this light engine is a complex LED assembly that mimics the light emission of the discharge tube of the sodium lamp, both in terms of the emitted total luminous flux and in terms of the spatial distribution of the emitted light. This LED assembly consists of several, identical PCB modules, carrying multiple LED packages. Thus, the corner stone of the design of this new light engine is the modelling such a PCB substrate populated with LED packages where the heat transfer from the packages towards the ambient takes place through the package footprints as thermal interfaces towards the PCB. The heat is dissipated by the LEDs then is transferred through the bottom side of the PCB substrates towards the light engine's cooling assembly from which the heat is finally transferred to the environment by convection. Now we restrict our discussion to the modelling and simulation of such a PCB module.

We set up a model in which we could easily vary the LED type, the number of LEDs used, and the properties of the substrate (e.g. different types of PCBs) during the product design optimisation phase. The properties of interest in such a system are the solder temperatures, the junction temperatures, and the "hot lumens" of each LED. "Hot lumens" is a sloppy expression widely used in the English technical language of solid-state lighting, to refer to the emitted total luminous flux of an LED or LED based light source at the operating junction temperature of the device under application conditions.

A fully parametrized model with arbitrary dimensions was created including a 3-layer (or less) PCB substrate with arbitrary dimensions and thermal conductivity, dummy LED footprints placed on top to provide the interface area for the compact models of the LED packages, through which LEDs' heat dissipation is conducted towards the ambient. The number of dummy LEDs can vary, potentially filling both x- and y-directions. The 3D model of the physical structure of the PCB substrate is shown in Figure 10a. Figure 10b illustrates the compact thermal model of such a substrate that has been used in the "design calculator tool" implemented as an Excel based application already used with success for a simpler design [5].

In our present example, the PCB substrate is considered as a single thermal component (described by a multi heat source compact model) and the luminaire body as a heat-sink is considered as another component (modelled by a single, equivalent thermal resistance to the ambient), such as shown Figure 10b. This way, libraries of substrates made of different materials and with different designs (copper coverage, footprint layout, etc) and luminaires with different thermal features can be created for subsequent system level design and optimization.

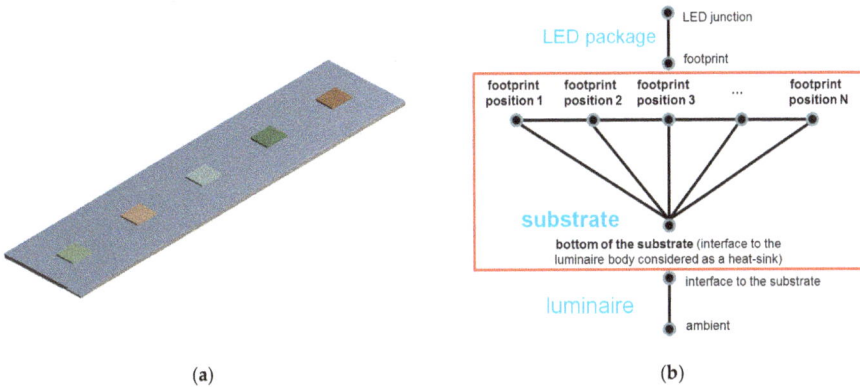

(a)   (b)

Figure 10. (a) an LED module with 5 LEDs; (b) the multi heat-source compact thermal model of the substrate (in red frame). The luminaire is represented in the overall system model as a single substrate-to-ambient thermal resistance, LED package thermal models are connected to the substrate model at the nodes corresponding to the LED footprint areas (coloured rectangles in Figure 6b).

For each LED module (a PCB populated with the chosen type and number of LED packages, considered as an elementary light source in the final light engine), following the method presented in [20] and [21] first the so called thermal characterisation matrix is created which is transformed into a Spice compatible thermal resistance network, as follows:

1. Run individual finite element model (FEM)-calculations for each LED-position assuming a unit heat dissipation at the footprint position and fixed temperature at PCB bottom interface towards the ambient (see Figure 10)

2. Generate a matrix of response temperatures at each footprint position for all cases. The resulting matrix is the so called $\mathbf{R}^*_{th}$ steady-state thermal characterisation matrix, that is specific to the given layout arrangement of the LED package footprints and represents the strength of the thermal coupling among the footprints and between each footprint and the ambient.

3. Using the algorithm described in [19] convert this $\mathbf{R}^*_{th}$ thermal characterisation matrix to a matrix of real thermal resistances that can also be exported as a real, Spice netlist of resistances. Complete this thermal resistance model with the junction to footprint thermal resistance of the chosen LED package type.

4. Convert the thermal resistance matrix corresponding to the above thermal resistance network to a solvable heat transfer coefficient (HTC) matrix. This matrix will represent the LED chips' 3D thermal environment.

5. This model is co-simulated with the chip level multi-domain LED model in a relaxation type scheme as follows:

    a. Matrix multiply the inverted HTC matrix with a power vector (containing all powers for each LED package footprint) to get the vector of temperature rises as a result.

    b. Execute the chip level multi-domain model for each LED instance, with the forward current and assumed ambient temperature. The model will calculate the LEDs' light output and heat dissipation.

    c. Take the new dissipation values as updates to the previous vector of dissipations and iterate until convergence is reached (i.e., the difference between the subsequent temperature/dissipation vectors shrinks below a given threshold).

6. When the convergence is reached, take the LEDs' calculated total radiant/luminous flux and junction temperature values and the footprint (solder joint) temperatures as final output.

2.5. The Overall Workflow

As seen in Figure 5, the entire workflow is divided into three major steps:

I.   Characterisation/measurement (using a JEDEC JESD51-51/51-52 and CIE 127:2007/225:2017 compliant, combined thermal, and radiometric/photometric LED test system) in order to gain the right set of isothermal current-voltage-flux characteristics of LED packages. This step is ideally performed by the LED vendors and the obtained LED characteristics are published in an electronic LED datasheet, in a machine processable format (see also Figure 4).

II. Extract parameters for the chip level multi-domain LED model as well as identify the thermal boundary condition independent dynamic compact thermal model of the LED package. The chip and package models combined represent the multi-domain behaviour of the LED package by means of a compact model that should part of the model library of an LED application level simulation system. Ideally these model parameter identification steps are also performed by LED vendors but in cases when such models are not yet available, LED testing laboratories should also be able to provide such models.

III. Create the compact models of the LED modules by combining LED package models taken from the above libraries (ideally provided by LED vendors or LED testing laboratories) with the design specific compact thermal models of the module substrates or luminaires (or both). The inputs for such substrate/luminaire models are the electronics and mechanical CAD models (PCB electrical layout models/3D solid models) of the modules and luminaires. A substrate/luminaire compact thermal modelling tool can convert these models into the right compact thermal models. For smaller designs the thermal characterisation matrix based method outlined above is feasible and is relatively easy to implement. For larger designs however, using model order reduction techniques might be considered [30–33].

The Delphi4LED approach provides a good handshake between the semiconductor industry (represented by LED vendors) and the lighting industry (represented by the luminaire designers/manufacturers). Each have a role to play in achieving more reliable and faster product designs: The LED vendors by providing the LED characteristics in the form of electronic datasheets and eventually the LED digital twins, the luminaire designers/manufacturers being able to use these data in their own, Industry 4.0 like design workflows and if the LED compact models are not provided, being able to extract them from the e-datasheets provided by the LED vendors.

In the third phase pre-characterisation of different module substrates and different heat-sink designs are performed by the luminaire manufacturers. This way their compact thermal models (thermal N-port models for the substrates, equivalent substrate-to-ambient thermal resistances for heat-sinks, see also Figure 10b) are also created and included in model libraries.

In the proof of concept implementation of the above workflow an Excel a spreadsheet application was created [5] into which both chip level multi-domain LED model parameter sets, LED package thermal resistances (as simplified LED package compact thermal models) as LED package model libraries and substrate/heatsink thermal models also as library items were included.

The user interface of this application allows setting major design goals (total luminous flux to achieve in this demo, see the blue field in Figure 11) and design constraints (foreseen ambient temperature, maximum allowed junction, and solder point temperatures; maximum allowed total forward voltage – the orange fields in Figure 11). The user (luminaire designer) can select the number of LEDs, substrate, and heat-sink types from pull-down menus and can set the efficiency of the luminaire optics and the total forward current provided the LED driver as input (light-green fields in Figure 11). (For the sake of simplicity it was assumed that all the LEDs of the module investigated were electrically connected in series, forming a string of LEDs.)

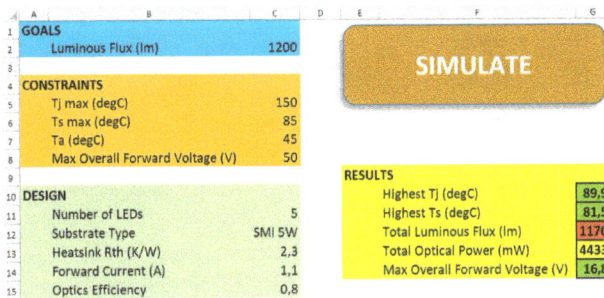

Figure 11. An Excel spreadsheet based application (a simple luminaire design tool) to support assessing basic luminaire design alternatives through virtual prototyping using Delphi4LED style compact models [3]. A "green" result means that the given target is met/constraint is not violated; "red" indicates failing a design target or violating a design constraint.

The complete digital twin of an LED luminaire is obtained by combining the luminaire (compact) thermal model with the digital twin of the LED package (chip level multi-domain model completed with the package DCTM). With such a digital twin of the complete luminaire as a virtual prototype, luminous flux calculations of different design versions can be performed under different environmental conditions that are close to the foreseen application conditions.

2.6. Implementation and Personas in the Overal Design Flow

Both within the Delphi4LED consortium and in real life, we identified two different company profiles where daily design practice significantly differs due to company cultures and available technical/financial resources. We distinguish between small and medium size companies (referred to as SMEs) and big, international companies (referred to as Majors).

2.6.1. From LED Physical Device to Multi-Domain Compact Model

LED vendors or certified expert labs play a key role in the characterisation and measurements of LEDs aimed as source of input for LED multi-domain modelling. The general vision is that these test date will be reported through electronic LED datasheets. Such e-datasheets, as vendor neutral ways of data exchange can be processed by appropriate software tools by these entities in order to extract the LED MDCM and store the models in a standard, LED vendor and software tool independent data exchange format (e.g., xml). Such LED models will be stored in a library of LED digital twins. Figure 12 shows the design process associating a set of libraries to a Luminaire Design Tool (LDT).

If a certain LED multi-domain digital twin cannot be obtained from an LED vendor or from an independent, certified/accredited LED testing laboratory, the luminaire designers/manufacturers will have to extract the necessary models themselves, taking the LED e-datasheet as input. This optional step is common to both SME and Majors.

2.6.2. From Luminaire Requirements to Final Application

In the overall Delphi4LED workflow of Figure 5 LED application designers (luminaire designers at luminaire manufacturers) are responsible for creating their own sets of libraries of heatsinks, optics, substrates and the matching (footprint) of the LED models (Figure 12).

The product designer will gather the luminaire requirements from the lighting designer matching the end application conditions (e.g. dimensions, emitted total luminous flux, light quality metrics such as correlated colour temperature, and/or colour rendering index, other optical requirements, such as light distribution pattern) and translates those requirements into critical to quality (CTQ) parameters.

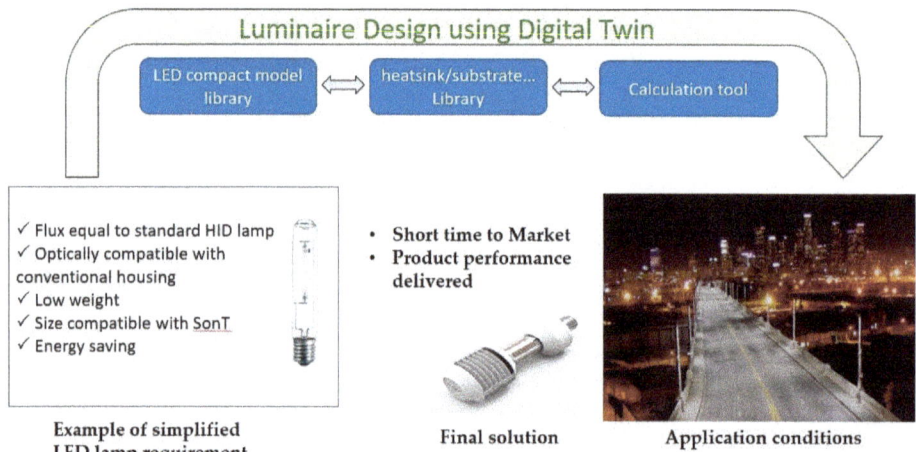

Figure 12. Computerized workflow: From luminaire requirement to final application with sets of model libraries.

Those parameters are split into three categories: Goals (equivalent to target performance), application constraints, and design choices (see Figure 13a). Goals set the target performance; such as luminous flux. Goals to achieve must fit the application constraints; which in turn are indicators for e.g., lifetime performance (maximum allowed temperature of the solder joints, junctions), limitations of the driver (maximum overall forward voltage), and ambient temperature. Design choices represent the degrees of freedom of the design. Parameters are typically LED type options (e.g., XPG3 and NF2L757DRT-V1) and corresponding forward current, substrates choices (e.g., FR4, MCPCB, and Cem3), optics efficiency, and cooling solution.

The luminaire calculation tools build up the system level model of a complete luminaire by mapping the design choices entered to items taken from the design libraries (Figure 12). The package and chip level model of the chosen LED is also taken from a library (ideally provided by LED manufacturers) and are combined with the other items, to create the internal, system level representation that is suitable for simulations (Figure 13b). This combined system model is executed when the "Simulate" button is pressed. If the user (LED application designer) changes a design choice, the luminaire calculation tools automatically updates the luminaire's system level model with the corresponding compact model taken from the model library.

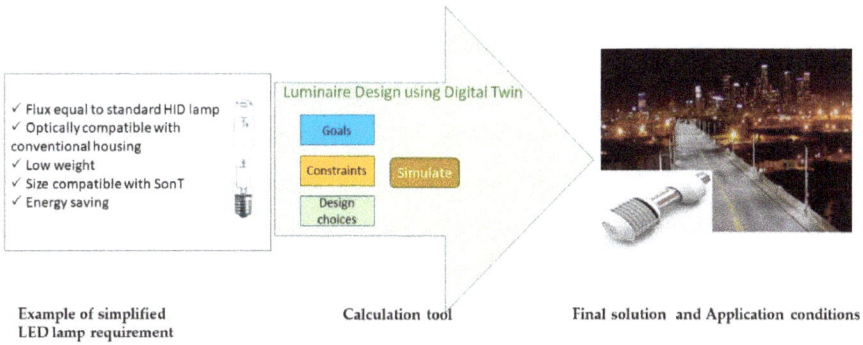

(a)

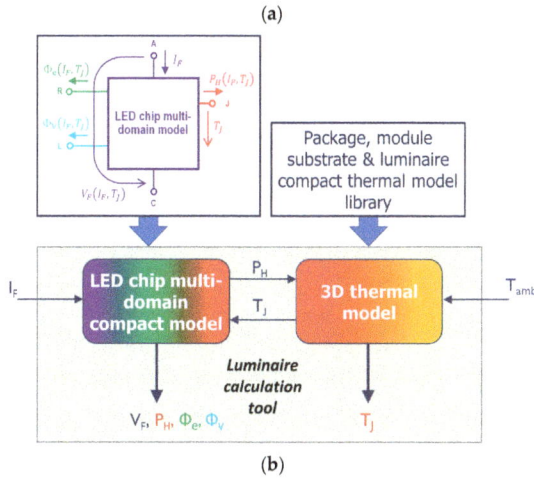

(b)

Figure 13. Computerized workflow: (**a**) from luminaire requirement to final application, (**b**) representation of the system in the luminaire calculation tool.

2.7. Case Studies

Two test cases are considered for assessing ease of use, performance of the "digital twins", impact of a large number of LEDs and to demonstrate that the approach of the Delphi4LED project is tool agnostic. Figure 14 shows the demonstrators. Demonstrator a) is used to compare SME versus Majors approach, while demonstrator b) is used to assess applicability to a large number of LEDs and its tool agnosticism.

(a)   (b)

Figure 14. Demonstration experiment proposed (not on scale): (**a**) "10 W outdoor LED spot" luminaire typology; (**b**) "70 W high-pressure sodium (HpS) replacement" LED lamp.

Experiment (A)

Based on this classification of foreseen users of the tools and methods developed within the Delphi4LED project, we set up two different demonstration experiments. One of the consortium members played the role of an SME and used the SME-type implementation of the overall workflow shown in Figure 5 (using the limited set of means usually available in small companies); another consortium member acted as a Major (i.e., large company) and used testing, modelling, and simulation tools typically available at a Major company. In this paper we present the SME style implementation of the proposed Delphi4LED workflow.

For the purpose of the study, two independent members of the consortium (Company A and B) played the role of developers and an independent industrial photometric test laboratory, respectively. Table 1 presents the four design styles assessed for SMEs and Majors. In this demonstration experiment, Company A executed the design and physical prototyping phase; and Company B executed the characterisation and final tests phase.

Table 1. Summary of design and development styles (workflows) followed during the project demonstration experiment of Delphi4LED.

Type of Design Flow	Concept/Development	Design and Prototyping	Characterisation and Final Tests
Majors	"Old process": tests/characterisation based optimization	Company A	Company B
Majors	"New process": compact model based optimization	Company A	Company B
SMEs	"Old process": prototyping and testing	Company A	Company B
SMEs	"New process": compact model based virtual prototyping	Company A	Company B

The assessment was done using Version 12.2 of the commercially available computer fluid dynamic (CFD) simulation tool Simcenter Flotherm software from *Mentor, a Siemens business* and a spreadsheet based luminaire calculation tool (realised in Visual Basic) and embedded in an Excel spreadsheet. The CFD tool was used to create the compact model library items corresponding to different substrate layouts and equivalent luminaire heatsink thermal resistances.

The design processes were monitored for each development style. The following performance indicators of the process were tracked: (1) Costs of development (time spent, incurred personal costs) for the realisation of virtual and physical prototypes samples, as well as (2) costs related to laboratory measurements and characterisation.

Experiment (B)

This experiment was performed to demonstrate how the Delphi4LED workflow is applicable to a real product development in a Major company. The chosen example was a high-intensity-discharge (HID) replacement lamp for street lighting application. The Delphi4LED virtual prototyping was used optimise the product by assessing different numbers and types of LEDs with which the total luminous flux target could be met. The purpose of this experiment was to demonstrate the ease of use when scaling the number of LEDs/applied forward current to meet the design specifications and to demonstrate the software tool independence of the proposed design flow.

In this experiment the models were created using the commercial finite element simulation tool ANSYS Version 17.2 (while in Experiment A the Simcenter Flotherm tool from Mentor was used).

The Ansys Parametric Design Language (ADPL) feature of ANSYS allowed the integration of both the LED package and the multi-domain chip level compact model to be solved in LTspice, in a similar relaxation type electro-thermal co-simulation scheme as depicted in Figure 13b. (Refer to [7] for details on the chip level multi-domain LED models aimed at different types of the implementation schemes of the electro-thermal co-simulation). This way the light output of the given LED configuration was obtained.

The middle part of the light engine of the HpS lamp replacement solution is composed of six symmetrically assembled LED modules. Due to the symmetry, it is enough to study only one single LED module. Figure 15 shows one of these six identical segments of the light engine of the lamp for two scenarios denoted by (x) and (y). The question was: Which combination of number of LEDs and forward currents for two different LED types meet the design goals and design constraints the best.

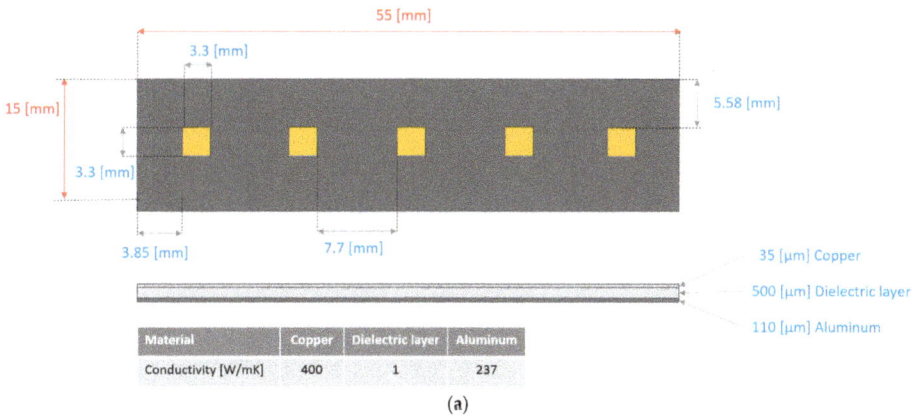

Material	Copper	Dielectric layer	Aluminum
Conductivity [W/mK]	400	1	237

(a)

Figure 15. *Cont.*

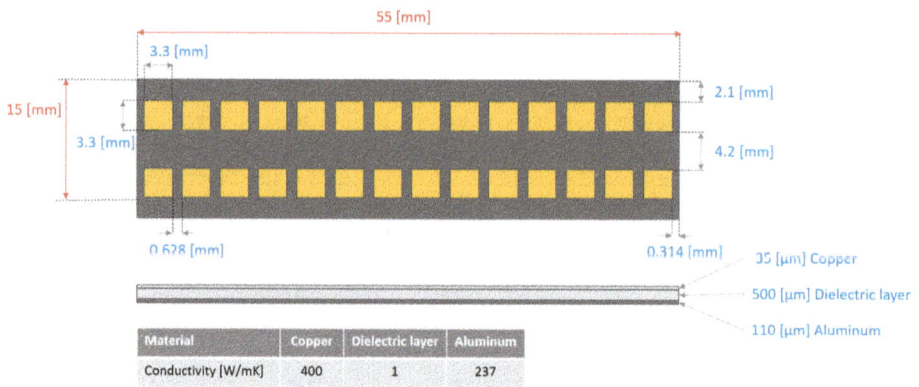

Material	Copper	Dielectric layer	Aluminum
Conductivity [W/mK]	400	1	237

(b)

Figure 15. One segment (not on scale) for Experiment b: (**a**) scenario (x) with details of the printed circuit board (PCB) for 5 LEDs powered by high forward current; (**b**) scenario (y) with details of the PCB for 22 LEDs powered by low forward current.

The designs were assessed with the following thermal boundary conditions:

1. Bottom temperature of the board at 85 °C.
2. Bottom temperature of the board at 45 °C.
3. Junction temperature of the board at 120 °C.

In scenario (y) with 22 LEDs is thermal coupling between adjacent LEDs was expected to be higher. For both scenario (x) and (y), the total luminous flux of the module and maximal power conversion efficiency were the design goals.

3. Results of Case Studies

This section presents the results of each key steps and related implementation in the overall Delphi4LED workflow of Figure 5.

3.1. Results of the Measurements and Modelling of LEDs

In this initial step, which is identical for both SMEs and Majors, the designer gathers input from LED suppliers and makes a pre-selection of sub-set of LEDs. In both demonstration experiments XPG3 LEDs from Cree were selected and used; in Figure 16 we show a few results of the measurements and modelling, performed by consortium members specialized in LED testing and modelling, mimicking the role of an LED supplier. (Such data are preferably input from LED suppliers or certified independent characterisation laboratories.)

	XPG3_01	XPG3_02	XPG3_03	XPG3_04	XPG3_05
Sample:					
Max Vf error:	0%	1%	0%	0%	0%
Max Fi_e error:	1%	1%	1%	1%	1%
Max Fi_v error:	1%	1%	1%	1%	1%
UT =	0,0296	0,0296	0,0296	0,0296	0,0296
I0 =	7,6395E-24	6,9812E-24	8,1736E-24	7,2375E-24	7,6335E-24
m =	1,7354	1,7349	1,7359	1,7353	1,7358
R =	0,1929	0,2141	0,197	0,2138	0,1973
I0_rad =	4,0317E-23	3,4889E-23	3,4027E-23	3,3826E-23	3,1585E-23
m_rad =	1,8150	1,8131	1,8075	1,8107	1,8089
R_rad =	0,0190	0,021001	0,020001	0,021001	0,019001
a_el =	-8,079E-06	-2,501E-06	-6,348E-06	1,973E-07	-3,155E-06
b_el =	2,153E-05	1,209E-05	1,762E-05	7,854E-06	1,475E-05
c_el =	-1,050E-06	2,362E-06	-5,003E-08	2,998E-06	-4,546E-07
d_el =	1,326E-03	5,207E-04	1,085E-03	1,150E-04	4,845E-04
e_el =	-4,104E-03	-2,919E-03	-3,586E-03	-2,243E-03	-2,982E-03
f_el =	-8,353E-04	-1,348E-03	-9,861E-04	-1,462E-03	-9,515E-04
a_rad =	-8,304E-06	-2,668E-06	-6,589E-06	2,394E-07	-3,053E-06
b_rad =	2,209E-05	1,261E-05	1,824E-05	8,137E-06	1,492E-05
c_rad =	-8,946E-07	2,563E-06	8,893E-08	3,154E-06	-1,960E-07
d_rad =	1,364E-03	5,481E-04	1,127E-03	1,034E-04	4,618E-04
e_rad =	-4,173E-03	-2,981E-03	-3,670E-03	-2,259E-03	-2,985E-03
f_rad =	-8,187E-04	-1,338E-03	-9,643E-04	-1,449E-03	-9,417E-04
a_Kap =	0,000	-0,002	-0,001	-0,002	-0,003
b_Kap =	-0,057	0,320	0,123	0,326	0,463
c_Kap =	2,306	-12,216	-5,028	-12,613	-17,489
d_Kap =	0,000	0,002	0,000	0,002	0,002
e_Kap =	0,081	-0,181	-0,028	-0,221	-0,324
f_Kap =	-6,896	3,296	-0,971	5,542	8,955
g_Kap =	0,000	0,000	0,000	0,000	0,000
h_Kap =	-0,066	-0,082	-0,088	-0,055	-0,057
i_Kap =	334,911	333,904	333,653	333,397	333,066

(a)

Thermal Conductivity	Calibrated value (W/mK)
Solder	36.7
Silicone+Phosphor	0.375
Ceramic body	103
MCPCB Dielectric	0.89

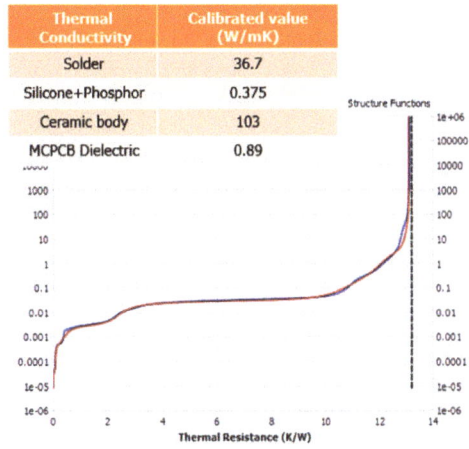

(b)

Figure 16. Model parameters and thermal test results for XPG3 type phosphor converted white LEDs from Cree: (**a**) Extracted sets of parameters used for a multi-domain LED chip level model [7]; (**b**) Typical representation of the junction-to-ambient thermal impedance through structure functions, used for package level modelling (model calibration based on test results).

3.2. Digitial Twins of Luminaires

From the thermal point of view, a luminaire is a multi-heat-source system where the heat-sources are the LED packages with their footprints. An obvious digital twin of an LED luminaire is its MCAD model (such as shown in Figure 17). The thermal model, as a thermal only digital twin of the luminaire can be extracted from this MCAD model. A luminaire compact thermal modelling tool (as indicated in Figure 5) may use different approaches to provide the compact thermal model of the luminaire.

(a) (b)

Figure 17. MCAD models of the demonstrator systems: (**a**) "10 W outdoor LED spot" luminaire typology; (**b**) "70 W HpS replacement" LED lamp.

One approach is to identify the thermal characterisation matrix of the luminaire and convert it into a Spice network model, through a series of CFD simulations, using the algorithm outlined in Section 2.4 (for details of the algorithm and for another application example refer to papers [20] and [21]). Figure 18 presents CFD simulation results of both demonstration examples. In Figure 19 CFD simulation result (temperature map) for a single run of the whole characterisation series is shown for the LED module according to scenario (x) of demonstrator b, together with the Spice netlist of the LED module. For creating the thermal compact model of the LED module substrates or complete luminaires, reduced order modelling techniques could also be used [32–35].

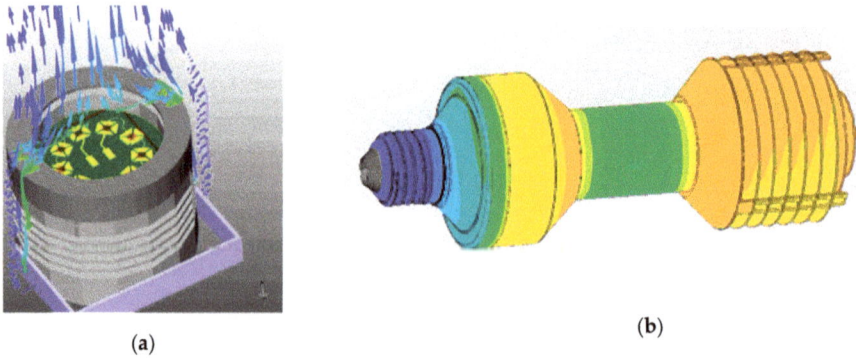

(a)

(b)

Figure 18. Computer fluid dynamic (CFD) simulation of the luminaires shown in Figure 14. The detailed simulation models were created from the MCAD models shown in Figure 17: (**a**) "10 W outdoor LED spot" luminaire typology; (**b**) "70 W HpS replacement" LED lamp.

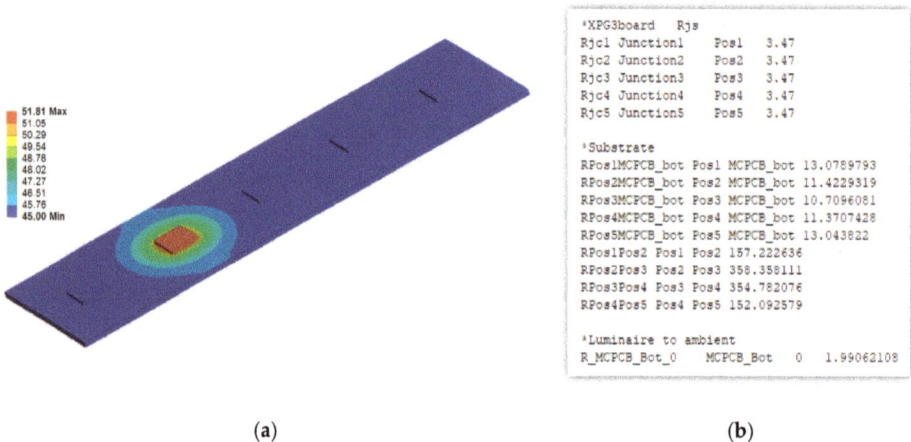

(a)

(b)

Figure 19. Thermal characterisation and compact thermal modelling of the LED module of the light engine of demonstrator B, according to scenario (x): (**a**) temperature map obtained by a CFD simulation in the series of simulations run to obtain the thermal characterisation matrix, $\mathbf{R}_{th}^*$; (**b**) Spice netlist of the steady-state compact model of the LED module (Substrate section).

3.3. Luminaire Level Results

The first results of Experiment A were reported in [5] together with the description of a pilot implementation of the overall Delphi4LED workflow. In this demo implementation, a set of models has been created and combined into an Excel spreadsheet application, aimed as a simplified design/virtual prototyping tool for SMEs. Based on prior characterisation of LEDs foreseen for the project demonstrators consortium members specialised in LED testing and modelling created a Spice-like chip level multi-domain models [5] and extracted the model parameters (see Figure 16a). The DCTM of the package was also created. In both cases variability among the different LED samples was also considered in a simplified way: Instead of creating Monte Carlo models from parameter distributions identified for a larger population of LEDs (as suggested in [26–29]) the model parameter sets obtained for the actual measured LED samples (five different ones, see Figure 16a) were randomly used.

In the subsequent paragraphs of this subsection we focus Experiment B and the version of the Delphi4LED workflow used by a Major company.

Figure 19 illustrates the thermal characterisation and compact modelling of the LED module of demonstrator b, according to scenario (x) (five LEDs driven by high current). Figure 19a is a temperature snapshot from the series of CFD simulations performed as part of the thermal characterisation of the module substrate (as outlined in Section 2.4 and detailed in papers [20] and [21]). Figure 19b presents the complete luminaire level Spice netlist that corresponds to the simplified thermal modelling approach shown in Figure 10b. The substrate model shown corresponds to the thermal boundary condition of fixing the bottom surface of the PCB substrate at 45 °C. For the geometric dimensions of the LED module in this example, see Figure 15a.

The results both for scenario (x)—5 LEDs operated at a forward current of 500 mA—and for scenario (y)—22 LEDs operated at a forward current of 100 mA—were obtained for two different LED types and different temperatures of the bottom surface of the PCB substrate. Tables 2 and 3 provide the simulation results for the Cree XPG3 type LEDs at 45 °C and 85 °C, respectively, and Tables 4 and 5 provide the results of for the Nichia NF2L757DRT-V1 type LEDs, also for 45 °C and 85 °C, respectively.

In both scenarios (x) and (y), the target of the emitted total luminous flux of 6000 lm for the entire lamp can be achieved. Note, that LED efficacies and junction temperatures are different which potentially leads to different energy consumption and different lifetime of the lamp, if one or the other design variant is realized. (Red colour in the tables indicates that the design target is failed.)

Table 2. Results for XPG3 4000K CRI 70 high power LEDs on PCB—temperature at the bottom of the PCB: 45 °C. (Green colour indicates that the design target is met.)

Calculated Property	5 LEDs; I_F = 500 mA	22 LEDs; I_F = 100 mA
Solder point temperature [°C]	51.6	46.9
Junction temperature [°C]	54.3	47.3
Light output per PCB [lm]	1056	1049
Total Light output [lm]	6336	6294
Efficacy [lm/W]	150	180

Table 3. Results for XPG3 4000K CRI 70 high power LEDs on PCB—temperature at the bottom of the PCB: 85 °C. (Green colour indicates that the design target is met.)

Calculated Property	5 LEDs; I_F = 500 mA	22 LEDs; I_F = 100 mA
Solder point temperature [°C]	91.7	86.9
Junction temperature [°C]	94.4	87.3
Light output per PCB [lm]	1000	1004
Total Light output [lm]	6001	6024
Efficacy [lm/W]	144	174

Table 4. Results for Nichia NF2L757DRT-V1 4000K CRI 80 mid-power LEDs on PCB—temperature at the bottom of the PCB: 45 °C. (Green colour indicates that the design target is met, red colour means target failed.)

Calculated Property	5 LEDs; I_F = 200 mA	22 LEDs; I_F = 60 mA
Solder point temperature [°C]	51.8	47.6
Junction temperature [°C]	61.1	49.7
Light output per PCB [lm]	664	1008
Total Light output [lm]	3984	6050
Efficacy [lm/W]	105	136

Table 5. Results for Nichia NF2L757DRT-V1 4000K CRI 80 mid-power LEDs on PCB—temperature at the bottom of the PCB: 85 °C. (Green colour indicates that the design target is met, red colour means target failed.)

Calculated Property	5 LEDs; I_F = 200 mA	22 LEDs; I_F = 60 mA
Solder point temperature [°C]	91.6	87.6
Junction temperature [°C]	100.8	89.7
Light output per PCB [lm]	818	951
Total Light output [lm]	3708	5707
Efficacy [lm/W]	101	131

4. Discussion

Benefits of using LED/luminaire digital twins in an early development phase were assessed considering the different working styles of SMEs and Major companies. The proposed, tool agnostic approaches were tested with multiple LED types in different demonstration experiments.

4.1. Benefit Observed in Experiment A

Table 6 compares the four design processes (SME "old process"/"new process" and Major "old process"/"new process") with comparison of relative costs of the development at the given type of company. Note, that the absolute total costs at SMEs and Majors are different and is also country specific, therefore fair comparison of design efforts can only be given in terms of relative costs where the basis is the total cost of development using the "old process" of product design (without virtual prototyping with digital twins).

Table 6. Cost comparison between the old process and the new Delphi4LED processes to design and manufacture the first prototype of a luminaire. The final total (relative) gains are indicated by boldface.

Main Design Costs	Major "Old Process"	Major "New Process"	Gain
Personal costs [1]	0.819	0.502	39%
Material costs [1]	0.055	0.028	48%
Testing [1]	0.126	0.045	65%
Total [1]	1.000	0.575	**42%**

Main Design Costs	SME "Old Process"	SME "New Process"	Gain
Personal costs [2]	0.896	0.633	29%
Material costs [2]	0.049	0.028	43%
Testing [2]	0.056	0.056	0%
Total [2]	1.000	0.717	**28%**

[1] Relative values compared to total costs of development in the "old way" at a Major company. [2] Relative values compared to total costs of development in the "old way" at an SME.

Based on the data gathered during our demonstration experiment roughly 30–40% overall design cost savings (see the exact values in boldface in Table 6) can be achieved by avoiding building and testing physical product prototypes and through increased efficiency of designers, using multi-domain simulations with compact models of LEDs combined with luminaires digital twin. The simulation based optimization led to a faster "time to pre-industrial sample", with a time reduction of about 30%.

4.2. Benefit observed in Experiment B

Analysing Table 2, Table 3, Table 4, and Table 5, an immediate observation is that 5 Nichia LEDs will not provide sufficient light output. Indeed, the using Cree's XPG3 LEDs assures to achieve the target luminous flux output of 6000 lm while keeping the junction temperature below 100 °C.

When usng Nichia NF2L757DRT-V1 4000K CRI 80 mid-power LEDs (considered for this case study), it can be seen, that with five LEDs of this Nichia mid-power LED type the target 6000 lm cannot be reached even when the LEDs are driven at the maximum allowed forward current of 200 mA. While in case of the modules with 22 pieces of Nichia's mid-power LEDs the LEDs are not driven at the maximum forward current: with only 60 mA of forward current the total luminous flux of the entire lamp close to the target of 6000 lm is reached.

For this specific case study, the XPG3 LEDs deliver in both scenarios (x) and (y) the targeted luminous flux at feasible junction temperatures and providing the highest system efficacy of both scenarios. The final decision between LED types, number of LEDs per module (5 LEDs, 22 LEDs, or any number of LEDs in between) will depend on the trade-off between optical performance, thermal limits, and cost.

The approach proposed allows the user, in a very short time frame, to develop very valuable insights in design choices. The case study highlighted the relation between the number of necessary LEDs and type of LED combination with system efficacy.

Finally, comparing demonstrations carried out in Experiment A (using Flotherm) with Experiment B (using ANSYS), the approach proved to be fast and accurate, independently of the software tool used: In a nutshell, tool agnostic.

4.3. Implementation of the New Digital Workflow in An Industrial Environment

Constraints of the actual industrial environment depend on the type of companies (SME versus Major). In both cases, the Delphi4LED overall workflow can be implemented in a way that best fits the company profile in question, providing ease of use by LED application designers and assuring accurate and fast results.

Experiment A was a proof of concept of the Delphi4LED methodology from LED device to luminaire design, both for SME and Major companies' industrial environment.

Experiments A and B show that results are accurate and fast, independently of the simulation tool chosen (simple Excel spreadsheet application ANSYS, Flotherm) confirming that the approach is tool agnostic.

Experiment B pushed the limits for a larger number of LEDs in a system. With the help of a parametric study, the results were obtained in less than an hour. The tedious part was to create the parametric model. Compared to the "old process", the Delphi4LED modelling and simulation provides answers to design choices early in the product development (when degrees of freedom are still high). These choice may include type and amount of LEDs, driving current, substrates, and layout configurations. In the dynamic market of lighting, this truly is a competitive advantage.

4.4. Impact on the Company Digital Footprint

Key to the success of this new LED design workflow is the availability of sets of libraries interacting with each other. The LED package libraries eventually should be provided obviously by the LED vendors, while other libraries such as module substrate, and heat-sink. Thermal model libraries, as indicated in Figure 12, are the new digital assets of a luminaire designer/manufacturer. This forms the future company digital footprint.

Figure 20 shows the input to those libraries from LED suppliers (LED e-datasheets or directly the LED digital twins) and customers (luminaire requirements). In case of SMEs, those databases are input for a spreadsheet based luminaire calculator tool. In case of Major companies, the model libraries can be directly linked to commercially available simulation tools (e.g., Flotherm, ANSYS).

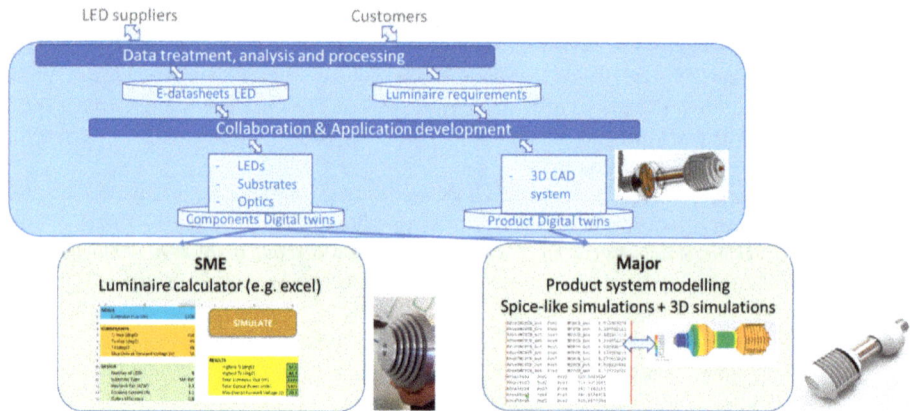

Figure 20. Digital footprint of an LED luminaire manufacturing company represented by the available design tools, libraries, and means of communications with LED suppliers and customers.

5. Conclusions

In the Delphi4LED project, new methodologies have been developed for compact modelling of LED products. The multi-domain nature of LED operation is treated on the chip level, using multi-domain LED models that can be implemented both in generic Spice circuit simulators and also as Visual Basic macros in a spreadsheet based luminaire calculation tool application. With an approach similar to the standard DELPHI methodology [25], boundary condition independent compact models of LED packages are created. These, combined with the chip level multi-domain model and the compact thermal models of module substrates and luminaires, allow a system level multi-domain simulation of an LED application. The combined chip + package + substrate + luminaire model is the "digital twin" or "virtual prototype" of a foreseen luminaire. This virtual prototype allows computer simulation assisted or computer simulation based optimization of an LED application. The methodology has been proven to be tool agnostic, tested with multiple LED systems.

Larger scale demonstrators are currently under study to validate the findings presented in this paper. Variability analysis applied to a luminaire level will also be further investigated.

Author Contributions: Conceptualization, G.M. and A.P.; methodology, G.M., A.P. and R.B.; software, R.B.; validation, J.Y. C.M., R.B.; formal analysis, J.Y., C.M., R.B. and G.O.; investigation, J.Y, G.O. R.B.; resources, J.Y. and C.M.; data curation, M.R.; writing—original draft preparation, G.M.; writing—review and editing, M.R., A.P and R.B.; visualization, G.M, A.P., C.M. and J.Y.; supervision, G.M.; project administration, G.M.; funding acquisition, G.M. and A.P.

Funding: This research has received funding from the European Union's Horizon 2020 research and innovation programme through the H2020 ECSEL project Delphi4LED (grant agreement number: 692465) (2016–2019). Co-financing of the Delphi4LED project by the national R&D funding organization of the participating countries. Additional information is available on: www.DELPHI4LED.eu.

Acknowledgments: Support from Delphi4LED project partners, especially from G. Farkas (Mentor Graphics), G., Hantos (BME), J. Hegedus (Mentor Graphics), L. Gaal (Mentor Graphics), D. Fournier (PISEO), E. Morard (Ecce'lectro), P.Zuidema (Signify), T.Merelle (PI-Lighting), and A.Alexeev (University of Technology of Eindhoven) are acknowledged.

Conflicts of Interest: The authors declare no conflict of interest. The funders had no role in the design of the study; in the collection, analyses, or interpretation of data; in the writing of the manuscript, or in the decision to publish the results.

Energies **2019**, *12*, 2389

Nomenclature

LED	Light Emitting Diode
PCB	Printed Circuit Board
DAL	Die Attach Layer
MDCM	Multi-domain compact model
LDT	Luminaire Design Tool
BCI	Boundary Condition Independent
DCTM	Dynamic Compact Thermal Model
HTC	Heat Transfer Coefficient
SME	Small and Medium Entreprise
MCAD	Mechanical Computer Aided Design
P_h	Total thermal power dissipated by an LED
P_{hJ}	Thermal power dissipated in a pn junction of an LED
P_{hC}	Thermal power dissipated on the cup reflector surface of an LED
P_{hD}	Thermal power dissipated in the dome (lens) volume of an LED
$\mathbf{R}^*_{th}$	thermal characterisation matrix

References

1. Delphi4LED Project Website. Available online: https://delphi4led.org (accessed on 27 March 2019).
2. Bornoff, R.; Hildenbrand, V.; Lungten, S.; Martin, G.; Marty, C.; Poppe, A.; Rencz, M.; Schilders, W.; Yu, J. Delphi4LED—From measurements to standardized multi-domain compact models of LEDs: A new european R&D project for predictive and efficient multi-domain modeling and simulation of LEDs at all integration levels along the SSL supply chain. In Proceedings of the 22nd International Workshop on THERMal INvestigation of ICs and Systems (THERMINIC'16), Budapest, Hungary, 21–23 September 2016; pp. 174–189. [CrossRef]
3. Schluse, M.; Rossmann, J. From simulation to experimental digital twins: Simulation-based development and operation of complex technical systems. In Proceedings of the 2016 IEEE International Symposium on Systems Engineering (ISSE), Edinburgh, UK, 3–5 October 2016. [CrossRef]
4. BITKOM (Bundesverband Informationswirtschaft Telekommunikation und neue Medien); VDMA (Verband Deutscher Maschinen- und Anlagenbau); ZVEI (Zentralverband Elektrotechnik- und Elektronikindustrie). *Umsetzungsstrategie Industrie 4.0. Ergebnisbericht der Plattform Industrie 4.0*; Bitkom Research GmbH: Berlin, Germany, 2015.
5. Marty, C.; Yu, J.; Martin, G.; Bornoff, R.; Poppe, A.; Fournier, D.; Morard, E. Design flow for the development of optimized LED luminaires using multi-domain compact model simulations. In Proceedings of the 24th International Workshop on THERMal INvestigation of ICs and Systems (THERMINIC'18), Stockholm, Sweden, 26–28 September 2018. [CrossRef]
6. Scope of the CIE TC2-84 Technical Committee. Available online: http://www.cie.co.at/technicalcommittees/recommendations-led-package-test-data-reporting (accessed on 27 March 2019).
7. Poppe, A.; Farkas, G.; Gaál, L.; Hantos, G.; Hegedüs, J.; Rencz, M. Multi-domain modelling of LEDs for supporting virtual prototyping of luminaires. *Energies* **2019**, *12*, 1909. [CrossRef]
8. Bornoff, R. Extraction of boundary condition independent dynamic compact thermal models of LEDs—A delphi4LED methodology. *Energies* **2019**, *12*, 1628. [CrossRef]
9. JEDEC JESD51-14 Standard. *Transient Dual Interface Test Method for the Measurement of the Thermal Resistance Junction-To-Case of Semiconductor Devices with Heat Flow through a Single Path*; JEDEC Solid State Technology Association: Arlington, VA, USA, 2010.
10. JEDEC JESD51-51 Standard. *Implementation of the Electrical Test Method for the Measurement of the Real Thermal Resistance and Impedance of Light-Emitting Diodes with Exposed Cooling Surface*; JEDEC Solid State Technology Association: Arlington, VA, USA, 2012.
11. JEDEC JESD51-52 Standard. *Guidelines for Combining CIE 127-2007 Total Flux Measurements With Thermal Measurements of LEDs with Exposed Cooling Surface*; JEDEC Solid State Technology Association: Arlington, VA, USA, 2012.

12. CIE Technical Report 225:2017. *Optical Measurement of High-Power LEDs*; CIE: Vienna, Austria, 2017; ISBN 978-3-902842-12-1. [CrossRef]

13. Farkas, G.; Poppe, A. Thermal testing of LEDs. In *Thermal Management for LED Applications*; Lasance, C.J.M., Poppe, A., Eds.; Springer: New York, NY, USA, 2014; pp. 73–165.

14. Poppe, A. Multi-Domain Compact Modeling of LEDs: An Overview of Models and Experimental Data. *Microelectron. J.* **2015**, *46*, 1138–1151. [CrossRef]

15. Poppe, A.; Hegedüs, J.; Szalai, A. Multi-domain modeling of power LEDs based on measured isothermal I-V-L characteristics. In Proceedings of the CIE Lighting Quality & Energy Efficiency Conference, Melbourne, Australia, 3–5 March 2016; pp. 318–327.

16. Hantos, G.; Hegedüs, J.; Bein, M.C.; Gaál, L.; Farkas, G.; Sárkány, Z.; Ress, S.; Poppe, A.; Rencz, M. Measurement issues in LED characterization for Delphi4LED style combined electrical-optical-thermal LED modeling. In Proceedings of the 19th IEEE Electronics Packaging Technology Conference (EPTC'17), Singapore, 6–9 December 2017. [CrossRef]

17. Farkas, G.; Gaál, L.; Bein, M.; Poppe, A.; Ress, S.; Rencz, M. LED characterization within the Delphi-4LED Project. In Proceedings of the 17th Intersociety Conference on Thermomechanical Phenomena in Electronic Systems (ITHERM'18), San Diego, CA, USA, 29 May–1 June 2018. [CrossRef]

18. Bein, M.C.; Hegedüs, J.; Hantos, G.; Gaál, L.; Farkas, G.; Rencz, M.; Poppe, A. Comparison of two alternative junction temperature setting methods aimed for thermal and optical testing of high power LEDs. In Proceedings of the 23rd International Workshop on THERMal INvestigation of ICs and Systems (THERMINIC'17), Amsterdam, The Netherlands, 27–29 September 2017. [CrossRef]

19. Onushkin, G.; Bosschaart, K.J.; Yu, J.; Van Aalderen, H.J.; Joly, J.; Martin, G.; Poppe, A. Assessment of isothermal electro-optical-thermal measurement procedures for LEDs. In Proceedings of the 23rd International Workshop on THERMal INvestigation of ICs and Systems (THERMINIC'17), Amsterdam, The Netherlands, 27–29 September 2017. [CrossRef]

20. Poppe, A.; Hegedüs, J.; Szalai, A.; Bornoff, R.; Dyson, J. Creating multi-port thermal network models of LED luminaires for application in system level multi-domain simulation using Spice-like solvers. In Proceedings of the 32nd IEEE Semiconductor Thermal Measurement and Management Symposium (SEMI-THERM'16), San Jose, CA, USA, 14–17 March 2016; pp. 44–49. [CrossRef]

21. Poppe, A. Simulation of LED based luminaires by using multi-domain compact models of LEDs and compact thermal models of their thermal environment. *Microelectron. Reliab.* **2017**, *72*, 65–74. [CrossRef]

22. Bornoff, R.; Farkas, G.; Gaál, L.; Rencz, M.; Poppe, A. LED 3D thermal model calibration against measurement. In Proceedings of the 19th International Conference on Thermal, Mechanical and Multiphysics Simulation and Experiments in Microelectronics and Microsystems (EuroSimE'18), Toulouse, France, 15–18 April 2018. [CrossRef]

23. JEDEC JESD15-4 Standard. *DELPHI Compact Thermal Model Guideline*; JEDEC Solid State Technology Association: Arlington, VA, USA, 2008.

24. Alexeev, A.; Martin, G.; Onushkin, G.; Linnartz, J.P. Thermal modeling of multiple heat sources and transient heat analysis of LEDs. *Energies* **2019**, *12*, 1860. [CrossRef]

25. Alexeev, A.; Martin, G.; Onushkin, G. Multiple heat path dynamic thermal compact modeling for silicone encapsulated LEDs. *Microelectron. Reliab.* **2018**, *87*, 89–96. [CrossRef]

26. Alexeev, A.; Martin, G.; Onushkin, G.; Linnartz, J.P. Accurate thermal transient measurements interpretation of monochromatic LEDs. In Proceedings of the 35th Semiconductor Thermal Measurement and Management Symposium (SEMI-THERM'19), San Jose, CA, USA, 18–22 March 2019; pp. 7–11.

27. Sari, J.; Mérelle, T.; Di Bucchianico, A.; Breton, D. Delphi4LED: LED measurements and variability analysis. In Proceedings of the 23rd International Workshop on THERMal INvestigation of ICs and Systems (THERMINIC'17), Amsterdam, The Netherlands, 27–29 September 2017. [CrossRef]

28. Mérelle, T.; Bornoff, R.; Onushkin, G.; Gaál, L.; Farkas, G.; Poppe, A.; Hantos, G.; Sari, J.; Di Bucchianico, A. Modeling and quantifying LED variability. In Proceedings of the 2018 LED Professional Symposium (LpS2018), Bregenz, Austria, 25–27 September 2018; pp. 194–206, ISBN 978-3-9503209-9-2.

29. Bornoff, R.; Mérelle, T.; Sari, J.; Di Bucchianico, A.; Farkas, G. Quantified insights into LED variability. In Proceedings of the 24th International Workshop on THERMal INvestigation of ICs and Systems (THERMINIC'18), Stockholm, Sweden, 26–28 September 2018. [CrossRef]

30. Bornoff, R.; Mérelle, T.; Sari, J.; Di Bucchianico, A.; Farkas, G. A Methodology to determine the sites of variability in an LED assembly. In Proceedings of the 35th Semiconductor Thermal Measurement and Management Symposium (SEMI-THERM'19), San Jose, CA, USA, 18–22 March 2019; pp. 1–6.

31. Mérelle, T.; Sari, J.; Di Bucchianico, A.; Onushkin, G.; Bornoff, R.; Farkas, G.; Gaál, L.; Hantos, G.; Hegedüs, J.; Martin, G.; et al. Does a single LED bin really represent a single LED type? In Proceedings of the CIE 2019 29th QUADRENNIAL SESSION, Washington, DC, USA, 14–22 June 2019. [CrossRef]

32. Lungten, S.; Bornoff, R.; Dyson, J.; Maubach, J.M.L.; Schilders, W.H.A.; Warner, M. Dynamic compact thermal model extraction for LED packages using model order reduction techniques. In Proceedings of the 23rd International Workshop on THERMal INvestigation of ICs and Systems (THERMINIC'17), Amsterdam, The Netherlands, 27–29 September 2017. [CrossRef]

33. Codecasa, L.; D'Alessandro, V.; Magnani, A.; Rinaldi, N. Structure-preserving approach to multi-port dynamic compact models of nonlinear heat conduction. *Microelectron. J.* **2015**, *46*, 1129–1137. [CrossRef]

34. Codecasa, L.; Magnani, A.; D'Alessandro, V.; Rinaldi, N.; Metzger, A.G.; Bornoff, R.; Parry, J. Novel MOR approach for extracting dynamic compact thermal models with massive numbers of heat sources. In Proceedings of the 32nd IEEE Semiconductor Thermal Measurement and Management Symposium (SEMI-THERM'16), San Jose, CA, USA, 14–17 March 2016; pp. 218–224. [CrossRef]

35. Codecasa, L.; D'Alessandro, V.; Magnani, A.; Rinaldi, N.; Zampardi, P.J. FAst novel thermal analysis simulation tool for integrated circuits (FANTASTIC). In Proceedings of the 20th International Workshop on THERMal INvestigation of ICs and Systems (THERMINIC'14), Greenwich, UK, 24–26 September 2014; pp. 1–6. [CrossRef]

energies

MDPI

Article

Multi-Domain Modelling of LEDs for Supporting Virtual Prototyping of Luminaires

András Poppe [1,2,*], Gábor Farkas [2], Lajos Gaál [2], Gusztáv Hantos [1], János Hegedüs [1] and Márta Rencz [1,2]

[1] Department of Electron Devices, Budapest University of Technology and Economics, Magyar tudósok körútja 2, bldg. Q, 1117 Budapest, Hungary; hantos@eet.bme.hu (G.H.); hegedus@eet.bme.hu (J.H.); rencz@eet.bme.hu or marta_rencz@mentor.com (M.R.)
[2] Mechanical Analysis Division, Mentor, a Siemens Business, Gábor Dénes utca 2, 1117 Budapest, Hungary; gabor_farkas@mentor.com (G.F.); lajos_gaal@mentor.com (L.G.)
*   Correspondence: poppe@eet.bme.hu or andras_poppe@mentor.com; Tel.: +36-1-463-2721

Received: 5 April 2019; Accepted: 14 May 2019; Published: 18 May 2019

Abstract: This paper presents our approaches to chip level multi-domain LED (light emitting diode) modelling, targeting luminaire design in the Industry 4.0 era, to support virtual prototyping of LED luminaires through luminaire level multi-domain simulations. The primary goal of such virtual prototypes is to predict the light output characteristics of LED luminaires under different operating conditions. The key component in such digital twins of a luminaire is an appropriate multi-domain model for packaged LED devices that captures the electrical, thermal, and light output characteristics and their mutual dependence simultaneously and consistently. We developed two such models with this goal in mind that are presented in detail in this paper. The first model is a semi analytical, quasi black-box model that can be implemented on the basis of the built-in diode models of spice-like circuit simulators and a few added controlled sources. Our second presented model is derived from the physics of the operation of today's power LEDs realized with multiple quantum well heterojunction structures. Both models have been implemented in the form of visual basic macros as well as circuit models suitable for usual spice circuit simulators. The primary test bench for the two circuit models was an LTspice simulation environment. Then, to support the design of different demonstrator luminaires of the Delphi4LED project, a spreadsheet application was developed, which ensured seamless integration of the two models with additional models representing the LED chips' thermal environment in a luminaire. The usability of our proposed models is demonstrated by real design case studies during which simulated light output characteristics (such as hot lumens) were confirmed by luminaire level physical tests.

Keywords: light emitting diodes; power LEDs; multi-domain modelling; LED luminaire design

1. Introduction

1.1. The Context of this Work

In previous decades, the design of light sources (incandescent, fluorescent) and luminaires were separated. The light sources were not sensitive to the ambient temperature, luminaire design was restricted mainly to the optics, and the electric and thermal coupling of the two worlds occurred at the socket, in both physical and conceptual sense.

With the advent of LED light sources, thermal aspects in luminaire design have become important because LEDs are very sensitive to heat. The "sockets" ceased to exist, and the close thermal coupling of luminaires and LEDs thus requires sophisticated electronic cooling design tools, such as CFD (computational fluid dynamics) based simulators.

Nowadays, the LED luminaire design is characterized by the presence of re-organized successors of previously known major players of the traditional lighting industry and also of many new, smaller start-up companies. These major companies have lots of expertise in the use of advanced CFD software for thermal-only simulations while start-ups have much less expertise and resources. The goal of the Delphi4LED project [1,2] is to develop new LED luminaire design workflows for both types of companies to support them in significantly reducing design costs and time-to-market [3,4], through virtual prototyping of luminaires, i.e., computer simulation of the luminaire. Figure 1 illustrates the process of creating simulation models at different integration levels, from the physical testing of LEDs, and creation of their simulation models to the use of LEDs at the higher level integration of a solid-state lighting product (e.g., a luminaire).

Inputs to such a digital design flow of LED luminaires are the overall lighting requirements, such as targeted minimal light output under application conditions, allowed maximal input electrical power, driver and optics efficiency, etc., to support the design of optimized LED luminaires through multi-domain simulations [3,4].

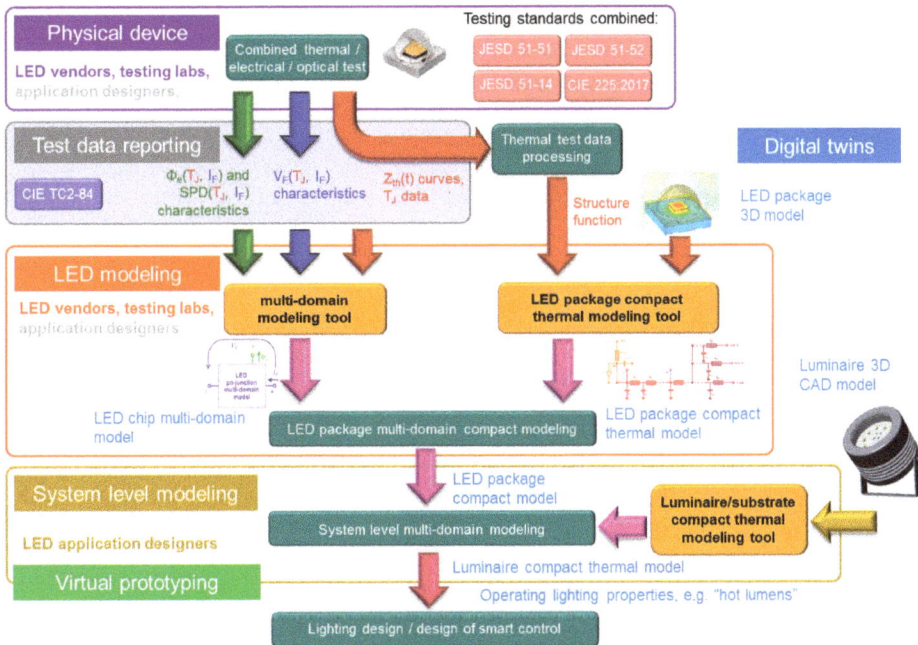

Figure 1. Delphi4LED workflow and design persona for the generation and application of digital twins (simulation models) of LED packages, modules, and luminaire with which luminaire level design optimization is performed in order to find the best solution to meet lighting design requirements (such as required light output properties) under application conditions.

In the bottom-up modular approach, the modelling starts from a multi-domain model of *LED chips* [5–7], followed by compact thermal models of the LED *packages, modules* (orange framed items in Figure 1), and finally, *luminaires* [8,9]. Details of the compact modelling of the 3D thermal environment of LED chips (package, substrate, luminaire with optics) are provided in other publications, such as [8,10,11]. In the "Industry 4.0" language, these simulation models are called "digital twins" to represent the relevant aspects of the physical samples of LED chips, LED packages, modules, etc. [3,4]. Thus, the ultimate goal is to be able to predict the total emitted luminous flux [12] of different versions of a luminaire under thermally different application conditions (also referred to by the sloppy English term "hot lumens").

1.2. The Wider Background

Many research teams have realized the importance of characterizing LEDs in multiple domains of operation. Mostly, these studies used empirical and sometimes rough approximations to obtain model parameters from measured data, see, e.g., [13–17]. In [13], an emphasis was placed on modelling LEDs' efficiency. The authors of [14] with their ambitious target design of retrofitted LED lamps, used the photo-electro-thermal theory, which in our view is a relatively rough, empirical way of describing LED behavior. The authors of [15–17] in their work relied on LED characterization tools that have been introduced to the SSL industry as a result of our team's pioneering work in multi-domain characterization [18,19]. In [15–17] (that are representative examples of the literature in the field), certain aspects of the junction temperature affected LED operating parameters were investigated, such as the electrical power [15], efficiency [16], and luminous flux [17]. However, none of these academic teams modelled LEDs using a holistic approach for daily use in industrial LED luminaire design flows, therefore, their studies were typically restricted to a few samples of certain LED types only.

Naturally, LED vendors have to support their customers. Therefore they usually publish different simulation models and other parts' data, which is needed for the electrical, mechanical, optical, and manufacturing design of LED modules/luminaires. Thus, their electronically published data include Spice diode model parameters for their LEDs as well, see, e.g., [20,21]. Some common properties of these parameter sets are that they target the modeling of only the electrical behavior of LEDs, though certain parameter sets also allow simulation of the I-V (current-voltage) characteristics at any assumed junction temperature in the range of the validity of the model. Other vendors provide model data allowing simulation only at a 25 °C junction temperature, strongly limiting the applicability of such a diode model applied to an LED. The problem of the usual Spice models of LED vendors is that the standard Spice diode model is not an electro-thermal one: It does not support a fully-coupled electro-thermal simulation. Spice network solvers—with a few exceptions like [22]—are not fully-coupled electro-thermal solvers either. Finally, in standard Spice, there are no built-in means for predicting the light output properties of LEDs consistently with the electrical characteristics and the junction temperature. Due to these limitations of the commercially available tools and model parameter sets provided by LED vendors, a few academic research teams (including ours as well) have devoted their efforts to LED multi-domain modeling.

Thus, there have been a few other academic teams who published their results roughly at the same time as we also published our first multi-domain LED model, which aimed for general application. The first author whom we have to mention is A. Keppens, who, in his PhD dissertation about LED modeling [23], provides a comprehensive overview about the current and temperature dependence of the major parameters for about 12 different power LEDs. His experimental results include the electrical characteristics of power LEDs from the microampere range up to the forward current ranges of practical operation (100 mA and above). Besides the I-V characteristic section corresponding to the classical Shockley model of pn-junctions, parasitic effects at very small currents (different leakage currents) and high currents (e.g., the influence of the diode internal series resistance) are also treated. This early work, however, did not attempt to model the multi-domain operation of LEDs, i.e., to describe the mutual dependence of the electrical and thermal operation and light emission.

The first macro-type LED models with a real multi-domain approach aimed for Spice solvers were published by C. Negrea et al. [24] and K. Górecki [25], independently. Following our early approach [19], they both intended to simulate LEDs' electrical, thermal, and optical properties simultaneously and consistently. Negrea's team relied on the built-in diode model of Spice. As discussed above, this model is not an electro-thermal capable one, so in [24], a relaxation type iteration is used to reach consistent electro-thermal simulation results. This iteration scheme also includes light output modelling. In contrast to Negrea's approach, in Górecki's works (such as [25–27]), a self-built Spice circuit macro is used to represent an LED, assuring consistency among the three operating domains consistently, without any need for an additional iteration loop external to the Spice solver. Over the last couple of years, a continuous evolution of Górecky's LED Spice model has been observed, resulting in an even newer model [28], which is still aimed at the calculation of the *illuminance* provided by the investigated LED. Illuminance, though, is not the property of the LEDs investigated, but it is a property that characterizes a given lighting situation, depending on the emitted total luminous flux of the light source and the geometrical arrangement of the lighting task. In their most recent model version [29] however, Gorecki and Ptak already calculate the primary light output properties: the radiant flux and the luminous flux.

These academic works did not have the ambition of developing a comprehensive multi-domain characterization and modelling methodology of power LEDs aimed at industrial application design flows, like those briefly discussed already and described in detail in [3,4]. Therefore, these studies were restricted to a few selected LED types, used as case studies only. In the work we present here, on the contrary, the models were applied to 5 to 20 samples of many different LED types, which are representative of today's industrial LED luminaire designs.

The results of our academic approach that was the starting point of our most recent developments are summarized in [5]. The steps of the evolution of our models from this academic study towards our present, industrial application oriented models were published in our recent conference papers [6,7]. In the subsequent sections of this paper, we report the latest achievements in chip level modelling of LEDs with a focus on the *thermally* influenced *electrical* and *optical* behavior treated in a self-consistent manner, with the needs of commercial implementation for industrial applications in mind, in a closer context as presented in Section 1.1.

1.3. Overall Considerations Regarding Multi-Domain Chip Level Modelling of LEDs Aimed at Use in Industrial Application Design Flows

Luminaires are characterized mainly in their stationary "hot" state, therefore, it is satisfactory to construct LED models for a constant current bias, I_F, and stabilized junction temperature, T_J. However, it must be noted that I_F is forced by the driving electronics, but T_J is determined by the ambient temperature, T_{amb}, and the actual thermal boundary, both in a simulation environment and in real life.

We have shown previously that thermal transient measurements can deduce the R_{thja} junction to ambient thermal resistance of the device along with the measurement system around it [30]. During physical tests of LEDs, R_{thja} is used in an iterative process to control T_{amb} such that a target T_J is reached. Thus, electric and radiometric properties can be measured at a programmed I_F, T_J operating point [31,32]. To obtain the thermal metrics, such as the junction-to-case thermal resistance (R_{thJC}), the applicable standard [33] prescribes thermal transient measurements at varying boundaries.

The Delphi4LED project targeted the elaboration of an automated process, which provides all modelling parameters in electrical, thermal, and optical domains, obtained in a single mechanical mounting of an LED or a compound of more LEDs in a combined thermal transient and radiometric measurement system [34].

In a round-robin test [35] (preceding measurement and modelling of dozens of LED types and hundreds of LED samples), the minimal number of I_F and T_J pairs was identified, which covered the foreseen design space of luminaires: The current range spanned from a few milliamps to amperes, and junction temperatures ranged from 15 °C to about 120 °C. Measurements also included the current region where the LEDs' efficiency peaks (see Figure 2). In all cases, we could observe the characteristic decay of the light output at high currents termed the "light droop".

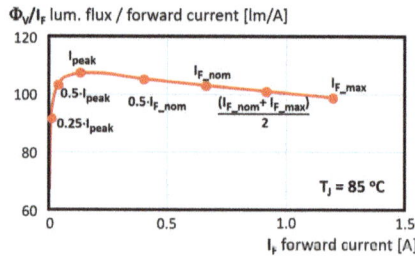

Figure 2. A typical LED efficacy diagram with forward current values suggested for measurement of isothermal IVL (current-voltage-light output) characteristics [35] to cover the LED operating domain relevant for LED luminaire design.

1.4. Expectations for a Chip Level Multi-Domain LED Model

An LED model aimed at the above outlined luminaire level simulations must be a functional one, and provide a set of simple analytic formulae. The same set should portray all power LEDs, each having a different complex material and geometry composition. A small number of numeric parameters must be assigned to the formulae to depict one distinctive LED type in all the three mutually dependent operating domains.

The proposed models provide the forward voltage, V_F, between the **A** (anode) and **C** (cathode) electrical terminals and the fluxes, leaving the LED as a function of a forced forward current, I_F, and junction temperature, T_J (Figure 3). The fluxes we are interested in are:

- The heat-flux (also known as the real heating power, P_H leaving the device through the thermal junction node denoted by **J**.
- The total emitted radiant flux, Φ_e [36] (also known as the optical power, P_{opt}), leaving the device through the optical port of the model denoted by **R**.
- The total emitted luminous flux, Φ_V [12], leaving the device through the optical port of the model denoted by **L**.

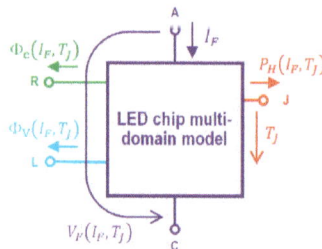

Figure 3. Functional chip level multi-domain LED model.

During the Delphi4LED project, we developed two different modeling approaches. Already, at the very beginning, a generic analytical model, which can be generated directly from the measured electrical and optical data with precise parameter fitting, primarily based on Shockley's model of semiconductor diodes, was required. The ultimate model version that we present in this paper can be used by "electrical only" circuit simulators, like the popular Spice program, freely available at many sites, such as LTspice, thus yielding accurate results in the typical forward current range of power LEDs (e.g., from 10 mA up to 1000 mA). Last, but not least, this model is a straightforward, natural extension of today's Spice models of LED vendors. Another advantage of this generic Spice circuit macro-based concept is its simplicity and speed. The extracted parameter sets can be promptly tested in the circuit simulator or the set of equations corresponding to the circuit macro can be directly used in Excel spreadsheet applications, as demonstrated recently [3].

As the project progressed, a semi-physical model was elaborated, which extends beyond the simple ideal diode equation, by considering some details of today's high efficiency, multiple quantum-well (MQW) hetero junction based LEDs, taking into account more aspects of the physics of the operation of such LEDs. The main driver for this work was luminaire designers' desire for accurate modelling of LED operations in the low forward current regime (mA range and below) as well. This satisfies their needs of being able to use the same LED type in all regions of its efficacy curve (Figure 2), allowing for more design trade-offs at the luminaire level. A slight drawback of this model is that it completely differs from the existing Spice models of LED vendors.

Both modelling approaches have been applied both to color LEDs, covering practically the major relevant wavelength regions of the visible range (blue, green, amber, red), and to different kinds of phosphor converted white LEDs. In both cases, samples originated from different leading LED vendors. In the case of every LED type, at least five individual samples were measured and modelled.

The targeted use of these functional models is in luminaire design. An approach to predict the operation of the complex LED + luminaire system is a relaxation type electro-thermal-optical simulation with a separate, but coupled thermal solver. In an iterative manner, the model of the LED chips passes the calculated P_H dissipation to the 3D thermal solver, which provides the updated values of the T_J LED junction temperatures back to the model, as illustrated in Figure 4.

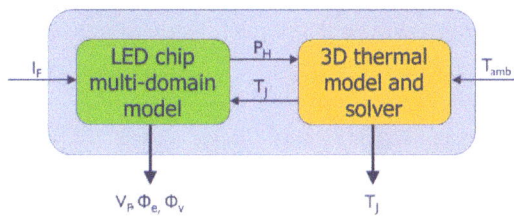

Figure 4. Application of a chip level multi-domain LED model in a relaxation type implementation of an electro-thermal-optical solver called an LED luminaire design calculator [3].

2. The Starting Point: Basic Spice Model

In nonlinear circuit simulators, the concept of generic circuit branches is used, theoretically represented by controlled sources (voltage/current controlled voltage/current sources). The relationship between the controlled property and the controlling property can be represented by any kind of (non)linear function. The most popular circuit simulators use the nodal method of solving Kirchhoff's equations for the investigated circuit. In this approach, the generic branch is a voltage controlled current source: The branch current is a function of any branch voltage in the circuit. Shockley's model of ideal diodes fits directly into this concept: In the forward region, its current depends on its own branch voltage in roughly an exponential manner. More realistic diode models are based on an equivalent model composed of two generic branches connected in series: An "ideal pn-junction model" and the diode's series electrical resistance (see Figure 5a).

An ideal pn-junction shows the exponential growth of the current when an increasing voltage is applied to its terminal nodes. However, in real diodes, this growth slows down when approaching the so called diffusion voltage of the junction. Hence, diodes are represented with the equivalent circuit of Figure 5a.

The physical considerations detailed in Section 4 imply that the internal junction in the scheme follows the Shockley equation over many orders of magnitude of current. In a V_F voltage controlled manner, it is formulated by Shockley's diode model:

$$I_F(V_{Fpn}) = I_0 \cdot \left[\exp\left(\frac{V_{Fpn}}{m \cdot V_T} \right) - 1 \right], \tag{1}$$

$$\text{where } m = const,$$

or expressed from the I_F forward current:

$$V_{Fpn} = m \cdot V_T \cdot \left[\ln\left(\frac{I_F}{I_0} \right) + 1 \right]. \tag{2}$$

Here, $V_T = k \cdot T_J / q$ is the thermal voltage, k is the Boltzmann constant, q is the elementary charge, so V_T is roughly 26 mV at room temperature. m is the ideality factor, representing how an actual overall diode characteristic deviates from an ideal, abrupt homojunction diode. Detailed discussion of the concept of the m ideality factor applied to today's multiple quantum well heterojunction based LED structures is provided in Section 4.

In contrast to the voltage driven concept of device models matched to the circuit solvers based on the method of nodal voltages, in practice, the LEDs used in luminaires are typically driven by current generators. Therefore, for the original voltage driven modelling approach (presented in [5]), it is more practical to use the current driven version of Shockley's diode model as given by Equation (2). After considering the electrical series resistance, R_S, the total diode forward voltage, V_F, is:

$$V_F = I_F \cdot R_S + V_{Fpn}. \tag{3}$$

In most Spice versions, thermal changes are not modeled, and all device equations are computed at a fixed reference temperature, T_{ref}, though T_{ref} is alterable as a run-time parameter.

Spice tacitly distinguishes between device parameters with a weak and strong temperature dependence. In the case of a weak temperature sensitivity of the parameter, Spice approximates parameter changes in a Taylor-series manner, using a polynomial function of the temperature change with respect to the reference temperature value at T_{ref}. For example, for the series resistance, R_S:

$$R_S(T_J) = R_{S0} + S_{RS1} \cdot (T_J - T_{ref}) + S_{RS2} \cdot (T_J - T_{ref})^2 \tag{4}$$
$$R_{S0} = R_s(T_{ref}).$$

Semiconductors have a strong temperature dependence; the parameters of Equation (2) cannot be recalculated just by a simple polynomial (quadratic) approximation. As shown in more detail in Section 4, the $V_T = k \cdot T_J / q$ formula for the thermal voltage, also I_0, has to be recalculated:

$$I_0(T_J) = I_{0ref} \cdot \left(\frac{T_J}{T_{ref}} \right)^{\frac{3}{m}} \cdot \exp\left[\frac{V_g(T_J)}{V_T} \cdot \frac{T_J - T_{ref}}{T_{ref}} \right], \tag{5}$$

where $I_{0ref} = I_0(T_{ref})$ and V_g denotes the bandgap voltage:

$$V_g = W_g / q. \tag{6}$$

The bandgap energy, W_g, can be determined in a quite complex way for standard diodes, from material parameters and doping levels. However, in the case of an LED, it can be calculated from the measured peak wavelength at a junction temperature. Assuming that most photons of the λ wavelength are generated by electrons jumping over the bandgap in a quantum well produces:

$$W_g = h{\cdot}v = h{\cdot}\frac{c}{\lambda_{peak}(T_J)},\tag{7}$$

where h is the Planck constant and c is the speed of light, resulting in $V_g = (h{\cdot}c)/(q\lambda_{peak})$.

The bandgap shrinks at growing temperatures, and both band edges move towards the intrinsic energy level. The empirical formula for this decrease is usually implemented in the different Spice versions by the widely known Varshni formula [37,38] as follows:

$$W_g(T) = W_{g0} - \frac{\alpha T^2}{\beta T}.\tag{8}$$

Most parameters can be gained from the measured I–V characteristics at a fixed T_{ref}. $I_{0_{ref}}$ and m can be determined from a logarithmic fitting at low currents. The approximate value of R_{S0} comes from the difference of the fitting and measured data at high currents. W_{g0} can be deduced from spectral measurements. The coefficients used in the modelling of temperature dependence (S_{RS1}, S_{RS2}, α, and β) can be concluded by repeating the measurements at other temperatures.

In some cases, the $3/m$ quotient in Equation (5) is replaced by an empirical **XTI**/m quotient for improved modeling of the temperature dependence of I_0 (**XTI** is a parameter of the built-in diode model of Spice, with a default value of 3). In Section 4, **XTI**/m is denoted as ϑ.

A severe practical problem is that the temperature dependence of I_0 is implemented in different ways in different versions of Spice. Therefore, if we want to model the thermally induced changes of LEDs' characteristics with the built-in diode model of Spice in a generic way, we have to "switch off" any temperature dependence in this model and we have to account for these changes ourselves.

3. An Analytical, Quasi Black-Box Multi-Domain Model of an LED Chip

3.1. The Basic Concept: Splitting the Total Forward Current into Two Components

When we model an LED, we can split its forward current into two components [5], where in each current component, the charge carriers are subject to one type of recombination process only. This way we can distinguish between a current component associated with radiative recombination processes, denoted by I_{rad}, giving rise to light emission, and a current component associated with non-radiative recombination processes, denoted by I_{dis}, resulting in heat-dissipation at the pn-junction, as illustrated in Figure 5b. Both the radiative and the dissipative current components can be described by the Shockley equation [5].

The total emitted radiant flux, Φ_e, leaves the device without adding to its heating. Accordingly, we calculate the heating power, P_H, of the LED by subtracting the radiant flux from the supplied electrical power:

$$P_H = I_F{\cdot}V_F - \Phi_e,\tag{9}$$

as prescribed also in the JEDEC JESD51-51 standard [39].

In the model represented by Figure 5b, either the region around the peak efficacy or the high current range was modeled at better precision. This was resolved by inserting a series resistance, R_R, in the "radiative branch" of the LED model as shown in Figure 5c [6,7].

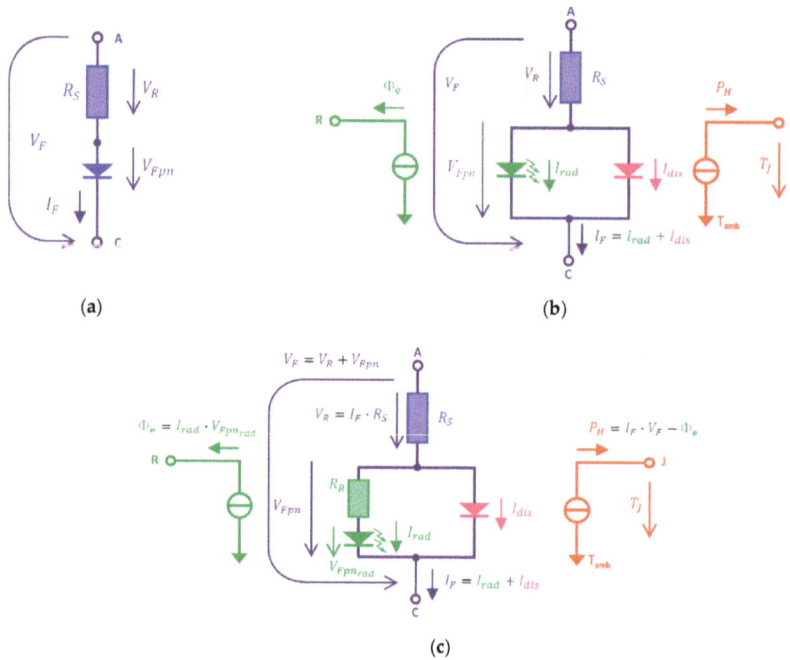

(a)

(b)

(c)

Figure 5. Evolution of the LED models: (**a**) basic diode model; (**b**) model with dissipating and radiative branches [5]; (**c**) LED model with added series resistance in the radiative branch, accounting for light decay at high currents [6,7].

3.2. Model Equations at a Fixed Reference Temperature, T_{ref}, in the Current Driven Mode

With the insertion of R_R, the model became an "implicit" one, and the currents and voltages cannot be expressed explicitly from I_F. However, a circuit simulator with an iterative solver inherently handles this.

For the electric domain, the Kirchhoff equations depict the topology of Figure 5c. That is, the voltages of components in series must be totaled, and the I_F current is split into I_{rad} and I_{dis} at the single internal node with a V_{Fpn} voltage. Regarding the branch equations, both diodes behave as dictated by Equation (2), with separate m_{rad}, $I_{0_{rad}}$ and m_{dis}, $I_{0_{dis}}$ parameters, respectively [5]. The emitted total radiant flux can be derived from the radiative branch:

$$V_{Fpn} = V_{Fpn_{rad}} + I_{rad} \cdot R_R = m_{rad} \cdot V_T \cdot \ln\left[\left(\frac{I_{rad}}{I_{0_{rad}}}\right) + 1\right] + I_{rad} \cdot R_R, \tag{10}$$

and:

$$\Phi_e = I_{rad} \cdot V_{Fpn_{rad}}. \tag{11}$$

A further stage in the evolution of our model is that the I_{dis} component is no longer calculated, since through the use of Equations (2) and (3), an LED as a "diode" is described and Equations (10) and (11) yield the radiant flux. From this, and using Equation (9), the total dissipated power heating the LED is calculated. This present, new approach has the advantage that one can apply a simpler parameter extraction procedure than with the previous approaches [5], thus enabling the entire LED model to be based on the built-in models of a generic Spice circuit simulator.

3.3. Implementation of the Temperature Dependence of the LED Chip Model

The second major evolutionary step in our LED chip multi-domain model is the way how the temperature dependences of the characteristics of an LED are modelled. Here, again, we consider the fact that in most applications, LEDs are driven by a constant DC forward current. In such a case, the electrical characteristics of the LEDs behave as illustrated in Figure 6a: The higher the junction temperature, the smaller the forward voltage drop is, when a constant forward current is forced through the LED. This can be understood by referring to Equation (5): The current coefficient, I_0, appearing in Equation (1) exhibits roughly an exponential temperature dependence (resulting in approximately 15% growth by a centigrade at room temperature). With this dependence, if the total forward current, I_F, is kept constant, Equation (1) can be satisfied only if the forward voltage across the diode shrinks as the junction temperature, T_J, increases. For a single LED, this is typically 1 mV. Furthermore, a 2 mV degree change on top of the 2 V to 3 V forward voltage is represented by a simple linear relationship. However, for an accurate model, the use of a quadratic relationship is advised, as revealed by recent measurements of multiple LED types [40]. Thus, the entire LED can be modelled as a temperature independent diode with a temperature controlled voltage source, $\Delta V_{F_{el}}$, connected in series, where this voltage source represents the temperature induced change of the forward voltage.

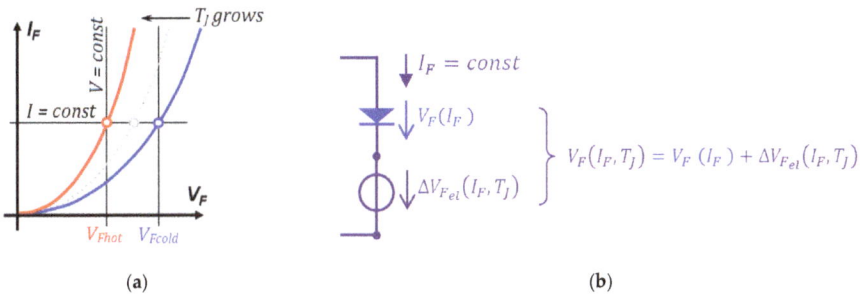

Figure 6. (a) Temperature dependence of the forward voltage of a diode at constant forward current; (b) Modelling the effect of self-heating for both the pn-junction and the series resistance by a single temperature dependent source.

In the complex model of Figure 5c, two resistors and two junctions are included. It would be impractical and hard to justify voluntarily distributing portions of the measured temperature related change of the entire LED to the four components.

It is more reasonable to start from a reference circuitry, fixing R_S and R_R at their value at T_{ref}, then adding diodes with $V_{F0}(I_F) = V_F(I_F, T_{ref})$ temperature independent equations (these can be easily implemented in a standard Spice macro model). After this, the temperature related variation of resistors and junctions can be cumulated into $\Delta V_{F_{el}}$ and $\Delta V_{F_{rad}}$ sources (as illustrated in Figure 6b), approximated by a quadratic temperature dependence:

$$\Delta V_{F_{el}} = \left(a_{el}{\cdot}I_F^2 + b_{el}{\cdot}I_F + c_{el}\right){\cdot}\left(T_J^2 - T_{ref}^2\right)+ \left(d_{el}{\cdot}I_F^2 + e_{el}{\cdot}I_F + f_{el}\right){\cdot}\left(T_J - T_{ref}\right),$$

(12)

and:

$$\Delta V_{F_{rad}} = \left(a_{rad}{\cdot}I_F^2 + b_{rad}{\cdot}I_F + c_{rad}\right){\cdot}\left(T_J^2 - T_{ref}^2\right)+ \left(d_{rad}{\cdot}I_F^2 + e_{rad}{\cdot}I_F + f_{rad}\right){\cdot}\left(T_J - T_{ref}\right).$$

(13)

These controlled sources can also be included in a Spice circuit macro model of an LED. This approach is similar to the approach used in a version of K. Górecki's LED model presented in [28], except that in [28], the additional temperature controlled voltage source represents the extra voltage drop in the series resistance only, approximated by a simple linear temperature dependence. Our approach presented here has the advantage that the model of $\Delta V_{F_{el}}$ can be directly derived from the results of the temperature calibration procedure required for the thermal testing of LEDs (as per the JEDEC JESD51-51 standard [39]).

With the above formula, for $\Delta V_{F_{rad}}$, the temperature dependence of the emitted radiant flux is calculated from the power of the radiative diode, as suggested earlier [5–7]:

$$\Phi_e = \left(V_{Fpn_{rad}} + \Delta V_{F_{rad}}\right) \cdot I_{rad}. \tag{14}$$

3.4. Modelling the Emitted Luminous Flux

A major target for luminaire design is the total emitted luminous flux, Φ_V, of an LED, which can be calculated from its emitted radiant flux, Φ_e.

The emitted luminous flux, Φ_V, of a light source is the quantity in which it is considered, how sensitive the human eye is to the emitted optical power at different wavelengths of the spectrum of the emitted light. This sensitivity is represented by the spectral luminous efficiency function, $V(\lambda)$. The ratio of the emitted total luminous flux and the emitted total radiant flux of a light source is thus characteristic of the spectral power distribution of the light source and is called the luminous efficacy of radiation [41], denoted by K. (The symbol of luminous efficacy of radiation is denoted by the capital letter Greek kappa that looks similar to the Latin letter K.) From the definition of K, it concludes that:

$$\Phi_V = \Phi_e \cdot K. \tag{15}$$

In the case of LEDs, the forward current and junction temperature dependence of K (caused by the current and temperature induced changes of their spectral power distributions) can be accurately described by a bi-quadratic empirical model:

$$K(I_F, T_J) = \left(a_{Kap} \cdot T_J^2 + b_{Kap} \cdot T_J + c_{Kap}\right) \cdot I_F^2 + \left(d_{Kap} \cdot T_J^2 + e_{Kap} \cdot T_J + f_{Kap}\right) \cdot I_F + \left(g_{Kap} \cdot T_J^2 + h_{Kap} \cdot T_J + i_{Kap}\right). \tag{16}$$

In Table 1, an overview of the model parameters is shown. The extraction of the parameters occurs in a heuristic way. At low currents, the R_S and R_R resistors can be neglected, thus we can assume that $V_F = V_{Fpn} = V_{Fpn_{rad}}$. I_F is forced, thus it is known. The current component I_{rad} can be determined from a radiometric measurement, so the I_{dis} current component can be produced as a difference.

Table 1. List of the parameters of the analytical quasi black-box model.

Input quantities: I_F, T_J			
Output quantities: V_F, P_H, Φ_e, Φ_V **Model parameters**			**Symbol in Spice diode model**
Thermal		T_{ref}, V_T	TNOM, UT
Electrical	entire LED	radiative branch only	
diode internal pn-junctions *at T_{ref}*	I_0, m	$I_{0_{rad}}$, m_{rad}	IS, N
resistors at T_{ref}	R_S	R_R	**RS** $= 0$ to switch of Spice's model of the series resistance
Coefficients of the model of *the "ΔV_F generators"*	a_{el}, b_{el}, c_{el}, d_{el}, e_{el}, f_{el}	a_{rad}, b_{rad}, c_{rad}, d_{rad}, e_{rad}, f_{rad}	-
Coefficients of the efficacy of *radiation model*	a_{Kap}, b_{Kap}, c_{Kap}, d_{Kap}, e_{Kap}, f_{Kap}, g_{Kap}, h_{Kap}, i_{Kap}		-

At a few low current points, the I_0, m, $I_{0_{rad}}$, and m_{rad} parameters can be calculated (similar to the way hinted later in Figure 10). The series resistances can be derived from a few V_F and Φ_e points at higher currents. Iterative trials can be conducted to have a better fit in the lower or higher current range for the optical ports. A better fit can be achieved with a global optimization scheme for all parameters on all measured V_F and Φ_e points. However, with this mathematic apparatus, better accuracy can be achieved with the concept presented in Section 4.

A Spice implementation of the LED model (without the luminous flux model) is shown as the LTspice circuit schematics in Figure 7.

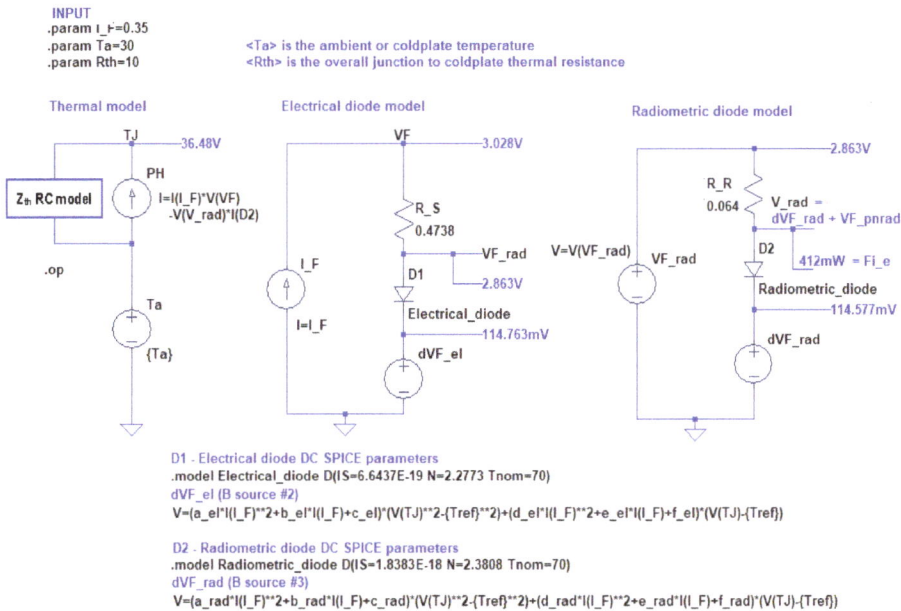

Figure 7. The generic Spice implementation of the chip level multi-domain LED model in LTspice shown as a schematic (resulting voltages for an actual parameter set and an actual setting of the forward current are also indicated in the circuit diagram). With an appropriate dynamic compact model of the environment (referred to as the Z_{th} RC model in the circuit diagram), the model also describes LED operation under alternating current driving conditions.

4. An LED Chip Multi-Domain Model Closer to Physics

The accuracy of the quasi-black box model (in terms of the radiant/luminous flux) for all the measured and modelled LED samples was good (within cca. 1%, as it will be shown later), but this accuracy was limited to about 1.5 decade of the forward current. Therefore, in order to better capture both sides of the maximum efficacy (shown in Figure 2), the range of this good fit had to be extended. In our quasi black-box model, a single ideality factor was assumed, while more realistic diode models use at least a second ideality factor for the regime of low forward currents. In the following subsections, the method of how our second multi-domain LED model was set up is presented. In our discussion, we rely on the basics of semiconductor physics that are detailed in well-known textbooks [37,38], also summarized in [42].

4.1. The Physical Roots

Up to date power LED devices approximately have the band diagram shown in Figure 8 (based on [38,43]). The active portion of the LED die consists of "bulky" p and n confinement layers, an electron blocking layer (EBL), and a sandwich of quantum barrier layers (QB) and quantum wells (QW). The bandgap in quantum wells defines the wavelength of the emitted light; a wider bandgap of the quantum barrier ensures low absorption of emitted photons and a high efficiency of the charge injection into the quantum wells.

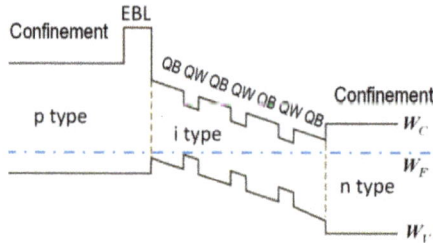

Figure 8. Band structure of a multi-quantum well (MQW) LED device (based on [38,43]).

In the stationary (DC) condition, the quantity of charge carriers in a material section is defined at a given temperature by the following factors:

- Number of generated electrons/holes and their recombination through different mechanisms within the section. These effects can be thoroughly treated by the collision theory, but a simplified quantity called the generation and recombination rate can be used in the treatment of the statistical mechanics, too.
- Injection from the adjacent sections, determined by the band structure near the interface.

The imbalance in these phenomena results in a current through all sections of the structure. The continuity equation from statistical mechanics dictates that this current is the same in all sections of the structure, but the actual carriers can be quite different. The exchange of the charge carriers occurs through generation and recombination effects. In the forward direction, the external driving force (voltage) diminishes the potential barriers in the band structure.

In "bulky" diodes, the diode characteristics in the forward direction can be calculated from the thermionic injection of the electrons confined in one side of the junction to the other similar side. This injection is driven by the diffusion of charge carriers that have a higher energy than the potential barrier separating them from the adjacent material section. As the distribution of the carriers between energy levels is approximately exponential, when an applied forward voltage, V_F, lowers the barrier, exponential growth in the diffusion current is experienced.

By solving the appropriate equations, one obtains a closed analytical formula under simplified conditions. These conditions are such that in the semiconductor material composed of adjoining ideal depleted and neutral zones, there is no particle generation and recombination in the depleted zone, and particles move until charge neutrality is reached in non-depleted zones.

The result is the well-known Shockley equation. By applying the forward voltage, V_F, on the diode, the forward current, I_F, is obtained as given in Equation (1), and reformatted for the forced forward current, I_F, in Equation (2). In Equation (3), $I_F \cdot R_S$ describes the additional voltage drop on the diode. R_S embodies the physical series resistance, but also some other effects, which do not follow the exponential equation.

The number of the movable particles quickly grows with the temperature. Solving the related equations of solid state physics, we can obtain that besides the thermal voltage, V_T, introduced earlier, the I_0 parameter is also temperature dependent:

$$I_0 = G \cdot T_j^\vartheta \cdot \exp\left(-\frac{W_g}{k \cdot T_j}\right), \tag{17}$$

where T_j (with a lowercase *j*) denotes the junction temperature in Kelvins, and W_g is the bandgap energy of the semiconductor introduced earlier. The coefficient, *G*, in Equation (17) and in further equations lumps appropriate constants that represent different material parameters. The value of ϑ for homojunctions is 3 (see, e.g., [37] or [42]). This equation is expressed in the same way as in Equation (5), but is now rewritten for an absolute T_j junction temperature, instead of a temperature change around T_{ref}.

When a 3D zone faces a 2D well, a complex injection mechanism can be identified: Thermionic emission for the particles above a potential barrier, tunneling into and over the barrier below the barrier height, etc. These equations can be solved numerically [43,44], and the results can be approximated with a value of ϑ, as shown in Equation (17), that is significantly higher than 3.

By solving the equations of the charge distribution for several sections of the semiconductor, such as the depletion layer at the pn interface, quasi-neutral layers around the depletion layer, and the quantum wells, several equations can be obtained:

$$I_{F_X}(V_F) = I_{0_X} \cdot \exp(V_F/(m_X \cdot V_T)). \tag{18}$$

Some dominating effects can be treated with the collision theory, for more detail, see, e.g., in [38]. In the following paragraphs, these effects are summarized briefly.

A. Monomolecular recombination, where the electrons are recombined at irregularities of the crystalline structure, which are present in a steady number. Such centers of recombination can be dislocations, section boundaries, etc.

The recombination rate can be written as $R = A \cdot n$, where *A* is the recombination probability and *n* is the electron density. The original Shockley-Read-Hall (SRH) recombination model treats the recombination at deep level traps in the band structure. The energy and momentum of the particle will be transferred to the partner in the recombination.

This kind of recombination produces current constituents of the $I_{F_A}(V_F) = I_{0_A} \cdot \exp(V_F/(2V_T))$ style ($m_A = 2$), and becomes saturated at high particle concentrations. Accordingly, in models of bulky diodes, the depleted region where the particle concentration is low is taken into consideration.

B. Bimolecular recombination, where the electrons and holes participate in a band to band recombination, in the same section. The recombination rate can be written as $R = B \cdot n \cdot p$.

This kind of recombination adds $I_{F_B}(V_F) = I_{0_B} \cdot \exp(V_F/V_T)$ constituents ($m_B = 1$). The energy difference between the electron and the hole will be emitted as a photon.

C. Auger recombination, an electron and a hole recombine in a band-to-band transition and give off the resulting energy to another electron (or hole).

Three particles have to be present. The related recombination rate is $R = C \cdot n \cdot n \cdot p$. The probability of this mechanism increases strongly with the carrier concentration.

The current component is not calculated as a closed analytic formula because at high carrier densities, the assumptions defined above do not apply, and the growth would be steeper than $I_{F_C}(V_F) = I_{0_C} \cdot \exp(V_F/V_T)$.

In all models, the Auger constituents can be lumped into the $I_F \cdot R_S$ term as they appear at high forward currents.

In the forward direction, the generation of carriers in a section can be neglected.

Further effects with the treatment of statistical mechanics are:

1. Diffusion of minority carriers at low concentrations, that is, in low numbers compared to the density of the majority carriers. This yields $I_{F_d}(V_F) = I_{0_d} \cdot \exp(V_F/V_T)$, corresponding to Equation (1) with $m = 1$.
2. Diffusion of both types of carriers at high concentrations, when the quasi-neutrality principle makes the density of both carriers equal. This yields a current constituent in an $I_{F_q}(V_F) = I_{0_q} \cdot \exp(V_F/2V_T)$ form.

We can observe that all effects contributing to the junction current correspond to the general form of Equation (18), where m_X is 1 or 2.

In the standard treatment of text books, disjoint locations in the structure are defined with dominant effects, which are then added up. As such, they claim the $\ln(I_F)$-(V_F) chart shown in Figure 9a, with sharp slope breaks at the change of the current forwarding mechanisms. However, real diodes do not expose this alteration of the slope (Figure 9b), producing a continuous slope over several orders of magnitude.

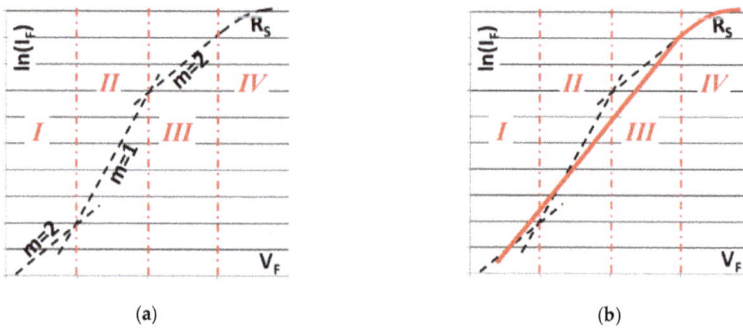

(a)   (b)

Figure 9. Current ranges of a pn-junction, I—recombination in the depletion layer, II—minority carrier diffusion, III—ambipolar diffusion, IV—serial resistance: (**a**) approximation of an LED I-V characteristic in the current ranges I–IV by means of theoretical values of the ideality factor m and an assumed series resistance R_S (**b**) real LED I-V characteristic covering the current ranges I–IV.

At this point, the text books typically make a sloppy step, stating that an average ideality factor, m, can be used without too much further examination. A more established statement outlining a step to consider separate ideal depleted and neutral zones, and then to add up the currents belonging to disjoint zones is simple, but not true.

In reality, there is more than a zero charge in the depletion zone, the frontier of it to the p and n side is fuzzy, and charge neutrality is questionable. In the treatment of collision theory, instead of statistical mechanics, proportionality would be experienced among different charge movement mechanisms, and one mechanism would establish a particle surplus in a zone (e.g., injection), which enhances the probability of another mechanism (e.g., recombination). Until one such mechanism gets exhausted (for example, SRH recombination decays when deep level traps are mostly filled), a product of the mechanisms is observed:

$$I_F(V_F) = I_{0_a} \cdot \exp[V_F/(m_a \cdot V_T)] \cdot I_{0_b} \cdot \exp[V_F/(m_b \cdot V_T)] \\ \cdot I_{0_c} \cdot \exp[V_F/(m_c \cdot V_T)] \dots = I_F \cdot \exp[V_F/(m \cdot V_T)], \tag{19}$$

where $I_0 = I_{0_a} \cdot I_{0_b} \cdot I_{0_c} \dots$ and $1/m = 1/m_a + 1/m_b + 1/m_c + \dots$.

4.2. Modeling the Electrical Characteristics

By using the proportionality concept, verified by the measurements on many actual LED samples (with peak wavelengths chosen from different ranges of the visible range), m appears in Equation (17):

$$I_0 = G \cdot T_j^\vartheta \cdot \exp\left(-\frac{W_g}{m \cdot k \cdot T_j}\right) = G \cdot T_j^\vartheta \cdot \exp\left(-\frac{V_g}{m \cdot V_T}\right). \tag{20}$$

Figure 10 shows the measured V_F of a royal blue power LED at several I_F currents, from 10 mA to 1 A at a fixed $T_J = 85\,°C$. A Shockley type operation can be observed at low currents. At higher currents, the R_S series resistance represents all losses which do not belong to the basic pn-junction operation.

Instead of just fitting a logarithmic trendline on a few points, the R_S value in Equation (2) can be optimized such that the $V_R = I_F \cdot R_S$ voltage at each measured I_F is subtracted from the measured V_F points (blue + markers) and the R_S value which yields the best logarithmic fit on the V_F–V_R relationship in the 10 mA to 1000 mA range is searched for. Herein, this optimization method is referred to as *OPT1*.

Applying *OPT1* on the measured points, the curve of the $V_{Fpn} = V_F - V_R = C \cdot \ln(I_F) + D$ format is obtained. The parameters, m and I_0, of Equation (2) can be calculated from C and D as $m = C/V_T$ and $I_0 = \exp(-D/C)$, respectively.

Repeating the *OPT1* procedure to all measured temperatures allows the temperature dependence of the parameters to be measured. First, the results of the method at various T_J junction temperatures were demonstrated, on a royal blue LED (Cree's XPE2 type).

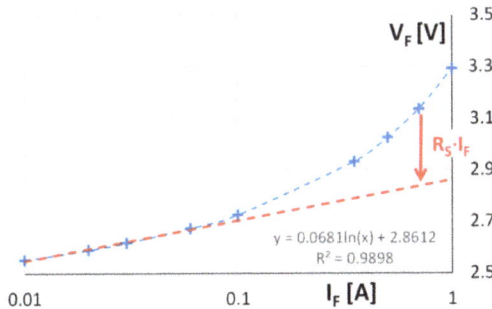

Figure 10. Measured forward characteristics of a power LED (+ markers) and the logarithmic equation of best fit.

The isothermal $V_F(I_F)$ characteristics of the LED were recorded at a number of current values, between 1 mA and 1000 mA, at four different T_J junction temperatures, namely at $T_J = 30, 50, 70$, and 85 °C.

The blue x markers in Figure 11a show the fitted R_S series resistance values generated by the *OPT1* method at the given temperatures. The temperature dependence in this range is rather "flat", so a good fit can be achieved by a second order polynomial approximation. For the model parameter, m, the optimization yields several values around $m = 2.5$ (the blue x markers in Figure 11b). Figure 12a presents the I_0 values for the best fit.

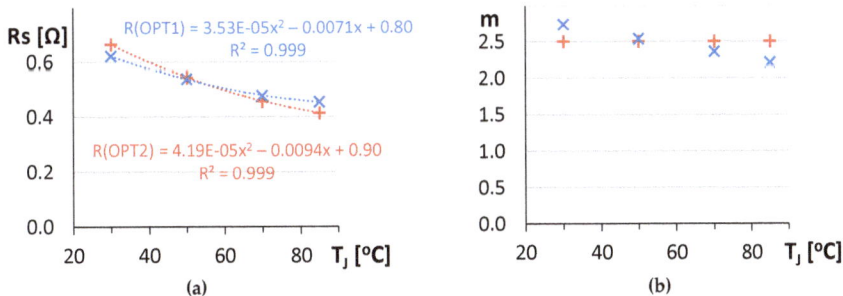

Figure 11. Results of the *OPT1* and *OPT2* optimization methods: (a) R_S series resistance; (b) for the m ideality factor—fixed at $m = 2.5$ in the case of the *OPT2* method.

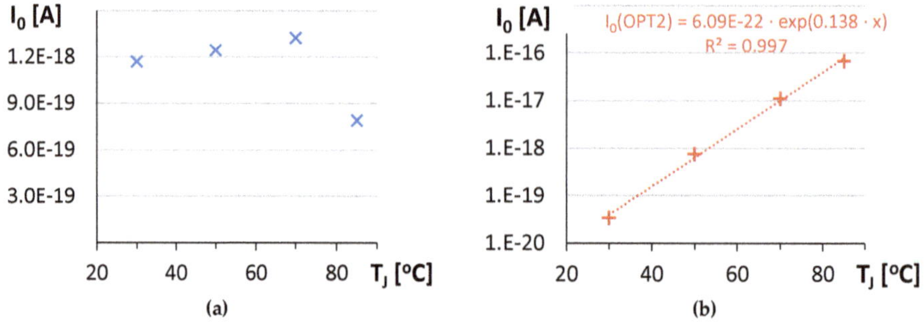

Figure 12. Results obtained for the I_0 current coefficient (saturation current): (**a**) obtained by the *OPT1* method; (**b**) obtained by the *OPT2* method.

In Equation (2), the forward voltage, V_F, depends on $m \cdot \ln(I_0)$, and accordingly small numeric perturbations in the calculated m values can cause unacceptable scatter in the I_0 values, often several orders of magnitude. Supposing that the distribution among different carrier transport mechanisms does not change much in the temperature range modeled, the m value can be fixed.

A similar optimization algorithm denoted by *OPT2* was also defined, where m is fixed, and the best (logarithmic) fit for R_S is targeted again.

Applying *OPT2* on the investigated XPE2 LED samples with $m = 2.5$ chosen, some accuracy of the fitting is lost, but the wild scatter of I_0 changes to demonstrate a reasonable trend (see the red + markers in Figure 11a,b, and Figure 12b).

The trend of I_0 in Figure 12b can be described by several mathematical formulae. The text books of solid state physics [37] offer several acceptable concepts.

The minimum expectation is that the formula should be valid in the relatively narrow temperature range of the measurement ($T_j = 300$ K to 360 K in absolute temperature).

The deduction in the related books can be found for the pn-junction in a bulk semiconductor, where $\vartheta = 3/m$. Thus, Equation (20) will be written in this case as:

$$I_0 = G \cdot T_j^{3/m} \cdot \exp\left(-\frac{V_g}{m \cdot V_T}\right),\tag{21}$$

and for a smaller range, ΔT_j, the change around T_j is:

$$\frac{d}{dT}[\ln(I_0)] = \frac{1}{I_0}\cdot\frac{dI_0}{dT} = \frac{3}{m \cdot T_j} + \frac{V_g}{m \cdot T_j \cdot V_T},\tag{22}$$

resulting in the following ratio of the saturation currents at a ΔT_j temperature shift:

$$\frac{I_{02}}{I_{01}} \approx \exp\left[\left(\frac{3}{m \cdot T_j} + \frac{V_g}{m \cdot T_j \cdot V_T}\right)\right]\cdot\Delta T_j.\tag{23}$$

According to this, the I_0 current coefficient plotted in Figure 12b can be approximated by:

$$I_0 = I_{01} \cdot \exp\left(a \cdot \Delta T_j\right),\tag{24}$$

where $I_{01} = 6.09 \times 10^{-22}$ and $a = 0.138$ in this particular example. The fit is correct in the given range with an R^2 determination coefficient of 0.9972.

A more ambitious attempt is through the direct use of Equation (20). By dividing the formula with the exponential part $\exp(-V_g/(m \cdot V_T))$, the following equation is obtained:

$$I_0 / \exp\left(-\frac{q \cdot V_g}{m \cdot k \cdot T_j}\right) = G \cdot T_j^\vartheta. \tag{25}$$

The *OPT2* optimization process already yields I_0 at several temperatures, but to construct the left side of Equation (25), the values of the V_g bandgap voltage are needed. As shown earlier by Equations (6) and (7), this can be calculated from the λ_{peak} peak wavelength values obtained from the LEDs' measured spectral power distributions. The temperature and current dependence of λ_{peak} can also be taken into consideration, as presented, e.g., in [5].

In Figure 13, this "power of ϑ" dependence of a royal blue XPE2 LED is illustrated, for which $\vartheta = 6.6531$ was obtained with an R^2 determination coefficient of 0.9966.

Two optimization steps were conducted on the measured data of several LED devices of different colors (in this particular example: Members of Cree's XPE2 family).

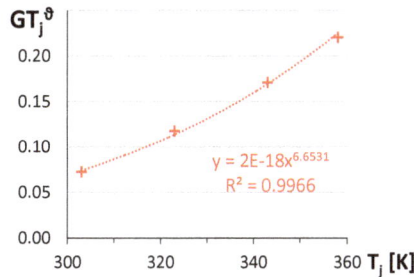

Figure 13. Modeling the ϑ parameter.

In Table 2, the measured λ_{peak} values of these and the calculated bandgap voltages, V_g, are listed. For the white devices, the peak wavelength of the blue part of their spectra was used.

In Table 3, the modeling parameters for all examined XPE2 devices are listed. To save space, only the minimum and maximum values of R_S in the given temperature range are presented. In the tables, a color identification code is assigned to the LEDs, which is used below to denote them.

Table 2. The color, identification code, peak wavelength, and bandgap voltage of the measured LEDs.

Color	Code	λ_{peak} (nm)	V_g (V)
Royal Blue, White	W, RB	450	2.76
Blue	B	470	2.64
Green	G	515	2.41
Amber	A	600	2.07
Red	R	640	1.94

Table 3. Model parameters for XPE2 LEDs obtained by the *OPT2* optimization process.

Code	R_S max (Ω)	R_S min (Ω)	m	ϑ
RB	0.662	0.413	2.5	6.6531
W	0.824	0.453	1.8	5.8421
B	0.610	0.277	3.0	7.7791
G	0.553	0.298	4.0	11.1289
A	0.508	0.525	1.5	8.8466
R	0.773	0.688	1.5	6.5855

4.3. Modeling the Light Output Characteristics

In the previous subsection, an explicit model for the description of the LED behavior in the electric domain was created, i.e., the applied forward current, I_F, directly determines the forward voltage, V_F, using closed analytic formulae. Besides providing an improved method to capture the phenomena related to the carrier transport and recombination processes in LEDs' active region, the use of explicit formulae is a clear advantage compared to the quasi black-box model described in Section 2, which requires that Equations (10) and (13) are solved iteratively (in the Spice circuit, macro implementations of the models of the circuit simulation algorithm are inherently iterative processes, so this difference between the two models is not relevant. However, in simulation environments, such as that depicted in Figure 4, avoiding internal iteration within the LED chip model is a clear advantage).

As hinted before, the optical domain is traditionally characterized by a bunch of current and temperature dependent parameters. The internal and external quantum efficiency (IQE and EQE) together yield the η_e energy conversion efficiency also known as the radiant efficiency or wall plug efficiency (η_e = WPE = IQE·EQE). Then, η_e leads to the emitted total radiant flux, Φ_e (also known as the emitted optical power), and the efficacy (η_V) or the luminous efficacy of radiation (K) metrics lead to the emitted total luminous flux, Φ_V. The chip level multi-domain LED model has to provide the Φ_e and Φ_V fluxes as output, therefore in the following, their modelling is provided.

Following the pattern of the calculation of the voltage drop of the "radiant diode" described in [5] (see also Figure 5), first, a new, virtual efficiency figure is introduced, which is called the radiant voltage:

$$V_{rad} = \frac{\Phi_e}{I_F}. \tag{26}$$

It will be later used in explicit models of the emitted total fluxes. The radiant voltage, V_{rad}, is tied to the energy conversion efficiency, η_e, because $\eta_e = \Phi_e / (I_F \cdot V_F) = V_{rad} / V_F$.

It must be noted that V_{rad} is merely a characterization tool, and ageing effects that degrade the external quantum efficiency, like graying of the lens, can significantly change it.

The measurements provided here prove that in a wide current range, where power LEDs are typically operated, V_{rad} can be approximated by a product of two Shockley style functions.

In the previous subsection, it was shown that a constant m parameter, over more orders of magnitude in current, expresses the proportionality between different charge transport mechanisms. Now, one blend of transport effects feeds the radiant recombination by injecting carriers into the quantum well, while other effects (such as the Auger recombination that enhances electron leakage over the electron blocking layer) deport the carriers before they can recombine. Detailed numeric analysis of the actual material compositions and geometries is broadly treated in the literature, like [43] or [44].

By maintaining the concept of proportionality that is underpinned by constant m in the forward current range of interest, whether the measurements support a:

$$V_{rad} = [a \cdot \ln(I_F / I_{0a})] \cdot [b \cdot \ln(I_F / I_{0b})], \tag{27}$$

type formula can be checked. The introduction of the notation, $z = \ln(I_F)$, $c = \ln(I_{0a})$, $d = \ln(I_{0b})$, allows Equation (27) to be rewritten as follows:

$$V_{rad} = a \cdot b \cdot (z - c) \cdot (z - d) = ab \cdot (z^2 - zc - zd + cd), \tag{28}$$

which can be further simplified to:

$$V_{rad} = Az^2 + Bz + C. \tag{29}$$

Figure 14 shows the V_F and V_{rad} values of the royal blue and phosphor converted white LEDs measured at different junction temperatures. It was found that the fitting of a quadratic polynomial with *A*, *B*, and *C* coefficients corresponding to Equation (29) obtained an excellent match on the 10 mA to 1000 mA region of the I_F total forward current, with an R^2 determination coefficient of around 0.99.

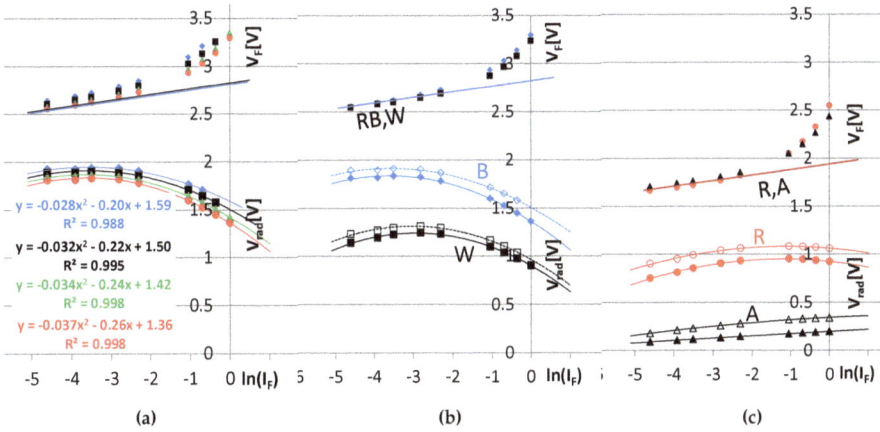

Figure 14. V_F and V_{rad} values of a royal blue LED sample measured at $T_J = 30, 50, 70$, and 85 °C: (a) Shockley approximation of V_F, log-quadratic fitting curves of V_{rad} and their determination coefficient are also shown; (b) comparison of a royal blue (RB) and white (W) LED sample measured at $T_J = 50$ and 85 °C; (c) comparison of a red (R) and an amber (A) LED measured at $T_J = 50$ and 85 °C.

The η_e energy conversion efficiency can be gained by dividing the V_{rad} values with the V_F forward voltage, which is also shown in the diagrams of Figure 14. This way, through V_{rad}, an explicit forward current dependent model for the radiant flux is given: $\Phi_e = V_{rad} \cdot I_F$.

Figure 14a,b show the quadratic fit on $\ln(I_F)$ for the royal blue and white devices, while Figure 14c shows the same for the red and amber devices.

The charts in Figure 15 present the temperature related changes of the *A*, *B*, and *C* parameters used in Equation (29), for all colors. The temperature dependence is rather flat again; it was found that a second order polynomial can be fitted at $R^2 > 0.99$.

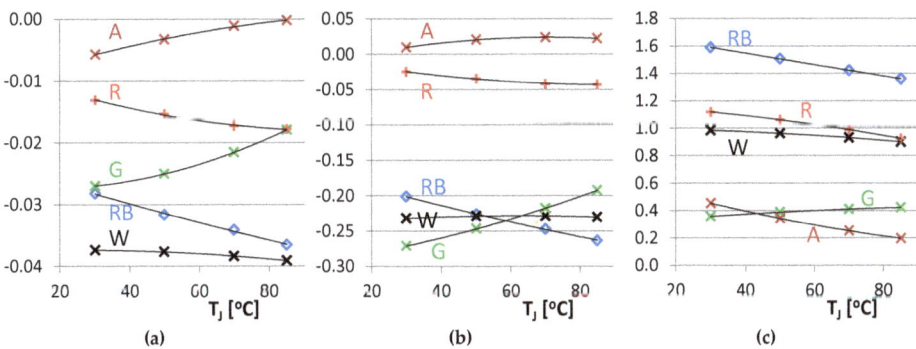

Figure 15. Temperature dependence of parameters *A*, *B*, and *C* used in Equation (29), for all LED colors: (a) parameter *A*; (b) parameter *B*; (c) parameter *C*.

4.4. The Predictive Power of the Physics Based Model

To prove that the models defined so far describe the electrical and optical domain in the current and temperature range of interest accurately the Cree XPE2 power LED family was characterized. All measurements were done in a combined thermal transient and optical measurement arrangement [34] that complies with CIE's and JEDEC's recent recommendations and standards on the optical and thermal testing of power LEDs [32,39,45].

Five different color LED types were measured, using multiple numbers of samples from each type. To make the modelling easier, the cold plate control mode was set to regulate towards a fixed T_J junction temperature as described in [31]. This ensured that the measured DC characteristics and the radiometric and photometric values corresponded exactly to the thermal boundary of a constant T_J, without further model correcting steps. Current levels were selected between 20 mA and 1 A, and T_J temperatures ranged from 30 to 85 °C.

In order to demonstrate the predictive power of the modeling concept, the red devices from Table 2 were selected because their η_e radiant efficiency has the largest temperature and current dependence (Figure 16b). Still, it is expected that the model extracted from a single median device accurately represents the devices in the measured binning class. (The median device is the LED sample, which is closest in terms of one if its selected parameters to the median value of the distribution of that parameter obtained for the entire LED population)

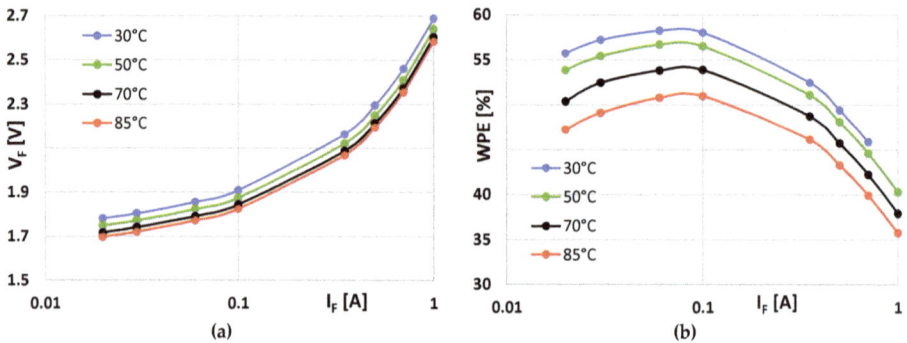

Figure 16. (a) Measured I_F–V_F characteristics; (b) measured radiant efficiency, η_e, of the median sample of a red LED population, at various junction temperatures, T_J.

The electrical model of red LEDs was characterized by just a few parameters in the previous subsection. The R_S values were approximated by a parabolic function with values around 0.7 Ω, m was set to 1.5, and for the "power of ϑ" dependence of the I_0 saturation, $\vartheta = 6.59$ with an R^2 of 0.99 was obtained.

Selecting the LED of median V_F value from the bin, the I_F–V_F plots at four T_J temperatures were measured as shown in Figure 16a.

In Figure 17a, a 3D plot of the relative error of the modeled V_F forward voltage is shown. The chart compares the calculated forward voltage, V_F, from the Shockley model of the median device to its own measured values from Figure 16a. The accuracy of the model is within ±0.5% on the whole current and temperature range.

The figure and also the subsequent similar figures show the 20 mA to 700 mA range, in which all T_J values were reached. The colors correspond to a ±4% range of error in the figures, representing an error of V_F and a ±8% range showing the error of Φ_e.

A "worst case" situation is shown in Figure 17b. Here, the measured forward voltage of the device which has the lowest V_F is compared to the model based on the median device. Still, the error remains in the 0% to 2% range.

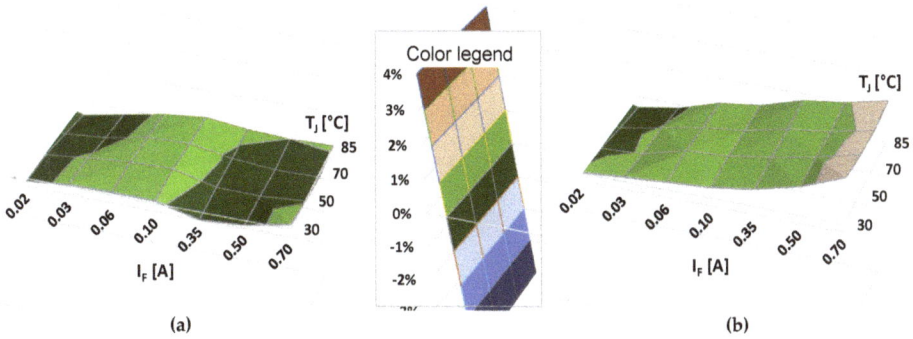

Figure 17. Relative error of the I_F–V_F model of red samples, (**a**) measured V_F of the median sample; (**b**) measured V_F of the sample with minimum V_F, both compared to the calculated V_F of the model, fitted on the parameters identified for the median sample.

In Figure 18 the modeled Φ_e radiant flux of the "median V_F" red sample to the measured Φ_e of the "median V_F" sample itself (Figure 18a) and to the measured Φ_e of the "lowest V_F" sample (Figure 18b) is compared. Again, the self-modeling is accurate to 1%, the scatter between samples is higher, and the Φ_e radiant flux of the least similar sample deviates a few percent from the median device.

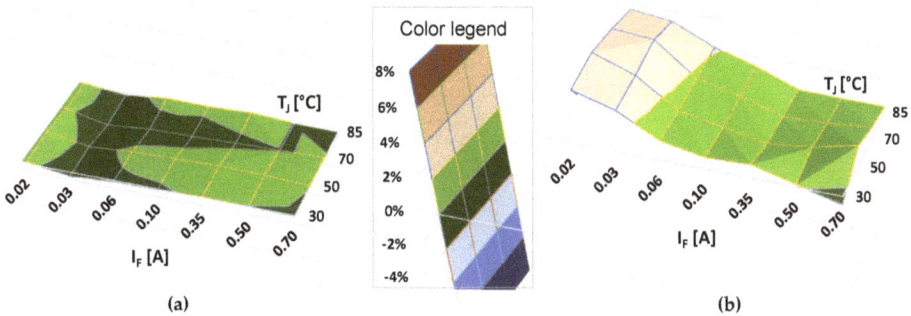

Figure 18. Relative error of the I_F–Φ_e model of red samples, (**a**) measured Φ_e of the median sample; (**b**) measured Φ_e of the sample with minimum V_F, both compared to the calculated Φ_e values of the model, fitted to the parameters identified for the median sample.

The test was also carried out on royal blue and phosphor converted white samples, composed of a similar royal blue chip and a yellow phosphor layer. It was found that the V_F scatter was very low—obviously the same as in the case of the royal blue devices—and the "median" white LED modeled its own radiant flux within 1% (Figure 19).

However, the white sample of the lowest V_F showed up to a −8% to +18% difference of Φ_e in the 30 mA to 1000 mA forward current range. This may have occurred due to the larger scatter in the phosphor quality and also indicates that the binning was performed at a single (I_{F_bin}, T_{J_bin}) operating point (Figure 20).

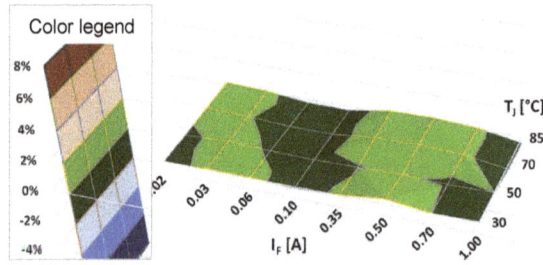

Figure 19. Predictive power of the I_F–Φ_e model for white LEDs: relative error, measured vs. modeled Φ_e, of the median sample.

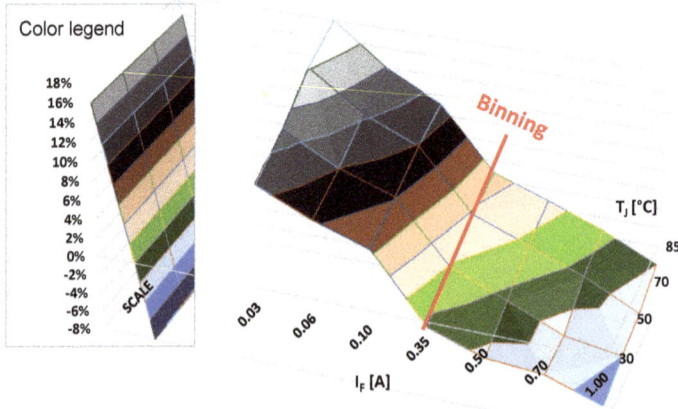

Figure 20. Predictive power of the I_F–Φ_e model for white LEDs: measured Φ_e of the minimum V_F sample vs. the Φ_e model generated from the median sample.

This finding corresponds to the results of the variability analysis of the thermal properties of the blue and white LED samples [46,47], where a significantly larger scatter of the properties of the white samples was found.

4.5. The Actual Implementation of the Physics Based Model

The physics based model differs from the ones hardcoded in all Spice program versions for semiconductor diodes. Therefore, a circuit macro in visual basic was created in which the controlled voltage or current sources directly represent the model equations.

It must be emphasized that this model is based on explicit formulae; I_F and T_J yield all output data directly, without internal iterations as implied by the circuit scheme in Figure 5c.

The I_F–V_F relationship in this model is coded as Equation (2). The R_S series resistance has the quadratic temperature dependence of Equation (4), as justified in Figure 11a. The temperature dependent I_0 saturation current was calculated based on Equation (20).

In all equations, the V_T thermal voltage and the V_g bandgap voltage were calculated for the actual T_j junction temperature (given in Kelvins). In the bandgap voltage formula, the peak wavelength was approximated by a relationship derived from Equations (7) and (8) for values of T_j close to 300 K:

$$\lambda_{peak} = L_{p1} \cdot T_j{}^2 + L_{p2}. \tag{30}$$

The emitted total radiant flux (optical power) was calculated with:

$$P_{opt} = \Phi_e = V_{rad} \cdot I_F. \tag{31}$$

As supported by Figure 15, the modest change in the temperature dependence of V_{rad} can be reflected by quadratic tuning equations:

$$
\begin{aligned}
V_{rad} = & \left(a\_A \cdot T_J^2 + b\_A \cdot T_J + c\_A\right) \cdot (\ln I_F)^2 + \\
& + \left(a\_B \cdot T_J^2 + b\_B \cdot T_J + c\_B\right) \cdot (\ln I_F) + \\
& + \left(a\_C \cdot T_J^2 + b\_C \cdot T_J + c\_C\right).
\end{aligned}
\tag{32}
$$

The heating power was calculated again, like in the quasi black-box model, using Equation (9). All the coefficients in these equations are model parameters that must be identified for a given LED type, given now by the systematic methodology of subsequent *OPT1* and *OPT2* operations.

For the Φ_V luminous flux, several approaches were tested, such as deriving it from Φ_e through the fitting of polynomials or searching for a polynomial on the raw I_F. Finally, it was found that best fidelity over a wide current range can be achieved with a polynomial on $\ln I_F$, Which was calculated similarly to Equation (32), with a different set of constants.

A Spice implementation of the proposed new model was created. A part of it representing the electrical characteristics by controlled sources is shown in Figure 21.

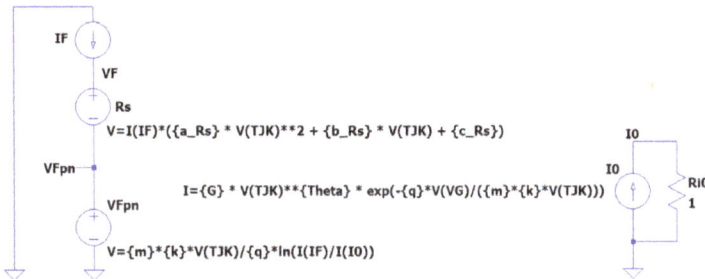

Figure 21. LTspice implementation of the electrical characteristics of LEDs based on our proposed new model, with controlled sources.

The T_J junction temperature can be computed by inserting the thermal model of the packaged LED into the thermal model of the environment. These models are deeply treated in [7,48].

A simple Foster-style equivalent thermal chain representing the device and the outer world is shown in Figure 22. The conversion from the T_J junction temperature in Celsius to T_j in Kelvin yields the TJK "voltage" for the model calculations in Figure 21.

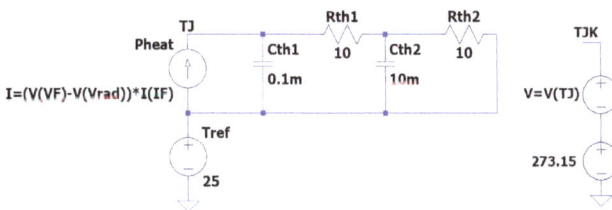

Figure 22. Spice implementation of the LED as a heat source and a simplified equivalent Foster chain representing the device and its environment in the thermal domain.

In Figure 23, the response of a blue XPE2 LED from Cree is monitored in its thermal environment represented by the simple compact thermal model. First, heating can be observed. The current grows linearly from 10 mA to 1 A in the first 1 μs (the current was multiplied by 10 in order to make it visible in the chart. The curvature corresponds to linear growth in the log-lin scale).

In the further two seconds, the increase of the temperature (25 to 70 °C) and the decay of the forward voltage and radiant flux can be observed. The "bumps" of the curves correspond to the thermal time constants in Figure 22, $\tau_1 = 1$ ms, $\tau_2 = 100$ ms.

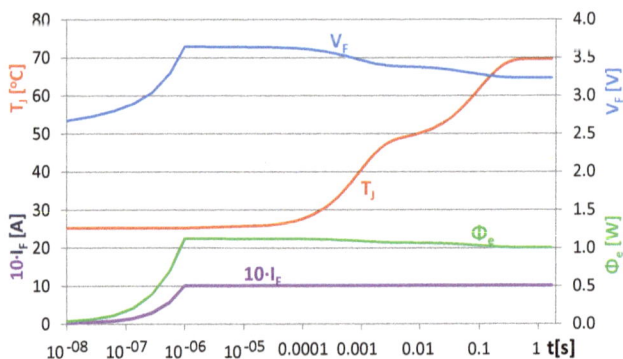

Figure 23. Transient simulation, LED switched on to a $I_{heat} = 1$ A heating current. Values of I_F, T_J, V_F, and Φ_e shown, simulated with the LTspice equivalent circuit implementations of the model equations and combined with the simple dynamic compact thermal model shown in in Figure 22.

5. Multi-Domain Modelling of LEDs for Real Design Tasks

5.1. Modeling an "LED type" and Parameter Extraction

Several approaches to represent an "LED type" based on the measurement results obtained for a given LED population exist. As mentioned earlier, in the Delphi4LED project, a specific task was devoted to study the sample-to-sample variations of LEDs and provide statistical models to describe these variations. Within this work, the properties of the junction-to-ambient heat-flow path of LED populations were investigated. The analysis was done on 11 royal blue and 10 phosphor converted white LEDs [46,47,49].

As one of the results of this study, the concept of the "median device" was established. The sample with a mechanical assembly providing the thermal resistance closest to the average of the entire population was selected as the "median device". The properties of the median device with their mean values and distributions are characteristic of the "LED type".

In terms of the IVL characteristics, nominating a median device is less straightforward [49]. In the study of the predictive power of the physics based multi-domain LED model, the median device was the one with an electrical characteristic closest to the average characteristic.

Another option is to use a combined metric, which lumps the "diode" property and the "light source" property of an LED. Such a metric could be the η_e energy conversion efficiency or the η_v efficacy. The "median device" could be the one which has an $\eta_e(I_F, T_J)$ efficiency or $\eta_v(I_F, T_J)$ efficacy surface closest to the average surface. Elaboration of the right method for identification of the "median" LED is still an ongoing activity. Some results obtained so far are reported elsewhere [49].

For demonstration purposes, however, a model parameter set for the measured LED populations was established, see Figure 24. In relation to the quasi black-box model presented in Section 2, a total of 75 LED samples of six different types were characterized in 35 operating points (the minimal number of operating points for the measurement of LEDs' isothermal IVL characteristics was defined in a round-robin test performed at the beginning of the project [34]). A global parameter fitting procedure was performed on all the individual measurement sets. The parameter sets obtained in this way represented the individual LED samples, rather than the given "LED type" or "bin of the LED type" to which the samples belonged to.

Then two average models for each LED type were set up with the intent to model the "LED type" instead of the individual samples: One by averaging the resulted model parameters of the same LED type and another one by averaging the corresponding measurement results first and performing the fitting procedure afterwards.

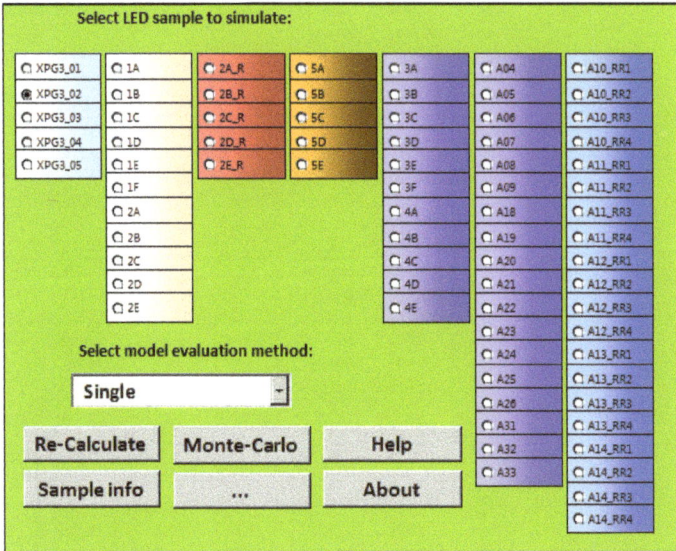

Figure 24. A snapshot of the user interface of the parameter identification tool developed for the quasi black-box multi-domain LED chip model, showing the LED families (two white, a red, an ember, and three blue LED types) and all individual LED samples for which isothermal IVL characteristics were measured. With this tool, parameter sets, as shown in Table 1, can be extracted for selected individual LED samples, or for average characteristics of the given LED population.

As an example, out of the 75 LED samples of six different LED types, Figure 25 compares the measured and modelled luminous flux values of a phosphor converted white and red high power LED sample, as a function of the forward current and junction temperature. The complete set of parameters and fitting errors in terms of the forward voltage and emitted fluxes obtained for five samples of white LEDs is shown in Figure 26.

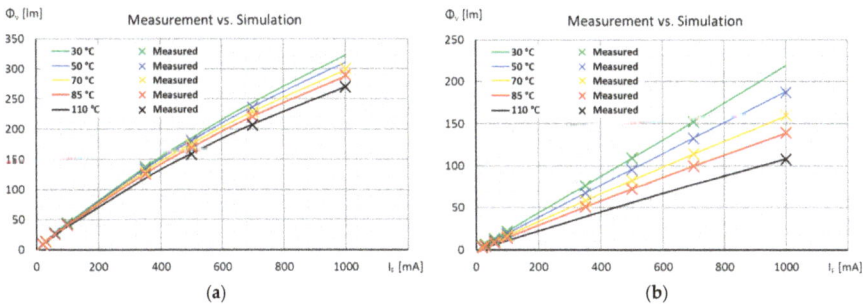

Figure 25. Luminous flux as the function of I_F and T_J, calculated by the Visual Basic macro implementation of the quasi black-box model: (**a**) a phosphor converted white LED; (**b**) a red LED.

A	B	C	D	E	F	G	H
15			4				
Sample:			XPG3_01	XPG3_02	XPG3_03	XPG3_04	XPG3_05
Max Vf error:			0%	1%	0%	0%	0%
Max Fi_e error:			1%	1%	1%	1%	1%
Max Fi_v error:			1%	1%	1%	1%	1%
	UT	=	0,0296	0,0296	0,0296	0,0296	0,0296
	I0	=	7,6395E-24	6,9812E-24	8,1736E-24	7,2375E-24	7,6335E-24
	m	=	1,7354	1,7349	1,7359	1,7353	1,7358
	R	=	0,1929	0,2141	0,197	0,2138	0,1973
	I0_rad	=	4,0317E-23	3,4889E-23	3,4027E-23	3,3826E-23	3,1585E-23
	m_rad	=	1,8150	1,8131	1,8075	1,8107	1,8089
	R_rad	=	0,0190	0,021001	0,020001	0,021001	0,019001
	a_el	=	-8,079E-06	-2,501E-06	-6,348E-06	1,973E-07	-3,155E-06
	b_el	=	2,153E-05	1,209E-05	1,762E-05	7,854E-06	1,475E-05
	c_el	=	-1,050E-06	2,362E-06	-5,003E-08	2,998E-06	-4,546E-07
	d_el	=	1,326E-03	5,207E-04	1,085E-03	1,150E-04	4,845E-04
	e_el	=	-4,104E-03	-2,919E-03	-3,586E-03	-2,243E-03	-2,982E-03
	f_el	=	-8,353E-04	-1,348E-04	-9,861E-04	-1,462E-03	-9,515E-04
	a_rad	=	-8,304E-06	-2,668E-06	-6,589E-06	2,394E-07	-3,053E-06
	b_rad	=	2,209E-05	1,261E-05	1,824E-05	8,137E-06	1,492E-05
	c_rad	=	-8,946E-07	2,563E-06	8,893E-08	3,154E-06	-1,960E-07
	d_rad	=	1,364E-03	5,481E-04	1,127E-03	1,034E-04	4,618E-04
	e_rad	=	-4,173E-03	-2,981E-03	-3,670E-03	-2,259E-03	-2,985E-03
	f_rad	=	-8,187E-04	-1,338E-04	-9,643E-04	-1,449E-03	-9,417E-04
	a_Kap	=	0,000	-0,002	-0,001	-0,002	-0,003
	b_Kap	=	-0,057	0,320	0,123	0,326	0,463
	c_Kap	=	2,306	-12,216	-5,028	-12,613	-17,489
	d_Kap	=	0,000	0,002	0,000	0,002	0,002
	e_Kap	=	0,081	-0,181	-0,028	-0,221	-0,324
	f_Kap	=	-6,896	3,296	-0,971	5,542	8,955
	g_Kap	=	0,000	0,000	0,000	0,000	0,000
	h_Kap	=	-0,066	-0,082	-0,088	-0,055	-0,057
	i_Kap	=	334,911	333,904	333,653	333,397	333,066

Figure 26. The complete parameter sets and fitting errors of the forward voltage and the emitted fluxes of the quasi black-box multi-domain LED model for a set of five samples of Cree XPG3 phosphor converted white LEDs. The physical prototypes of the luminaire aimed for project demonstration were built by using this LED type [3,4].

XPG3 white LEDs from Cree were used as a device demonstrator in the project. As an ultimate proven method for the obtainment of a model parameter set with statistical models of their distributions was lacking, the above mentioned averaging of the measured characteristics was applied before the actual parameter fitting procedure.

5.2. Applying the Quasi Black-Box Model for a Real Luminaire Design Task

The same set of visual basic macros that were used for parameter identification were also built into a spreadsheet application based on the LED luminaire design calculator [3]. This way the model applied to Cree XPG3 LEDs was used in two implementations of the foreseen Delphi4LED design processes [3,4] to predict the performance of different versions of an "outdoor 10 W LED spot" type luminaire used as the first project demonstrator.

The luminaire design calculator is equipped with a simple user interface, as shown in Figure 27. The designer has to provide the major lighting design goal in terms of the total emitted luminous flux of the luminaire and has to provide the design constraints, such as the maximum allowed temperatures and maximum total electrical input power. The third group of input parameters describes the major properties of a foreseen luminaire design variant, such as the total number of LEDs (in a series electrical configuration) to be used in the luminaire, foreseen input electric current (the common I_F forward current of the LED string), driver and optics efficiencies, and the choice of the LED module substrate and luminaire heatsink.

Once all these inputs are provided, the spreadsheet application creates the compact thermal model of the LEDs' 3D environment and calculates the LEDs' operating points corresponding to the specified ambient temperature and forward current through a relaxation type iterative solution process as illustrated in Figure 4.

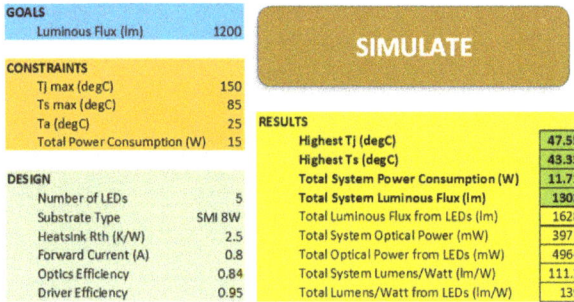

Figure 27. The user interface of the luminaire design calculator Excel spreadsheet application with the design input settings and calculated results of the final, optimized version of the first Delphi4LED demonstrator design.

In Figure 27, the calculated results for the SME-version of the first project demonstrator, the "10 W outdoor LED spot", are shown. The prototype of the luminaire has also been manufactured with the chosen number of LEDs, substrate material, layout arrangement, and heatsink.

At an independent testing laboratory, good agreement between the measured and simulated major operating parameters was found, as shown in Table 4. In terms of the total emitted luminous flux, the mismatch between the simulated and measured value was cca. 2.7% while in the case of the total input electric power, the mismatch was cca. 8%. Considering the fact that the luminaire design calculator uses a rough thermal model of the luminaire as a heat-sink, these results are satisfactory. Work is ongoing for the application of a second, physics related multi-domain LED model to other demonstrator systems of the project.

Table 4. Simulated and measured major parameters of the variant of the first Delphi4LED project demonstrator performed in the design style of an SME (small and medium sized enterprise).

Property	Simulated	Measured
Total input electric power	11.71 W	10.7 W
Total emitted luminous flux	1302 lm	1339 lm

6. Conclusions

In this paper, we first presented an implementation of the latest version of the improved chip level multi-domain LED model proposed earlier for the Delphi4LED project [6,7] as a quasi black-box model of LEDs. This model was based on formulae widely used in textbooks and built-in diode models of the Spice-like circuit simulator. We also presented a new multi-domain LED model, closer to the physics of LED operation. Both models were implemented in the form of visual basic macros and as circuit models suitable for usual "electrical only" Spice circuit simulators. In the Spice implementations, crucial elements are the different controlled sources, which allowed us to directly represent our model equations.

Both models were included in a spreadsheet application to support the design of different demonstrator luminaires of the Delphi4LED project and both models were also tested with simple LTspice simulation case studies.

The advantage of the quasi black-box model is that it can be considered as an extension of LED vendors' present Spice models of LEDs, thus, from a user's perspective, they can be used in a similar way, with a Spice solver, to the current existing Spice models of LEDs. This model calculates the luminous flux from the radiant flux, using an empirical model of the efficacy of radiation of LEDs. The disadvantage of this model is that for forward currents below the current where the efficacy peak occurs, it overestimates the light output. Another possible disadvantage of this model is that one of its equations is an implicit one, requiring the model to perform internal iterations. In the case of the design complexities encountered in the Delphi4LED project demonstrators, however, this did not result in unacceptable response times of the Excel spreadsheet application in which the model was used.

Our new proposed model, more closely related to the actual physics of the operation of modern power LEDs, is largely different from the usual textbook models. Therefore, the Spice implementation of this model cannot rely on the standard built-in models of semiconductor diodes. The major advantage of this model is that it solely uses explicit formulae to describe both the electrical and light output characteristics of LEDs. This model calculates the luminous flux directly from the forward current and the junction temperature.

Unlike the previous academic versions of Spice models for LEDs, with the present implementations of our new chip level multi-domain LED models, we targeted real life, industrial design flows aimed at LED luminaire design, such as that proposed by the Delphi4LED project. An important aspect of our approach is the strictly modular, hierarchical modelling, according to which the chip level behavior of LEDs is completely separated from the compact model of the chips' thermal environment, thus even the compact thermal models of the LED packages are external to our proposed chip level LED multi-domain models.

LED modeling is not finished here. A couple of issues still must be dealt with, such as the necessary number of LED samples to be characterized for the extraction of the parameter set that is representative of the given LED type; how to represent the sample-to-sample variations of the model parameters; how to include typical LED product parameters, such as the binning value of the forward voltage, etc. These are questions that are still the subject of research within the Delphi4LED project [49].

Author Contributions: Conceptualization, A.P. and G.F.; Data curation, L.G., G.H. and J.H.; Formal analysis, A.P., G.F. and M.R.; Funding acquisition, A.P.; Investigation, G.F., L.G., G.H. and J.H.; Methodology, L.G., G.H. and J.H.; Project administration, A.P. and M.R.; Software, L.G., G.H. and J.H.; Supervision, A.P. and M.R.; Writing—original draft, A.P., G.F., G.H. and J.H.; Writing—review and editing, A.P., G.F. and M.R.

Funding: This research received funding from the European Union's Horizon 2020 research and innovation program through the H2020 ECSEL project Delphi4LED (grant agreement 692465). Co-financing of the Delphi4LED project by the Hungarian government through the NEMZ_16-1-2017-0002 grant of the National Research, Development and Innovation Fund is also acknowledged. The modelling work of LEDs performed at BME was also funded by the K 128315 grant of the National Research, Development and Innovation Fund and was also supported by the Higher Education Excellence Program of the Ministry of Human Capacities in the frame of Artificial Intelligence research area of Budapest University of Technology and Economics (BME FIKP-MI/SC).

Acknowledgments: Support from Delphi4LED project partners, especially from G. Martin (Signify), Ch. Marty (Ingelux), R. Bornoff (Mentor, a Siemens business), D. Fournier (PISEO) and E. Morard (Ecce'lectro) is acknowledged.

Conflicts of Interest: The authors declare no conflict of interest. The funders had no role in the design of the study; in the collection, analyses, or interpretation of data; in the writing of the manuscript, or in the decision to publish the results.

References

1. Martin, G.; Poppe, A.; Lungten, S.; Heikkinen, V.; Yu, J.; Rencz, M.; Bornoff, R. Delphi4LED—From measurements to standardized multi-domain compact models of light emitting diodes (LED). *Electronics Cooling Magazine*, 4 August 2017; pp. 20–23.
2. Delphi4LED Project Website. Available online: https://delphi4led.org (accessed on 27 March 2019).

3. Marty, C.; Yu, J.; Martin, G.; Bornoff, R.; Poppe, A.; Fournier, D.; Morard, E. Design flow for the development of optimized LED luminaires using multi-domain compact model simulations. In Proceedings of the 24th International Workshop on Thermal Investigation of ICs and Systems (THERMINIC'18), Stockholm, Sweden, 26–28 September 2018. [CrossRef]

4. Martin, G.; Marty, C.; Bornoff, R.; Poppe, A.; Onushkin, G.; Rencz, M.; Yu, J. Luminaire Digital Design Flow with Multi-Domain Digital Twins of LEDs. *Energies* **2019**. submitted to the Special Issue on Thermal and Electro-thermal System Simulation.

5. Poppe, A. Multi-Domain Compact Modeling of LEDs: An Overview of Models and Experimental Data. *Microelectron. J.* **2015**, *46*, 1138–1151. [CrossRef]

6. Hantos, G.; Hegedüs, J.; Bein, M.C.; Gaál, L.; Farkas, G.; Sárkány, Z.; Ress, S.; Poppe, A.; Rencz, M. Measurement issues in LED characterization for Delphi4LED style combined electrical-optical-thermal LED modeling. In Proceedings of the 19th IEEE Electronics Packaging Technology Conference (EPTC'17), Singapore, 6–9 December 2017. [CrossRef]

7. Farkas, G.; Gaál, L.; Bein, M.; Poppe, A.; Ress, S.; Rencz, M. LED characterization within the Delphi-4LED Project. In Proceedings of the 17th Intersociety Conference on Thermomechanical Phenomena in Electronic Systems (ITHERM'18), San Diego, CA, USA, 29 May–1 June 2018. [CrossRef]

8. Poppe, A. Simulation of LED Based Luminaires by Using Multi-Domain Compact Models of LEDs and Compact Thermal Models of their Thermal Environment. *Microelectron. Reliab.* **2017**, *72*, 65–74. [CrossRef]

9. Poppe, A.; Hegedüs, J.; Szalai, A.; Bornoff, R.; Dyson, J. Creating multi-port thermal network models of LED luminaires for application in system level multi-domain simulation using Spice-like solvers. In Proceedings of the 32nd IEEE Semiconductor Thermal Measurement and Management Symposium (SEMI-THERM'16), San Jose, CA, USA, 14–17 March 2016; pp. 44–49. [CrossRef]

10. Bornoff, R.; Farkas, G.; Gaál, L.; Rencz, M.; Poppe, A. LED 3D Thermal Model Calibration against Measurement. In Proceedings of the 19th International Conference on Thermal, Mechanical and Multiphysics Simulation and Experiments in Microelectronics and Microsystems (EuroSimE'18), Toulouse, France, 15–18 April 2018. [CrossRef]

11. Bornoff, R. Extraction of Boundary Condition Independent Dynamic Compact Thermal Models of LEDs—A Delphi4LED Methodology. *Energies* **2019**, *12*, 1628. [CrossRef]

12. CIE e-ILV Term 17-738 (Luminous Flux). Available online: http://eilv.cie.co.at/term/738 (accessed on 25 April 2019).

13. Titkov, I.E.; Karpov, S.Y.; Yadav, A.; Zerova, V.L.; Zulonas, M.; Galler, B.; Strassburg, M.; Pietzonka, I.; Lugauer, H.-J.; Rafailov, E.U. Temperature-Dependent Internal Quantum Efficiency of Blue High-Brightness Light-Emitting Diodes. *IEEE J. Quantum Electron.* **2014**, *50*, 911–920. [CrossRef]

14. Chies, L.; Dalla Costa, M.A.; Bender, V.C. Improved design methodology for LED Lamps. In Proceedings of the 2015 IEEE 24th International Symposium on Industrial Electronics (ISIE), Buzios, Brazil, 3–5 June 2015; pp. 1196–1201. [CrossRef]

15. Tao, X. Study of Junction Temperature Effect on Electrical Power of Light-Emitting Diode (LED) Devices. In Proceedings of the 2015 IEEE 22nd International Symposium on the Physical and Failure Analysis of Integrated Circuits, Hsinchu, Taiwan, 29 June–2 July 2015; pp. 430–433. [CrossRef]

16. Tao, X.; Yang, B. An Estimation Method for Efficiency of Light-Emitting Diode (LED) Devices. *J. Power Electron.* **2016**, *16*, 815–822. [CrossRef]

17. Raypah, M.E.; Sodipo, B.K.; Devarajan, M.; Sulaiman, F. Estimation of Luminous Flux and Luminous Efficacy of Low-Power SMD LED as a Function of Injection Current and Ambient Temperature. *IEEE Trans. Electron. Devices* **2016**, *63*, 2790–2795. [CrossRef]

18. Farkas, G.; van Voorst Vader, Q.; Poppe, A.; Bognár, G.Y. Thermal Investigation of High Power Optical Devices by Transient Testing. *IEEE Trans. Compon. Packag. Technol.* **2005**, *28*, 45–50. [CrossRef]

19. Poppe, A.; Farkas, G.; Székely, V.; Horváth, G.Y.; Rencz, M. Multi-domain simulation and measurement of power LED-s and power LED assemblies. In Proceedings of the 22nd IEEE Semiconductor Thermal Measurement and Management Symposium (SEMI-THERM'06), Dallas, TX, USA, 14–16 March 2006; pp. 191–198. [CrossRef]

20. Osram LED PSpice Libraries. Available online: https://www.osram.com/apps/downloadcenter/os/?path=/os-files/Electrical+Simulation/LED/PSpice+Libraries/ (accessed on 25 February 2019).

21. Lumileds LWS LTSpice Libraries. Available online: https://www.lumileds.com/support/design-resources/electrical (accessed on 25 February 2019).
22. Raynaud, P. Single Kernel Electro-Thermal IC Simulator. In Proceedings of the 19th International Workshop on Thermal Investigation of ICs and Systems (THERMINIC'13), Berlin, Germany, 25–27 September 2013; pp. 356–358. [CrossRef]
23. Keppens, A. Modeling and Evaluation of High-Power Light-Emitting Diodes for General Lighting. Ph.D. Thesis, Katholeieke Universiteit Leuven, Leuven, Belgium, 2010; D/2010/7515/9, ISBN 978-94-6018-256-3. Available online: https://lirias.kuleuven.be/bitstream/123456789/274568/1/PhD+text+AK.pdf (accessed on 25 April 2019).
24. Negrea, C.; Svasta, P.; Rangu, M. Electro-Thermal Modeling of Power LED Using Spice Circuit Solver. In Proceedings of the 35th International Spring Seminar on Electronics Technology (ISSE 2012), Bad Aussee, Austria, 9–13 May 2012; pp. 329–334. [CrossRef]
25. Górecki, K. Electrothermal model of a power LED for SPICE. *Int. J. Numer. Model.* **2012**, *25*, 39–45. [CrossRef]
26. Górecki, K. The Influence of Mutual Thermal Interactions Between Power LEDs on Their Characteristics. In Proceedings of the 20th International Workshop on Thermal Investigation of ICs and Systems (THERMINIC'13), Berlin, Germany, 25–27 September 2013; pp. 188–193. [CrossRef]
27. Górecki, K. Modelling mutual thermal interactions between power LEDs in SPICE. *Microelectron. Reliab.* **2015**, *55*, 389–395. [CrossRef]
28. Górecki, K.; Ptak, P. Modelling Power LEDs with Thermal Phenomena Taken into Account. In Proceedings of the 22nd International Conference on Mixed Design of Integrated Circuits and Systems, Toruń, Poland, 25–27 June 2015; pp. 432–435. [CrossRef]
29. Górecki, K.; Ptak, P. New dynamic electro-thermo-optical model of power LEDs. *Microelectron. Reliab.* **2018**, *91*, 1–7. [CrossRef]
30. Farkas, G.; Poppe, A. Thermal testing of LEDs. In *Thermal Management for LED Applications*; Lasance, C.J.M., Poppe, A., Eds.; Springer: New York, NY, USA; Heidelberg, Germany; Dordrecht, The Netherlands; London, UK, 2014; pp. 73–165. [CrossRef]
31. Bein, M.C.; Hegedüs, J.; Hantos, G.; Gaál, L.; Farkas, G.; Rencz, M.; Poppe, A. Comparison of two alternative junction temperature setting methods aimed for thermal and optical testing of high power LEDs. In Proceedings of the 23rd International Workshop on Thermal Investigation of ICs and Systems (THERMINIC'17), Amsterdam, The Netherlands, 27–29 September 2017. [CrossRef]
32. *CIE Technical Report 225: 2017. Optical Measurement of High-Power LEDs*; CIE: Vienna, Austria, 2017; ISBN 978-3-902842-12-1. [CrossRef]
33. JEDEC Standard JESD51-14. *Transient Dual Interface Test Method for the Measurement of the Thermal Resistance Junction-To-Case of Semiconductor Devices with Heat Flow Through a Single Path*; JEDEC: Arlington, VA, USA, 2010.
34. T3Ster-TeraLED Product Website. Available online: https://www.mentor.com/products/mechanical/micred/teraled/ (accessed on 27 March 2019).
35. Poppe, A.; Farkas, G.; Szabó, F.; Joly, J.; Thomé, J.; Yu, J.; Bosschaartl, K.; Juntunen, E.; Vaumorin, E.; di Bucchianico, A.; et al. Inter Laboratory Comparison of LED Measurements Aimed as Input for Multi-Domain Compact Model Development within a European-wide R&D Project. In Proceedings of the Conference on "Smarter Lighting for Better Life" at the CIE Midterm Meeting 2017, Jeju, Korea, 23–25 October 2017; pp. 569–579. [CrossRef]
36. CIE e-ILV Terms 17-1025 (Radiant Flux) and 17-1027 (Radiant Power). Available online: http://eilv.cie.co.at/term/1025 and http://eilv.cie.co.at/term/1027, respectively (accessed on 25 April 2019).
37. Sze, S.M.; Ng, K.K. *Physics of Semiconductor Devices*, 3rd ed.; John Wiley & Sons: Hoboken, NJ, USA, 2007; ISBN 0-471-14323-5.
38. Schubert, E.F. *Light-Emitting Diodes*, 2nd ed.; Cambridge University Press: Cambridge, UK, 2006; ISBN 0-511-34476-7.
39. JEDEC JESD51-51 Standard. *Implementation of the Electrical Test Method for the Measurement of the Real Thermal Resistance and Impedance of Light-Emitting Diodes with Exposed Cooling Surface*; JEDEC: Arlington, VA, USA, 2012.

40. Hantos, G.; Hegedüs, J.; Poppe, A. Different questions of today's LED thermal testing procedures. In Proceedings of the 34th IEEE Semiconductor Thermal Management Symposium (SEMI-THERM'18), San Jose, CA, USA, 19–23 March 2018; pp. 63–70. [CrossRef]

41. CIE e-ILV Term 17-730 (Luminous Efficacy of Radiation). Available online: http://eilv.cie.co.at/term/730 (accessed on 27 March 2019).

42. Van Zeghbroeck, B. Principles of Semiconductor Devices. Available online: http://ecee.colorado.edu/~bart/book (accessed on 27 March 2019).

43. Tansu, N.; Mawst, L.J. Current injection efficiency of InGaAsN quantum-well lasers. *J. Appl. Phys.* **2005**, *97*, 054502. [CrossRef]

44. Zhao, H.; Liu, G.; Zhang, J.; Arif, R.A.; Tansu, N. Analysis of Internal Quantum Efficiency and Current Injection Efficiency in III-Nitride Light-Emitting Diodes. *J. Disp. Technol.* **2013**, *9*, 212–225. [CrossRef]

45. JEDEC JESD51-52 Standard. *Guidelines for Combining CIE 127-2007 Total Flux Measurements with Thermal Measurements of LEDs with Exposed Cooling Surface*; JEDEC: Arlington, VA, USA, 2012.

46. Bornoff, R.; Mérelle, T.; Sari, J.; Di Bucchianico, A.; Farkas, G. Quantified Insights into LED Variability. In Proceedings of the 24th International Workshop on Thermal Investigation of ICs and Systems (THERMINIC'18), Stockholm, Sweden, 26–28 September 2018. [CrossRef]

47. Mérelle, T.; Bornoff, R.; Onushkin, G.; Gaál, L.; Farkas, G.; Poppe, A.; Hantos, G.; Sari, J.; Di Bucchianico, A. Modeling and quantifying LED variability. In Proceedings of the 2018 LED Professional Symposium (LpS2018), Bregenz, Austria, 25–27 September 2018; pp. 194–206, ISBN 978-3-9503209-9-2.

48. Farkas, G.; Poppe, A.; Gaál, L.; Hantos, G.; Berényi, C.S.; Rencz, M. Structural analysis and modelling of packaged light emitting devices by thermal transient measurements at multiple boundaries. In Proceedings of the 24th International Workshop on Thermal Investigation of ICs and Systems (THERMINIC'18), Stockholm, Sweden, 26–28 September 2018. [CrossRef]

49. Mérelle, T.; Sari, J.; Di Bucchianico, A.; Onushkin, G.; Bornoff, R.; Farkas, G.; Gaál, L.; Hantos, G.; Hegedüs, J.; Martin, G.; et al. Does a single LED bin really represent a single LED type? In Proceedings of the CIE 2019 29th Quadrennial Session, Washington, DC, USA, 14–22 June 2019.

![energies logo] *energies*

MDPI

Article

Multiple Heat Source Thermal Modeling and Transient Analysis of LEDs

Anton Alexeev [1,*], Grigory Onushkin [2], Jean-Paul Linnartz [1,2] and Genevieve Martin [2]

[1]   Department of Electrical Engineering, Eindhoven University of Technology, 5612 AZ Eindhoven,
      The Netherlands; j.p.linnartz@signify.com
[2]   Research, Signify, High Tech Campus 7, 5656 AE Eindhoven, The Netherlands;
      grigory.onushkin@signify.com (G.O.); genevieve.martin@signify.com (G.M.)
*   Correspondence: a.alexeev@tue.nl

Received: 1 April 2019; Accepted: 9 May 2019; Published: 15 May 2019

Abstract: Thermal transient testing is widely used for LED characterization, derivation of compact models, and calibration of 3D finite element models. The traditional analysis of transient thermal measurements yields a thermal model for a single heat source. However, it appears that secondary heat sources are typically present in LED packages and significantly limit the model's precision. In this paper, we reveal inaccuracies of thermal transient measurements interpretation associated with the secondary heat sources related to the light trapped in an optical encapsulant and phosphor light conversion losses. We show that both have a significant impact on the transient response for mid-power LED packages. We present a novel methodology of a derivation and calibration of thermal models for LEDs with multiple heat sources. It can be applied not only to monochromatic LEDs but particularly also to LEDs with phosphor light conversion. The methodology enables a separate characterization of the primary pn junction thermal power source and the secondary heat sources in an LED package.

Keywords: dynamic thermal compact model; LED; silicone dome; phosphor light conversion; structure function; thermal transient analysis; thermal characterization; multiple heat source; secondary heat path

---

1. Introduction

The market of solid-state lighting has rapidly expanded over the past decades. Due to commoditization of light emitting diodes (LEDs), the expansion sets a need for accurate thermal modeling of LEDs to ensure highly reliable end products. Compact thermal modeling is an approach that enables fast and reliable thermal simulations of the devices of interest without disclosing confidential information from LED suppliers. Compact thermal models (CTMs) are widely used in modern production and in optimization processes (particularly in the semiconductor industry) [1–3]. One of the goals of our European Union project Delphi4LED [4,5] is the standardization of the CTMs for LEDs.

CTMs are traditionally either directly generated based on measurement data of thermal transients or are extracted by model order reduction [1] from detailed full-3D finite element thermal models [6]. CTMs derived from a 3D model can predict more thermal parameters of the original device. Yet a calibration of the original 3D model with thermal transient measurements is required. The calibration includes tuning of the thermal properties of the materials by the model's thermal transient response or structure function (SF) alignment with the data measured from the corresponding physical sample [7]. Geometry tuning and a proper choice of the right physical phenomena to be modeled should also be carefully executed.

The geometry of the 3D models can be extracted precisely from x-ray scans of a physical device [8]. The choice of the physics is complicated as multiple phenomena define the heat generation inside an

LED package. Firstly, the LED chip generates heat due to internal optical losses (non-ideal internal quantum efficiency *IQE*), and electrical losses [9,10]. Secondly, the light emitted by the LEDs chip can be trapped in the encapsulating optical dome, partially re-absorbed inside the LED chip and on the walls of the LED package [11]. The impact of the heat loss caused by trapped light is often underestimated. We will show that these losses can have a significant influence on the LEDs transient and steady-state thermal behavior. Thirdly, if the LED has a silicone/phosphor layer, Stokes losses occur in the light conversion and cause thermal heating [12]. The phosphor material can cover the entire LED chip, be distantly located within the LED package, or be placed remotely on a secondary optic diffuser [13,14]. In the first or the second case, phosphor power losses can significantly affect the thermal performance of the LEDs [15].

Various approaches to calibrate multiple heat source LED thermal models are demonstrated in scientific works. A well-known impedance matrix approach enables the thermal characterization of multi-heat source multichip LEDs [16–18]. Yet this method cannot be applied for the characterization of the secondary heat sources: they are bounded to the main ones and cannot be activated separately. This makes the required measurements of transfer impedance impossible. The secondary heat sources can be characterized by ray tracing simulations [19–21]. Yet this approach requires access to proprietary information, such as phosphor composite particles spatial distribution, excitation and emission spectra, reflection properties, and the angular light distribution of the LED die. Another way to characterize secondary heat sources is comparative analysis of the thermal transient and optical measurements of an original bare chip LED and the same LED with a dispensed dome [13,22]. Yet, this method requires access to the bare chip LED packages, which are typically not available on the market and phosphor/silicone mixtures used by the LED manufacturers. Thus, we lack methods to characterize the LEDs secondary heat sources [23]. As a result, the trapped light and phosphor light conversion losses are typically coarsely approximated during the calibration of 3D models, e.g., Bornoff et al. assume 75% phosphor conversion efficiency without direct measurements when demonstrating their calibration procedure [7]. However, we show that the secondary heat sources have a significant impact on thermal transient and, therefore, these sources have to be estimated based on measurement data for proper calibration of 3D models. This work is filling the methodological gap by describing a procedure of multiple heat source LED thermal model calibration by analysis of LED package transient response.

We begin with familiarizing the reader with the considered LED package architecture, the transient analysis method, and the thermal SF concept. Next, we describe the LED's finite element analysis (FEA) thermal model. Then, we demonstrate an analytical estimation of the impact of such LED parameters as internal quantum efficiency (*IQE*), external quantum efficiency (*EQE*), light extraction efficiency (*LEE*), and dome geometry on the dome light extraction losses. Afterwards, we propose the novel thermal transient analysis methodology for multiple heat sources LED characterization. Then, we validate the proposed methodology and demonstrate the impact of the secondary heat sources on the interpretation of the transient measurements with our FEA model. We experimentally demonstrate the importance of the secondary thermal sources consideration for the thermal transient analysis of LEDs. Finally, we discuss the topologies of thermal resistor networks for physical-based modeling an LED. A list of abbreviations and variables can be found in the end of the paper.

2. Materials and Methods

2.1. LEDs Architecture

In this section, we give an overview of the LED packages of interest, their structures, characteristic dimensions and typical materials. We focus on mid-power (MP) surface mounted packages with lateral LED chips. MP LEDs are currently the most commonly used type in the lighting industry. Figure 1 represents a sketch of a typical MP LED package mounted on a metal core printed circuit board (MCPCB). MP LEDs typically contain one to three separate dies interconnected with wire bonds. Each die is attached to a thermal pad by a die attach layer (DAL). For sake of simplicity, we consider a one-die LED package. Multichip LED packages with a larger number of dies can also be characterized by the approach presented in this paper. As shown in [24–26], the thermal transient measurements can also be applied to such packages. Yet, the interpretation of the results for CTM calibration must be done with careful approach. Numerous pn junctions in such packages may have different thermal behavior.

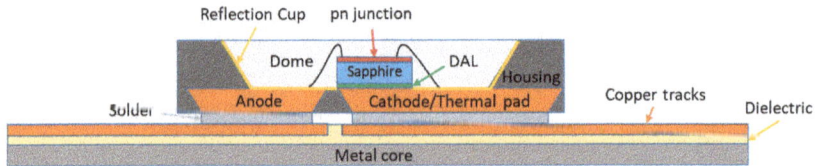

Figure 1. Sketch of a typical MP LED package soldered on an MCPCB.

A typical MP LEDs with square size of 3 mm × 3 mm (3030 package) have a characteristic reflection cup with a radius of approximately 1.2 mm. The representative height and width of the die are chosen as 0.2 mm and 0.8 mm, respectively. Highly efficient GaN LEDs are grown on c-plane sapphire substrate [27]. Despite the fact that sapphire crystal's thermal conductivity is anisotropic, the known crystal orientation enables high accuracy of sapphire thermal parameter estimation. The encapsulating dome above an LED die is typically fabricated out of either transparent silicone or silicone mixed with phosphor particles. LEDs' anodes, cathodes, and the thermal pads are traditionally made of copper.

2.2. Thermal Transient Analysis

A well-established method to characterize and extract a CTM of an LED is thermal transient analysis [28–30]. It requires sampling of device under test (DUT) junction temperature T_j transient response data to an applied thermal power step P_h. The power step P_h is derived as a difference between the measured supplied electrical power P_{el} and the measured emitted optical power P_{opt}:

$$P_h = P_{el} - P_{opt} \tag{1}$$

A one-dimensional thermal RC Cauer network with a step response identical to T_j is synthesized next. If the DUT heat path is sufficiently one-dimensional, the derived RC Cauer network genially represents its thermal properties. The details of the LED's transient characterization are described in the JESD51-5x series of standards [31–35].

A thermal RC Cauer network can be graphically represented as an SF. An SF is a plot of the cumulative R_{th} values versus the cumulative C_{th} values along the DUT heat path starting from the pn junction. An SF can be converted to a differential SF by taking the derivative of C_{th} with respect to R_{th}. The peaks of differential SFs usually indicate new regions of the heat flow path [36]. We relate the thermal structures of the LED presented in Figure 1 to the particular regions of the correspondent SF and differential SF plots shown in Figure 2 according to the methods demonstrated in [28,37]. The pn junction is always located at the beginning of the plot. Thus, the initial increase of the thermal capacitance is related to the first element of the thermal path, the sapphire crystal. The subsequent shelf represents the DAL that has high partial R_{th} and low partial C_{th}. A sharp increase of the cumulative thermal capacitance observed after the DAL shelf is related to the massive copper thermal pad. The thermal pad is attached to the copper tracks of the printed circuit board (PCB), a region with significant radial heat spreading. The SF image of it looks like a tilted straight line [28]. We use the differential SF peak related to the thermal pad to characterize the total R_{th} of the die and the DAL. The peak is marked with a red arrow in Figure 2. Nevertheless, as will be shown further, the die and the DAL SF regions can be significantly distorted if the secondary heat sources are present. Thus, the R_{th} value estimated with this method may be inaccurate. We analyze the factors contributing to the SF distortion, the related physics and the impact of the secondary heat sources on thermal transient processes in this work. We use simulated SFs to quantify the impact.

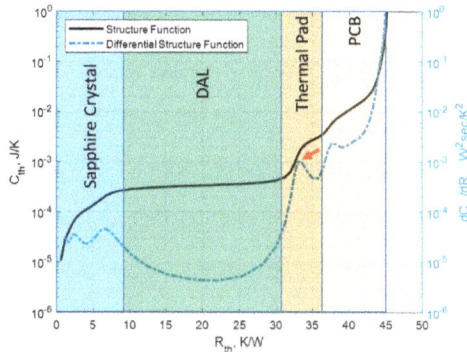

Figure 2. A characteristic structure function and a correspondent differential structure function of a typical one die MP LED and its relation to the LED's thermal heat path structures. The peak of the differential structure function related to the thermal pad is indicated.

The DAL is the most significant contributor to total R_{th} of LED packages [38,39]. The partial R_{th} of a DAL can reach up to tens of K/W for a single die MP LEDs package (Figure 2), while the partial R_{th} values of other heat path structures as sapphire crystal and thermal pad are below 10 K/W. Thus, estimation of DAL properties is crucial for LEDs reliability prediction and calibration of FEA thermal models. The R_{th} of a DAL is the major contributor to junction to thermal pad thermal resistance R_{th_J2T}. For our further analysis of the SF distortion we compare the R_{th_J2T} values derived with two different methods: SF analysis and FEA. The correspondent values are named $R_{th\_J2T\_SF}$ and $R_{th\_J2T\_FEA}$, respectively.

As shown before, the location of the correspondent differential SF peak enables a straight-forward identification of the $R_{th\_J2T\_SF}$ value. Yet, the peak might become blurred and impossible to identify by this method for some LEDs configurations. In this case the $R_{th\_J2T\_SF}$ value is determined by a method inspired by [40] as cumulative R_{th} value correspondent to characteristic C_{th_J2T} value which is defined as:

$$C_{th\_J2T} = C_{th\_crystal} + 0.5 C_{th\_thermal\ pad} \qquad (2)$$

$R_{th\_J2T\_SF}$ values derived by these methods are in excellent agreement.

We use $R_{th\_J2T\_FEA}$ to verify $R_{th\_J2T\_SF}$ values and determine the accuracy of the thermal transient measurements interpretation. The $R_{th\_J2T\_FEA}$ derivation method is based on a direct evaluation of steady-state thermal FEA results. It requires calculation of $\overline{T_j}$ and $\overline{T_{DAL}}$, the average temperatures of the pn junction and the bottom of the DAL finite element model nodes, respectively. According to the definition of the thermal resistance:

$$R_{th\_J2T\_FEA} = \frac{\overline{T_j} - \overline{T_{DAL}}}{P_{hJ}} \qquad (3)$$

The difference between the $R_{th\_J2T\_SF}$ and the $R_{th\_J2T\_FEA}$ defines the accuracy of SF representation of the LED's main heat path. We define the relative error of $R_{th\_J2T\_SF}$ derivation as:

$$R_{th\_J2T\_err} = \frac{R_{th\_J2T\_FEA} - R_{th\_J2T\_SF}}{R_{th\_J2T\_FEA}} \qquad (4)$$

2.3. Thermal FEA Modeling

We use a modification of the FEA model demonstrated in our previous publication [41]. The model was built with the MATLAB (2018b, MathWorks, Natick, MA, USA) partial differential equation toolbox. We aim to determine the effects of the secondary heat sources on the accuracy of the transient analysis, in particular, on the R_{th_J2T} value. We use a generalized axisymmetric geometry of an LED package mockup. Our model contains only the major thermal elements of an LED: pn junction, sapphire crystal, DAL, thermal pad, and dome. The fine details of the LED package are omitted since their effect on the T_j transient response and SF is negligible and unique for each LED package. The model can simulate characteristic thermal behavior of MP LEDs packages of various size by adjusting the characteristic dimensions of the thermal structures. Figure 3a depicts the geometry of the FEA thermal model and defines key characteristic dimensions. The geometrical and thermal parameters of the model are presented in Table 1.

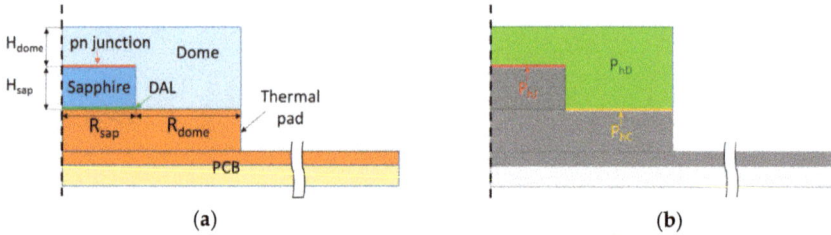

Figure 3. (a) Cross-sectional view of the axisymmetric FEA LED package mockup and its dimensions. (b) Thermal power dissipation regions.

Table 1. Geometrical and thermal properties of the LEDs FEA model structures.

Element	Material	Radius (mm)	Height (mm)	ρ (g/mm³)	C (J/gK)	k (W/mK)
Crystal	Sapphire	0.4	0.2	3.98	0.85	32
DAL	-	0.4	0.002	-	-	0.1
Thermal pad	Copper	1.2	0.25	8.93	0.39	380
Dome	Silicone phosphor	0.8	0.2	1.10*	1.15*	0.2*
MCPCB tracks	Copper	10	0.07	8.93	0.39	380
MCPCB dielectric	FR4	10	0.1	1.90	1.2	0.2

* The data are given for silicone. Silicone/phosphor composite thermal properties are defined in Equations (22)–(24).

Plastic housing of LEDs is not explicitly considered in the thermal simulations because its thermal conductivity is considerably lower than that of the copper thermal pad. Moreover, the housing is separated from the junction by a dome made of an extremely low thermal conductive material. Thus, we assume the housing thermal effect on the T_j response to be small. Nevertheless, as shown in [42], the housing geometry significantly affects the *LEE* and the trapped light thermal losses. The trapped light thermal losses related to the plastic housing walls (cup reflector surface) are added to the thermal losses on top of the thermal pad surface.

The model contains three heat sources. The sources are defined in Figure 3b and related to the pn junction P_{hJ}, dome P_{hD}, and cup reflector surface P_{hC}. P_{hJ} and P_{hC} are homogeneously distributed over the corresponding junction and cup surfaces. P_{hD} is uniformly distributed within the dome volume. The sum of the considered heat sources is equal to the total thermal power losses P_h of the LED:

$$P_h = P_{hJ} + P_{hD} + P_{hC} \tag{5}$$

We define the sum of the parasitic secondary heat losses as P'_h, with:

$$P'_h = P_{hD} + P_{hC} \tag{6}$$

We also evaluate heat transfer through the copper tracks and the dielectric layer of the MCPCB in our model. A constant temperature boundary condition is set at the bottom of the PCB's dielectric layer to simulate the upper surface of the PCB's highly conductive metal core. We set uniform initial temperature condition. Convection and radiation heat transfers from the outer surfaces of the LED package are ignored. A number of works has previously demonstrated that a negligible fraction of heat leaves an LED via these mechanisms compared to conduction via the main heat path [43,44]. Thus, adiabatic boundary conditions are applied to all outer LED surfaces. Therefore, the numerical problem is linear and the system temperature response, normalized with applied thermal power, does not depend neither on the absolute value of the initial temperature nor on the applied power. The heat sources are activated at $\tau = 0$. Transient response of the averaged pn junction temperature is evaluated. For calculation of the $R_{th\_J2T\_FEA}$ the steady-state averaged temperature of the bottom of the DAL finite element model nodes is used.

2.4. Experimental Setup and Physical Specimens

To confirm the impact of secondary heat sources, experiments using two samples of MP 3030 LED with a bare non-encapsulated chip have been conducted. We made various modifications for the dome encapsulant of each LED. The 1st LED got a flat silicone dome as an encapsulant filling the package cup. The 2nd LED was initially modified with a clear silicone rim layer around the sapphire chip and then a silicone/phosphor mixture as a light conversion layer covering the whole open surface in the package cup. OE-6650 resin material (Dow Corning, Midland, Michigan, United States) was used as the silicone for these layers. The resulting LED configurations are presented in Figure 4. Transient testing of each of them was performed separately. Comparative analysis of the results correspondent to each LED enabled identification of the impact of the trapped light and the phosphor secondary power sources.

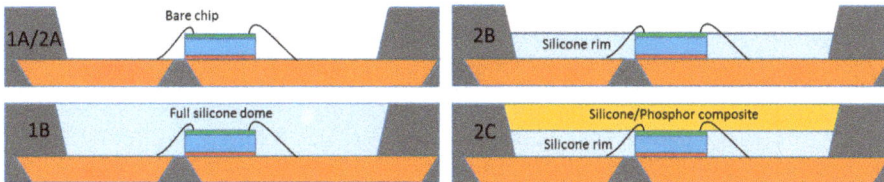

Figure 4. The physical LED specimens' configurations.

To perform the thermal transient measurements, the LEDs were soldered on an MCPCB which was mounted on a heat sink. T3Ster (Mentor Graphics, Wilsonville, Oregon, United States) thermal tester was used to preform transient testing. An on/off forward voltage response was measured while the heat sink temperature was kept 50 °C. We used 150 mA driving and 10 mA measurement currents. The dependence of the LEDs' forward voltage on T_j was calibrated in the range of 25–85 °C while the LEDs were mounted on a coldplate and biased with the measurement current. We use a quadratic approximation of the T_j dependence on the forward voltage to increase the accuracy, since it was shown in the literature that use of a linear approximation may lead to a significant error of the junction temperature evaluation for LEDs [45,46].

We used a square root initial correction to substitute the initial electrical transient. The correction was based on T_j response data sampled in the 50–400 µsec interval. The optical fluxes were measured by an integration sphere to calculate the total thermal power P_h dissipated by each LED. The LEDs were horizontally mounted on a heat sink during the optical measurements in order to minimize the convention heat flux losses that may be caused by the non-isothermal environmental conditions at open interface with integration sphere [45]. The LEDs' forward current and the heat sink temperature were kept the same during all thermal transient measurements.

2.5. Multiple Heat Source Characterization

In this chapter, we perform an analysis of the fraction of the secondary thermal power sources in the total power dissipation for MP LED packages. First, we analytically estimate the power losses related to the trapped light P_{hC}. We demonstrate a method to estimate P_{hC} based on the values of *IQE*,

EQE, and *LEE* parameters and on the geometries of the LED chip and the package. Next, we estimate the thermal losses caused by phosphor light conversion P_{hD}. Finally, we propose an experimental method that enables separation of the secondary heat sources power P'_h from the total thermal power P_h. The method is based on a revised analytical solution of an initial thermal transient response. We analyze the applicability of the proposed method for LEDs with transparent domes and for LEDs with phosphor light conversion.

2.5.1. Secondary Heat Sources

In this subchapter, we aim to estimate the fraction that secondary heat sources contributing to the total heat dissipation. First, we analyze LEDs without phosphor light conversion. In this case the secondary heat sources are originated from the light trapped in the encapsulating dome layer due to total inner reflections (TIRs), (see Figure 5). This light is partially absorbed by the reflector cup walls, the reflective metal contact pads, and the pn junction when light re-enters the LED crystal [47].

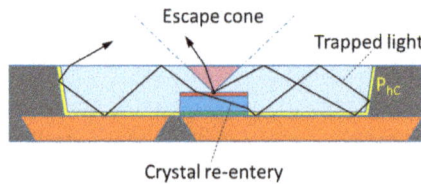

Figure 5. Illustration of a trapped light ray emitted by the pn junction experiencing multiple TIRs before leaving a flat silicone dome of an LED.

In general, evaluation of the trapped light losses requires sophisticated ray tracing modeling and extensive LED characterization. Yet, it is possible to make a coarse analytical estimation employing approximations of the *LEE* and *IQE* coefficients and other LED parameters, such as cup reflectivity, dome curvature, die and package dimensions, etc. First, we define the total thermal power P_h dependence on the applied electrical power. The total *LEE* of an LED package is a product of the chip-to-dome light extraction coefficient LEE_{chip} and the dome-to-air light extraction coefficient LEE_{dome}, thusly:

$$LEE = LEE_{chip} \cdot LEE_{dome} \tag{7}$$

EQE is a product of *IQE* and *LEE*, that is:

$$EQE = IQE(I) \cdot LEE_{chip} \cdot LEE_{dome} \tag{8}$$

Unlike LEE_{chip} and LEE_{dome}, *IQE* is dependent on the forward current I. The total thermal power dissipated in an LED is:

$$P_h = P_{el} \cdot \left(1 - IQE(I) \cdot LEE_{chip} \cdot LEE_{dome}\right) \tag{9}$$

The fraction of the light initially left the die and trapped on the reflection cup walls is:

$$P_{hC} \cong IQE(I) \cdot LEE_{chip} \cdot (1 - LEE_{dome}) \cdot \Lambda \tag{10}$$

where coefficient Λ provides a correction for the crystal light re-absorption. Appendix A analyzes Λ dependence on the LED's package geometry and other parameters.

The LEE_{dome} coefficient is difficult to measure directly without manufacturing packages of custom calibration LEDs and without advanced measurement setups. Therefore, we use the results of ray tracing simulation presented by Tran et al. [42]. The results enable estimation of the LEE_{dome} for MP LED packages. The authors define the LEE_{dome} of an LED, dependence on the dome curvature, and angle of the reflector cup. The simulation results evidence that the LEE_{dome} coefficient for conventional dome designs varies from 0.65–0.92. The LEE_{dome} coefficient for conventional multiple-chip LEDs with flat light-emitting surface (LES) and the absence of special light extraction enhancement structures is around 60% [48–50]. Optimization of the LEDs packages by using a gradient refractive index encapsulant, roughened or patterned lead-frame substrates, and the scattering effect of phosphor particles can increase the LEE_{dome} up to 85%.

Now we estimate the fraction of P_{hC} in the total power dissipated by an LED:

$$\frac{P_{hC}}{P_h} \cong \frac{IQE(I) \cdot LEE_{chip} \cdot (1 - LEE_{dome})}{\left(1 - IQE(I) \cdot LEE_{chip} \cdot LEE_{dome}\right)} \cdot \Lambda \tag{11}$$

We then estimate the P_{hC}/P_h ratio for blue GaN LEDs. Modern state-of-the-art high brightness blue LEDs can have a LEE_{chip} coefficient of 85% for double-side textured-crystals [51]. The reflectance coefficient of the cup is set to 93%. The ratio between the areas of the cup and the crystal S_{cup}/S_{cry} plays an important role. It defines the probability of crystals re-entry by the trapped light. We consider two S_{cup}/S_{cry} ratios: 7 and 2. The first corresponds to the MP LED architecture presented in Table 1. The second represents a smaller single die or a multiple die LED package. Indeed, multiple-die packages with low S_{cup}/S_{cry} ratios have increased shielding of the light by the neighboring dies (denser chip placement leads to an increased light re-absorption by the LED package chips) [52]. The results are presented in Figure 6.

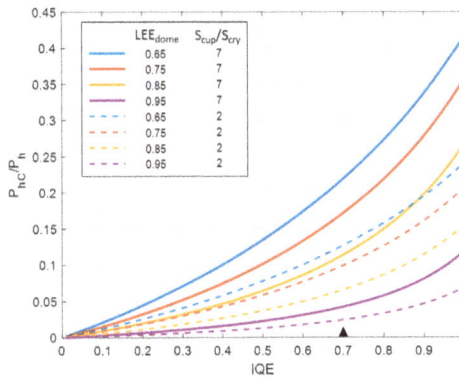

Figure 6. P_{hC}/P_h ratio as a function of IQE for a range of LEE_{dome} coefficient.

Figure 6 plots evidence that LED chips with high IQE in low LEE_{dome} packages have the highest relative P_{hC} thermal losses. The fraction of P_{hC} rapidly increases with increase of IQE. A decrease of the S_{cup}/S_{cry} value leads to a decrease of the P_{hC} fraction due to the shielding effect.

The IQE of blue GaN LEDs approaches its theoretical limit of 95%. Nevertheless, in the majority of high-power applications, blue LEDs are driven in the droop regime when the IQE is approximately 70% (marked at the plot). These parameters indicate that approximately up to 25% of the total thermal power P_h can be dissipated on the LED cup walls due to TIR.

LEDs with a phosphor light conversion layer always have extra heat losses due to Stokes effect. This emphasizes the fact that the secondary heat sources are significant. Next, we propose a methodology of their experimental estimation.

2.5.2. Estimation of the Secondary Heat Sources

Modern LED transient testing methods determine the total thermal power P_h as a difference between the applied electrical power P_{el} and the emitted radiant flux P_{opt} [32–35]. It is impossible to separate the secondary heat sources P'_h with this approach. In this section, we propose a method of experimentally estimate P'_h. The method is based on a revised solution of the initial T_j transient response.

In practice, it is challenging to measure the initial T_j response during the first tens of milliseconds due to the electrical transient processes in the pn junction, the connecting wires and the transient measurement equipment. Therefore, correction methods were developed to restore this data. One of these methods is a square root correction. The method is based on the analytical solution of heat propagation into a semi-infinite material from a homogeneous surface heat source: the pn junction substrate can be often approximated as a surface heat source and a semi-infinite body in the beginning of thermal transient. The initial T_j transient response can be approximated with the following equation [53,54]:

$$T_j(t) = T_0 + \frac{P_{hJ}}{S_{pn}} K \sqrt{t} \tag{12}$$

The form of the equation evidences that the initial T_j transience response plotted versus a square root of time is linear. The coefficient K bounds the slope of this plot with the junction dissipated thermal power P_{hJ} and the chip surface S_{pn}. An example of measured LED data and of the applied square root initial transient correction based on 50–400 μsec interval is shown in Figure 7.

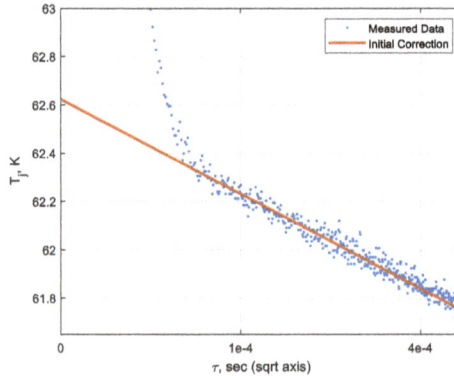

Figure 7. An example of the square root initial transient correction applied to measured data of an on/off transient.

The correction method assumes a unilateral one-dimensional heat propagation in the beginning of the transient. While LEDs chips are typically encapsulated in a dome to enhance light extraction and shape the light beam. Thus, the heat dissipated by the pn junction propagates bilaterally, both to the substrate and the dome. We solve a problem of the bilateral heat transfer into two semi-infinite bodies representing the sapphire crystal substrate and the encapsulating dome. We solve a problem of the bilateral heat transfer into two semi-infinite bodies representing the sapphire crystal substrate and the dome. The solution is presented in Appendix B. The resulting equation retains the square root time-dependence. After substituting the power density q with P_{hJ}/S_{pn} in Equation (A17) we obtain the coefficient K_{bi_lat} characterizing T_j initial response for the case of bilateral heat propagation. The original coefficient K_{uni_lat} and the derived K_{bi_lat} are presented below:

$$K_{uni\_lat} = \frac{2}{\sqrt{\pi k_s \rho_s C_s}} \tag{13}$$

$$K_{bi\_lat} = \frac{2}{\sqrt{\pi}\left(\sqrt{k_s \rho_s C_s} + \sqrt{k_d \rho_d C_d}\right)} \tag{14}$$

The similar form of the bilateral heat propagation solution justifies the application of the square root initial correction for LEDs. Nevertheless, the bilateral solution bounds quite accurately the initial T_j response to the thermal power dissipated by the junction P_{hJ}.

We determine the values of the heat flows toward the dome and the substrate during the initial thermal transient by substituting the correspondent thermal properties in Appendix B Equations (A15) and (A16). We derive that approximately 5% of the total heat is dissipated by the junction propagates to the dome. This corresponds to approximately 5% systematic error for P_{hJ} power evaluation with the classic method when no heat propagation into the dome is considered.

The T_j response follows the square root time dependency only for a finite amount of time while the assumption of one-dimensional heat propagation is valid on both sides of the active region. On the one hand, it is limited by the characteristic time constant of the substrate τ_s. On the other hand, by the requirement of sufficiently one-dimensional heat propagation into the dome.

The substrate time constant τ_s can be found as a product of the partial thermal resistance and thermal capacitance of the die crystal, which can be expressed with sapphire thermal properties and the crystal height H_{sap}:

$$\tau_s = R_{th\_sap}C_{th\_sap} = \frac{H_{sap}^2 \rho_{sap} C_{sap}}{k_{sap}} \tag{15}$$

The heat propagation into the dome can be considered sufficiently one-dimensional if the characteristic width of the pn junction l'_{pn} is much larger than the characteristic depth of heat propagation into the dome x'_d:

$$\frac{x'_d}{l'_{pn}} \ll 1 \tag{16}$$

We estimate l'_{pn} as a one half of the minimal dimension of the top of the sapphire crystal. Carslaw and Jaeger [55] have derived a closed-form solution of the time-dependent temperature profile for heat propagation into a semi-infinite media:

$$T(x,t) = T_0 + \frac{2q}{\sqrt{\pi k \rho C_p}} \sqrt{t} \exp\left(-\frac{x^2}{4\tau}\frac{\rho C}{k}\right) - \frac{qx}{k} erfc\left(\sqrt{\frac{x^2}{4\tau}\frac{\rho C}{k}}\right) \tag{17}$$

The form of the exponential term of the Equation (17) yields the x'_d dependence on time:

$$x'_d \approx 2\sqrt{\tau \frac{k_d}{\rho_d C_d}} \tag{18}$$

Thus, the characters time τ_d at which $x'_d \approx l'_{pn}$ is:

$$\tau_d \approx \frac{1}{4} \frac{l'_{pn}{}^2 \rho_d C_d}{k_d} \tag{19}$$

Therefore, the initial heat propagation is sufficiently one-dimensional on both sides of the active region if:

$$\tau \ll \min(\tau_s, \tau_d) \tag{20}$$

τ_s and τ_d are estimated as 3.8 msec and 250 msec, respectively, considering the data of Table 1. Estimated value of τ_s is significantly smaller than τ_d. Thus, we conclude that the heat propagation into the silicone dome is always sufficiently one dimensional until the heat flux has not reached the DAL via sapphire substrate. Therefore, P_{hJ} can be reliably extracted by Equations (12) and (14). Then, P'_h can be found as a difference between P_h and P_{hJ}.

2.5.3. Applicability of the Approach for LEDs with Phosphor Light Conversion

In this section, we investigate the applicability of the proposed method of P_{hJ} estimation for LEDs with silicone/phosphor composite domes. The heat generation by phosphor particles may disturb the initial thermal transient which can affect the accuracy. We investigate the impact of the presence of phosphor on the accuracy of the proposed P_{hJ} extraction method.

Firstly, we perform estimation of the thermal properties of the phosphor/silicone composite material. We use the phosphor filler volume fraction f parameter defined with the volumes of the silicone V_{sil} and the phosphor V_{pho} fractions as:

$$f = \frac{V_{pho}}{V_{pho} + V_{sil}} \tag{21}$$

Numerous models have been proposed to model the effective thermal conductivity of this type of composites [56–58]. These models are typically derived for a certain range of the phosphor volume fraction. We use a high volume fraction limit model proposed by Every [59]. The effective thermal conductivity k_d of the silicone/phosphor composite material dome is expressed as:

$$k_d = \frac{k_{sil}}{(1 - f)^{3(1-\alpha)(1+2\alpha)}}. \tag{22}$$

Here, k_{sil} is the thermal conductivity of the silicone and α is a nondimensional parameter bounding the particle size and particle-composite matrix interface effect.

Zhang et al. [60] have fitted the model of the equation to experimental measurements of the typical silicone/phosphor composite used in LEDs They used and Ce^{3+} doped YAG ($Y_3Al_5O_{12}$) phosphor particles of 13.0 ± 2.0 μm diameter encapsulated in high optical transparency silicone. They achieved an excellent agreement with the experimental results for high volume concentrations f from 20–40%. It was found that α is 0.004. This fitting result slightly overestimate the thermal conductivity for the composites with low phosphor volume fractions.

The YAG phosphor density ρ_{pho} and the specific heat C_{pho} are 4.56 g/cm^3 and 0.6 J/(g K), respectively [61]. The density ρ_d and specific heat C_d of the dome composite are estimated based on the volume fraction of the phosphor particles f as:

$$\rho_d = \rho_{sil}(1 - f) + \rho_{pho}f \tag{23}$$

$$C_d = C_{sil}(1 - f) + C_{pho}f \tag{24}$$

The phosphor dome light conversion efficiency was shown to be dependent on the phosphor particles type, concentration, temperature, experienced thermal stress [13,44,62,63]. Yet, we assume the silicone/phosphor composite properties to be constant during the fast initial transient processes due to small temperature variations. Consequently, the heat transfer problem remains linear. Therefore, we address the multiple heat source initial thermal transient analysis using the principle of superposition. We analyze heating of the LED package with P_{hD} and P_{hJ} heat sources separately and compare the heating rates.

We start with the analysis of the silicone/phosphor composite dome. The data presented by Chung [64] shows the evidence that the heating time constant of the remote phosphor layers is significantly slower than the pn junction heating time constant (1 min vs. 0.02 sec). However, the phosphor layers deposited over the pn junction may have significant higher heating rates due to the higher optical power density. Lou et al. [20,65] determined the phosphor energy conversion efficiency both experimentally and numerically. For warm white LEDs with high phosphor volume fractions, up to 45% of blue light optical power can be dissipated as heat during light conversion [66]. If we assume that $WPE_{chip} = 70\%$ for the LED chip (a typical value for modern blue LEDs under typical operational conditions), then it will mean that up to half of the total thermal power P_h for white LED package can be related to the phosphor thermal losses P_{hD}. Similar power ratio results were previously determined in the literature [19,22].

The rate at which the temperature of the silicone/phosphor composite dome increases during the first hundreds of milliseconds after turning an LED on is linear. It is determined mainly by the capacitive thermal effects due to the low thermal conductivity of the silicone/phosphor composite. Thus, we estimate the composite dome temperature increase ΔT_d at times $\tau \ll \tau_d$ as:

$$\Delta T_d(\tau) = \frac{P_{hD}\tau}{\rho_d C_d V_d} \qquad (25)$$

Here V_d is the characteristic dome volume per die. Multiple die LEDs have higher dome optical power density than the one-die LEDs. This leads to a faster rate of silicone/phosphor composite heating. To consider a worst-case scenario we perform an estimation for a case of two-die MP 3030 white LED package. Thus, we chose V_d as a half of the correspondent LED dome volume.

We compare the characteristic dome temperature increase with the junction temperature increase $\Delta T_j(\tau)$ estimated by Equation (12) and K_{bi_lat} coefficient. The comparison yields evidence that for τ below 400 µsec (which is a typical upper time limit used for transient correction), ΔT_d is less than $0.03 \cdot \Delta T_j$. Thus, the initial heating rate of the silicone/phosphor composite is significantly slower than one of the pn junction. Moreover, the thermal conductivity of the silicone/phosphor composite is considerably lower than one of the sapphire crystal. All these factors indicate that the phosphor-related thermal losses have insignificant impact on the initial transients, as will be confirmed in the next chapter.

The unilateral coefficient K_{uni_lat} and the derived dependence of the bilateral coefficient K_{bi_lat} as a function of the phosphor fraction f are shown in Figure 8. The data confirms 5–15% K_{uni_lat} relative error if compared with more precise K_{bi_lat} values. The same error will have P_{hJ} when estimated by Equation (12) under the classical unilateral heat propagation assumption.

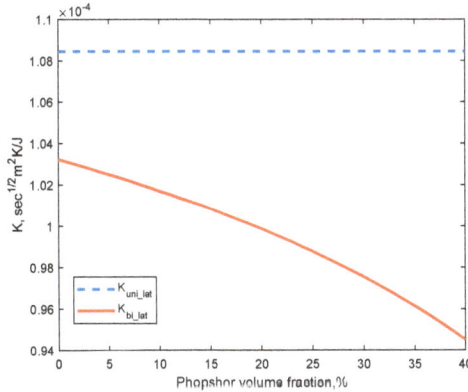

Figure 8. Comparison of K_{uni_lat} and K_{bi_lat} coefficients.

3. Results

3.1. Numerical Simulations

In this section, we determine the impact of the power P'_h for secondary heat sources on the transient analysis results with our FEA model. The analytical estimations presented above show evidence that P_{hC} can reach up to 25% of total P_h for LEDs with inefficient dome design. Phosphor-related thermal losses P_{hD} can be estimated as high as 50% of P_h if we consider an LED that has 65% phosphor light conversion efficiency [14] driven with current correspondent to 75% IQE. We aim to determine general patterns and estimate the thermal transient analysis inaccuracies associated with the secondary heat sources.

3.1.1. Secondary Heat Sources Impact on Junction Thermal Transient

First, we analyze the logarithmic time derivative and the initial transient of T_j response of the FEA model (Figure 9). We consider two extreme cases: $P'_h = P_{hD}$ and $P'_h = P_{hC}$.

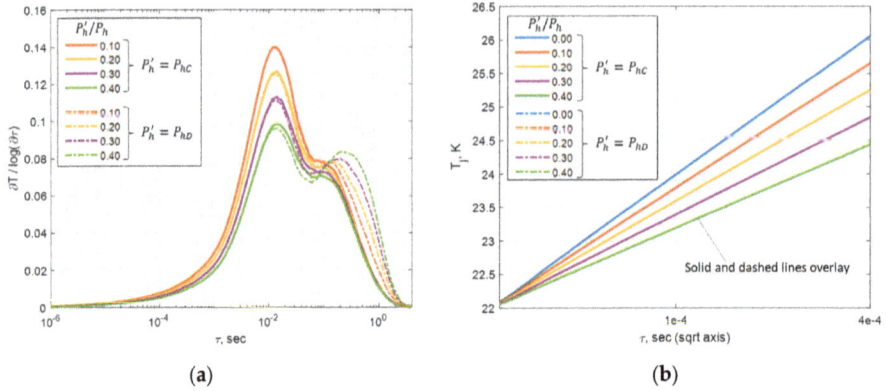

(a)

(b)

Figure 9. (a) Log time derivative of T_j response. The solid lines represent LEDs with $P'_h = P_{hC}$. The dashed lines are correspondent to LEDs with $P'_h = P_{hD}$. (b) Initial T_j response vs. The square root of time. No considerable difference between the $P'_h = P_{hD}$ and $P'_h = P_{hC}$ cases is observed, the dashed lines are merged with the solid ones.

Figure 9a shows that the initial thermal responses are similar up to 10 msec. They linearly scale with the P'_h/P_h fraction. A significant effect of the phosphor heat generation is observed for $\tau > 10$ msec. Figure 9b shows that the initial T_j transient responses plotted versus square root of time are linear up to 400μsec. As estimated in Section 2.5.3, phosphor thermal power generation P_{hD} has no influence on the initial transient in the considered time interval. Thus, the numerical modeling confirms that the K_{bi_lat} coefficient extracted from the initial thermal transient is weakly affected by the heat generation in the encapsulating dome. This confirms the applicability of the P_{hJ} estimation from initial transient response analysis even for LEDs with phosphor light conversion layer.

3.1.2. Secondary Heat Sources Estimation Verification

To verify the proposed method of P'_h and P_{hJ} evaluation, we determine the increase of the ΔT_j at $\tau = 400$ μsec by deriving K_{bi_lat} and K_{uni_lat} coefficients using Equations (12)–(14). We compare the obtained ΔT_j values against the simulation results in Table 2. The data show that the proposed bilateral initial heat propagation model predicts initial thermal transient response with significantly higher accuracy as compared to the classical unilateral model. Thus, the K_{bi_lat} coefficient helps to decrease significantly the P'_h and P_{hJ} evaluation errors.

Table 2. Initial ΔT_j response predicted by unilateral and bilateral analytical models at $\tau = 400$ μsec.

Phosphor Volume Fraction	ΔT_j (K)		
	FEA Reference	Bilateral Estimation	Unilateral Estimation
0	4.1	4.1	4.3
0.4	3.7	3.7	4.3

3.1.3. Secondary Heat Sources Impact on Structure Functions

For numerical simulations of the LED thermal structure functions we consider two extreme cases: $P'_h = P_{hC}$ and $P'_h = P_{hD}$. We change the secondary heat sources P'_h/P_h fraction from 0 to 0.4.

We use P_h as an input power for thermal transient analysis. The resulting simulated SFs are presented in Figure 10a. SF-derived $R_{th\_J2T\_SF}$ and reference $R_{th\_J2T\_FEA}$ thermal resistances are indicated on the graph. The $R_{th\_J2T\_FEA}$ is found to have an extremely weak dependence on the secondary heat sources presence, thus, $R_{th\_J2T\_FEA}$ is considered to be independent of the secondary heat source's power fraction. The blue SF represents the LED without secondary heat sources. The DAL SF based estimation error $R_{th\_J2T\_err}$ defined by Equation (4) is plotted in Figure 10b.

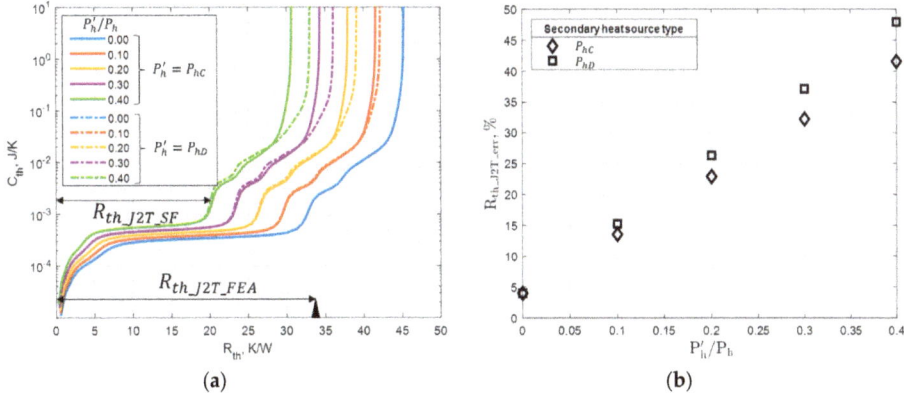

Figure 10. (a) Influence of the secondary heat sources on SFs. The solid structure functions represent LEDs with a heat source imitating the dome trapped light losses. The dashed structure functions are related to LEDs with heat source imitating the phosphor light conversion losses. The reference thermal resistance $R_{th\_J2T\_FEA}$ is indicated. (b) Error of the $R_{th\_J2T\_SF}$ part relative to the $R_{th\_J2T\_FEA}$.

We notice that the error is approximately linearly proportional to the P'_h/P_h ratio for both considered cases:

$$R_{th\_J2T\_err} \simeq P'_h/P_h \tag{26}$$

Any actual configuration of secondary thermal sources within an LED is formed as a superposition of these extreme cases. Since the considered problem is linear, thus, our proposed $R_{th\_J2T\_err}$ estimation is valid for both types of encapsulated LEDs, with and without phosphor light conversion.

$R_{th\_J2T\_err}$ is slightly higher for the $P'_h = P_{hD}$ than for $P'_h = P_{hC}$ case. The difference is explained by the fact that in steady-state conditions a fraction of the heat generated by the phosphor containing encapsulating dome goes through the sapphire crystal and through the DAL, while the heat generated by the trapped light in clear encapsulant is dissipated directly on the thermal pad, bypassing the die and the DAL. In phosphor converted LED package, the die and the DAL experience additional heat flow caused by the phosphor losses in particles located close to the die crystal. An important remark is that this heat flow is significantly delayed due to low thermal diffusivity of silicone/phosphor composite. Figure 9a confirms the "slow" impact of the delayed phosphor heat flow on the pn junction temperature transient. These results allow us to conclude that the phosphor heat generation have no significant impact on the initial thermal transient, yet it partially increases the thermal flux flow through the die and the DAL during steady state.

It should be noted, that the dome phosphor-related heat sources P_{hD} significantly impact the "tails" of the SFs, in Figure 10a. Indeed, as shown above, the phosphor dome heat generation P_{hD} produces an additional delayed thermal flow that affects the slow part of the T_j response. Based on this observation, we conclude that the presence of the distributed heat sources in the dome leads to the distortion of the regions of SFs correspondent to structures with relatively "slow" time constants (e.g. PCB and the further thermal path). Therefore, application of such techniques as transient dual interface measurements [67] may be limited in cases of LEDs with phosphor light conversion.

The analysis presented in this chapter shows that the value of total thermal resistance of an LED (as derived from the thermal transient measurement) decreases when the fraction of the secondary heat sources increases. This decrease of thermal resistance is associated with the fact that a significant part of the thermal power from secondary heat sources is distributed over the LED package volume. Thus, a part of the total heat losses used to calculate the LED's thermal resistance value bypasses the pn junction, the die, and the DAL. This effectively decreases the junction temperate comparing to the case when all the thermal power is generated only by the pn junction.

3.1.4. Die and DAL Thermal Characterization

We notice a "scaling" effect of the SFs presented in Figure 10a: the higher the P'_h/P_h fraction the lower the R_{th} and the higher the C_{th} values. At the same time, the thermal properties of the main heat path remain constant in all the numerical experiments. Moreover, we have noticed the characteristic dependence of $R_{th\_J2T\_err}$ on P'_h/P_h, and assumed that the initial heat propagation through the die and DAL is not dependent on the secondary heat sources P'_h. In order to verify this hypothesis, we use P_{hJ} instead of P_h as the power step for thermal transient analysis. The resulted new SFs are presented in Figure 11.

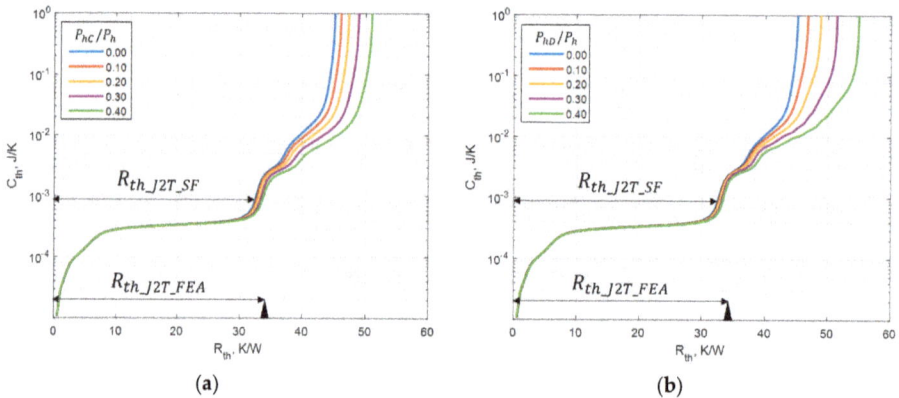

Figure 11. FEA derived SFs, when P_{hJ} power step is used for thermal transient analysis. Junction to thermal pad resistances derived by the SF analysis and a reference FEA are indicated. (**a**) when only trapped light losses are considered as the secondary heat source; and (**b**) when only phosphor dome light conversion losses are considered as the secondary heat source.

We observe a cancelation of the "scaling" effect up to the thermal pad step regardless of the nature of the secondary heat source. The derived $R_{th\_J2T\_SF}$ values are almost identical to the reference $R_{th\_J2T\_FEA}$. The $R_{th\_J2T\_err}$ error is reduced to 5% and does not depend on the P'_h/P_h ratio anymore. Thus, we conclude that our hypothesis is valid: the distributed secondary heat sources are not influencing the initial heat propagation though the die and the DAL. We conclude that the die and the DAL thermal properties can be extracted from thermal transient measurements quite accurately if the secondary heat sources are subtracted from the total thermal power.

The properties of the other elements in the thermal path and the total thermal resistance of the LED become distorted if one corrects SF for the power from the secondary heat sources. This can be seen as a spread of the SFs shapes after the thermal pad step. These results show that this method of LED thermal transient analysis can be applied only to characterize the thermal properties of the die and the DAL.

3.2. Experimental Verification

We evaluate the thermal power dissipated at the junction P_{hJ} and the secondary heat losses P'_h for each physical LED configuration. To evaluate these parameters we measure P_{el} and P_{opt} and perform initial transient analysis by Equations (12) and (14) in the range of 50–400 μsec. The obtained thermal power distributions are presented in Table 3.

Table 3. Thermal power distribution.

LED Sample	Dome Configuration	P_{el} (W)	P_{opt} (W)	P_h (W)	P_{hJ} (W)	P'_h/P_h
1A	Bare chip	0.444	0.189	0.255	0.250	0.02
1B	Flat silicone	0.445	0.122	0.323	0.229	0.29
2A	Bare chip	0.441	0.199	0.242	0.221	0.09
2B	Rim dome	0.442	0.159	0.283	0.203	0.28
2C	Phosphor top	0.441	0.091	0.350	0.233	0.33

First, we analyze the log time derivative of the measured junction temperature T_J response. The measurement data are presented in Figure 12.

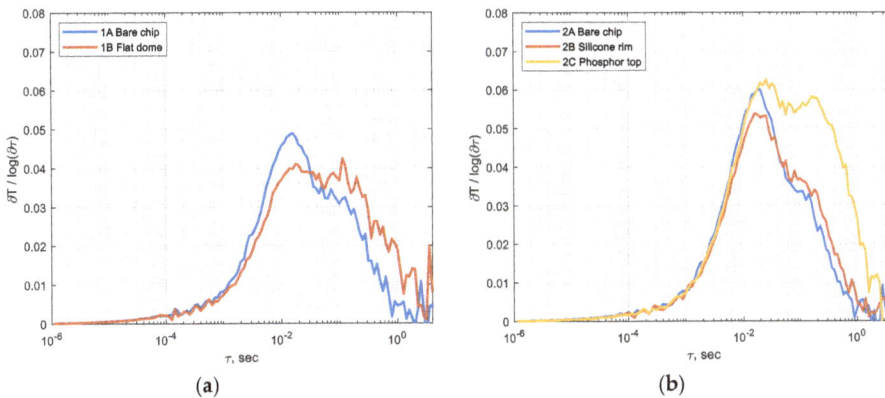

Figure 12. Experimentally derived log time derivative of T_j response: (a) 1st LED; and (b) 2nd LED.

Despite a significant variation of the measured total thermal power P_h values, we observe an expected small difference for the initial thermal transient response between both LEDs in our set of experiments. Indeed, alterations of the dome configuration mainly change the power of the secondary heat sources P'_h and weakly affect the pn junction heat power P_{hJ}. The initial T_J increase rate is dependent on P_{hJ} according to Equation (12). Moreover, we confirm the predicted impact of the phosphor heat generation on the slow time constants (Figure 12b). Additionally, we notice that the measurement data noise patterns up to 10 msec look similar. This may be an evidence for a presence of a systematic error that is most likely caused by the measurement equipment or a raw data processing algorithms.

The results of the thermal transient testing of the 1st LED are presented below. We employ the experimentally derived P_h and P_{hJ} values as power inputs for the transient analysis.

The comparison of the SFs presented in Figure 13a exhibits the "scaling" effect demonstrated with the FEA modeling. The 1B flat silicone dome configuration has a slightly higher fraction of the secondary thermal losses. Our reasoning for this is:

- The silicone dome enhances the light extraction from the chip into the encapsulant;
- The enhanced light extraction from the chip leads to a reduction of P_{hJ} (Table 3); and

- The light is trapped in silicone dome due to TIR and is absorbed on the reflector surfaces, this effect increases P'_h.

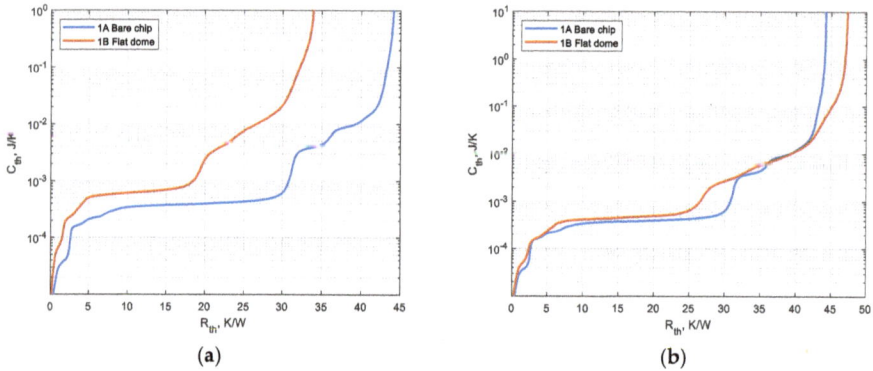

Figure 13. SFs of 1st LED: (**a**) Total thermal power P_h is used for the transient analysis; and (**b**) only junction thermal power P_{hJ} is used for the transient analysis.

Now we analyze the SFs when P_{hJ} is used for the transient analysis. The resulting SFs are presented in Figure 13b. The SFs are aligned up to the 25 K/W. The scaling effect is significantly reduced. Yet, considerable discrepancy is observable. The SF of the 1B configuration is significantly "smoother" compared to the bare chip one. We have shown in our previous works [41,43] that this effect is related to the heat storage in the dome encapsulating material.

Analyzed SFs of the 2nd LED are presented in Figure 14. Again, we observe the "scaling" effect when the total thermal power P_h is used for thermal transient analysis (Figure 14a): the higher the P'_h/P_h ratio the more distorted the initial SF regions representing the sapphire chip and the DAL. Like the dome of 1B configuration, the rim of the 2B configuration enhances the light extraction from the crystal, and it increases the fraction of the secondary thermal losses P'_h/P_h. Introduction of the phosphor top layer increases the P'_h/P_h ratio even more.

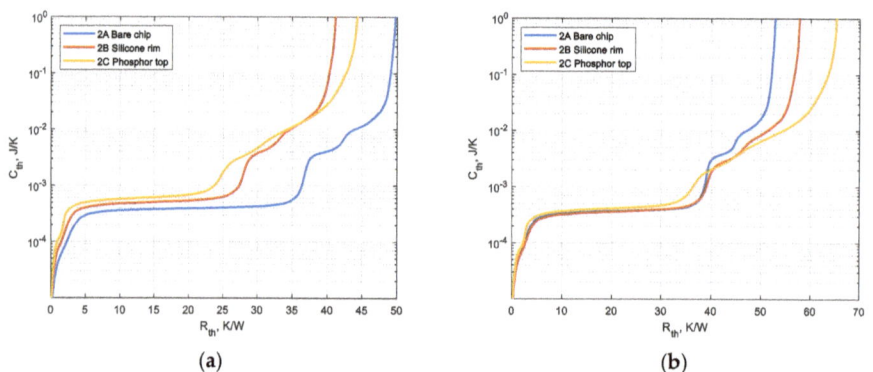

Figure 14. SFs of 2nd LED: (**a**) Total thermal power P_h is used for the transient analysis; and (**b**) only junction thermal power P_{hJ} is used for the transient analysis.

Figure 14b demonstrates that usage of the P_{hJ} power for transient analysis suppresses the scaling effect. The resulting 2A and 2B configurations' SFs perfectly overlay each other up to the thermal pad step. The silicone rim does not create a significant parallel capacitive thermal heat path, unlike the full silicone dome, thus, 2B SF is not blurred. Yet, the 2C configuration SF is blurred due to the secondary heat path created by the phosphor top layer. The derived sapphire chip thermal capacitance is almost identical for all three SFs. The SF of the 2C configuration has a significantly expanded tail compared to the pure silicone dome configuration 2B. This is caused by the phosphor heat generation impact demonstrated with the numerical simulations.

The error of the junction to thermal pad thermal resistance evaluation with thermal transient measurements without the correction for the secondary heat sources is 35% for configuration 1B, 30% for configuration 2B and 37% for 2C. The error is proportional to P'_h/P_h ratio. If the pn junction is completely encapsulated into the dome the error increases by approximately 5% due to the SF distortion caused by heat storage in the dome.

4. Discussion

We have numerically and experimentally proven the applicability of the proposed secondary heat sources separation method. Moreover, we have demonstrated that, firstly, the die and the DAL thermal resistance can be accurately derived from a SF only if the correction for the secondary heat sources is done and the thermal power dissipated exclusively by the pn junction is used for the transient analysis. Secondly, we have shown that the trapped light and the phosphor heat generation have no significant influence on the initial transient. Thirdly, we have demonstrated that the phosphor heat generation is affecting the slow time constant region of the transient spectrum and, as a result, it changes the parts of SFs correspondent to such bulky elements as the LED package and assembly elements (e.g. PCB).

Next, we analyze the topologies of various thermal models and calibration procedures used by other authors based on the obtained knowledge. Works [22] and [68] use a bidirectional thermal resistance network to model the thermal behavior of LEDs. This bidirectional model is presented in Figure 15a. It contains two power sources modeling the active region and the dome heat generation. The nodes connected by thermal resistances represent the phosphor dome, the junction and the case of an LED. To derive the model, the authors compare and analyze SFs of bare chip, silicone and silicone/phosphor composite dome packages of the same LED design. The SFs have a noticeable decrease of the cumulative thermal resistance and an increase of the cumulative thermal capacitance of the die and DAL regions evidencing the "scaling" effect demonstrated in Figure 10a. The authors explain the junction to ambient thermal resistances decrease by the enhancement of the heat conduction through the dome towards ambient (e.g., a decrease of R_{th_D2J} and R_{th_D2A}). Nevertheless, as shown in other scientific works, the steady-state convection and radiation heat transfer from the top of an MP LED package is negligible comparing to the conduction heat transfer via the main heat path [43,44,69] ($R_{th\_D2A} \gg R_{th\_J2C}$). Thus, the bidirectional model proposed by the authors [22] and [68] does not provide a physics-based explanation of the observed phenomena.

Juntunen et al. also developed a thermal resistance network model that considers the secondary heat sources [13]. The model is shown in Figure 15b. The model contains a thermal pad node correspondent to the temperature of the top of the substrate on which an LED die is mounted. The authors consider a one-dimensional heat path but do not separate the junction and the phosphor heat sources in their model. This may not accurately capture all relevant physical phenomena. The SF data presented in their paper also exhibits the "scaling" effect. Yet, they do not consider distributed secondary thermal sources. They explain this effect only by an additional parallel heat path (represented as R_{th_shunt}) formed by the dome that shunts the die and the DAL thermal resistance R_{th_J2T}.

Based on the analysis of the SFs we propose a thermal resistance network model that eliminates the above mentioned flaws. The topology of the model is shown in Figure 15c. It contains three power sources responsible for heat generation in the pn junction, the dome and on the reflective cup surface. The dome heat source is connected both to the pn junction and the thermal pad. The trapped light heat source is located at the thermal pad node. The secondary heat sources are spread over the package according to their actual placement. Thus, a significant part of the heat generated by them is dissipated through the thermal pad bypassing the die and the DAL. This better explains the main reason of the total R_{th} decrease for the LEDs with secondary heat sources rather than assumed increase of the heat flow through the top of the dome, or the shunting heat path for the die and the DAL. We propose a possible extraction procedure for the parameters of this compact thermal model in Appendix C.

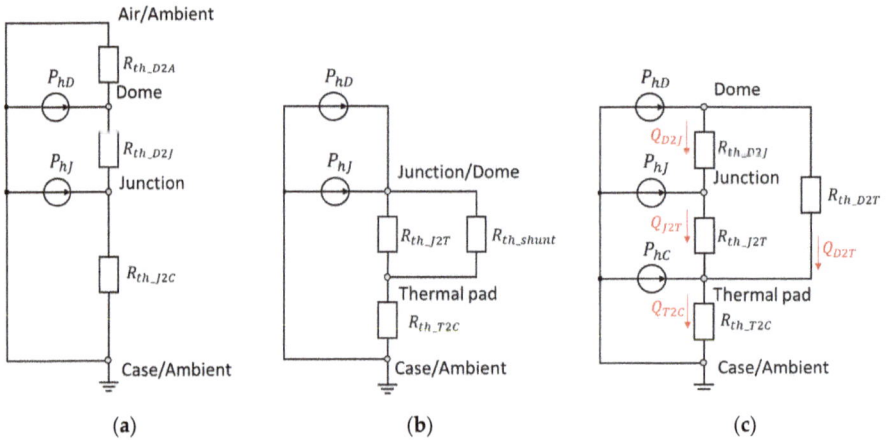

Figure 15. LED package thermal resistance models. (**a**) Bidirectional model presented in works [22,68]; (**b**) the model presented in work of Juntunen et al. [13]; and (**c**) the model proposed by us. Heat flows are indicated with red.

The discussed impact of the phosphor secondary heat source on the "tails" of SFs is noticed in the experimental and numerical simulations data of works [13,15,22]. Like demonstrated in Figure 10a, the presence of phosphor losses leads to an extension of the SFs' thermal pad and PCB regions' SF images. As discussed earlier, this effect is caused by the relatively low thermal diffusivity of silicone/phosphor composite and associated alteration of the slow time constant spectrum, as demonstrated in Figure 9a. This knowledge coupled with the demonstrated procedure of heat source separation could enhance the accuracy of the CTM calibration procedure demonstrated in the work of Bornoff et al. [7] and similar ones.

5. Conclusions

We have shown that the secondary heat sources have a significant impact on the accuracy of the thermal transient analysis results, in particular, they distort the die and the DAL regions of SF causing a "scaling" effect. We have numerically modeled and experimentally confirmed the distortion of the SF's die and the DAL regions associated with the presence of secondary heat sources. We have experimentally confirmed that the related error of the die and the DAL thermal resistance evaluation can be as high as 35%. Moreover, our estimations indicate that this error can reach up to 50% for LEDs with phosphor light conversion. We have shown that this error is proportional to the fraction of the secondary heat sources in the total power dissipation.

We have proposed a novel method of separation of the main and the secondary heat sources based on the thermal transient analysis and radiant flux measurements. The method is confirmed both numerically and experimentally. The method can significantly increase the accuracy of calibration procedures for 3D FEA thermal models.

We have proposed a technique that can be used to determine actual thermal properties of the die and the DAL. The technique enables suppression of the distortion effects in SFs caused by the secondary heat sources. Using the new SF analysis insights obtained with this technique, we have analyzed current thermal resistor models for multiple heat source LEDs. We notice that these models are partially derived using incorrect SF interpretations. Thus, we propose a new model topology that enables physically accurate thermal modeling of multiple heat source LEDs.

The characteristic impact of the heat generation by phosphor particles on the transient response was demonstrated. We have found that the heat caused by the phosphor light conversion losses affects "slow" thermal time constants. These time constants are often related to the heat transfer through the PCB. This is an important observation for enhancing calibration accuracy of 3D FEA thermal models of phosphor converted LEDs against thermal transient results.

We have revised and improved the analytical solution for the initial transient response, by considering a case of bilateral heat propagation. Our approach increases the accuracy of the junction thermal power estimation up to 15% for LEDs with phosphor light conversion and up to 5% for the LEDs with clear silicone encapsulant.

In a general case, only the total power of the secondary heat sources can be derived with the proposed secondary heat sources estimation method. It appeared not yet possible to separate the phosphor heat generation and the reflection losses. The sensitivity analysis of the proposed method to the accuracy of the k-factor calibration and the measurement noise, as well as additional cross-verification with other methods of the secondary heat source evaluations should be performed.

Moreover, the heat generation profile in a silicone/phosphor composite dome is not uniform. It depends on the phosphor particle spatial distribution around the die, the dome shape and the reflection cup geometry. Thus, the secondary heat sources power density distribution within an LED package is difficult to evaluate. There is no technique yet to estimate the part of the heat dissipated by the phosphor particles that bypasses the die and DAL. As a result, the proposed power separation method still cannot guarantee a precise calibration if used alone without FEA. Thus, we see room for future research.

Author Contributions: Methodology, software, validation, writing—original draft preparation: A.A.; conceptualization: A.A. and G.O.; supervision, writing—review and editing: J.-P.L. and G.M.; project administration, and funding acquisition: G.M.

Funding: This research has received funding from the European Union's Horizon 2020 research and innovation program through the H2020 ECSEL project Delphi4LED (grant agreement number: 692465) (2016–2019). Co-financing of the Delphi4LED project by the national R&D funding organization of the participating countries. Additional information is available on: www.DELPHI4LED.eu.

Acknowledgments: Support from Delphi4LED project partners, especially from Robin Bornoff (Mentor), Andras Poppe (Mentor), Marta Rencz (Mentor) and Gabor Farkas (Mentor) is acknowledged.

Conflicts of Interest: The authors declare no conflict of interest.

Nomenclature

CTM	Compact thermal model
DAL	Die attach layer
DUT	Device Under Test
FEA	Finite Element Analysis
LED	Light-emitting diode
PCB	Printed circuit board
MCPCB	Metal core PCB
MP	Mid-power
SF	Structure function
TIR	Total internal reflection
EQE	External quantum efficiency
IQE	Internal quantum efficiency
I	Current
a	Nondimensional parameter characterizing composite particles
f	Phosphor particles volume fraction
Λ	Crystal light re-absorption correction coefficient
P_{el}	Electric power

P_{opt}	Radiant flux
P_h	Total thermal power
P'_h	Combined secondary heat sources thermal power
P_{hJ}	Thermal power dissipated in a pn junction
P_{h_re}	Thermal power dissipated in a crystal during the trapped light re-entries
P_{hC}	Thermal power dissipated on the cup reflector surface
P_{hD}	Thermal power dissipated in the dome volume
C_{th}	Thermal capacitance
R_{th_y}	Thermal resistance
Q_y	Heat flow
y	suffix made of two capital letters separated by "2". The letters are can be "J", "D", "T", and "C", denoting pn junction, and the LEDs' dome, top of the thermal pad and case. The suffix denotes elements connecting two entities. The suffix can be followed by "FEA" which designates value derived directly from finite element analysis, "SF" which designates value derived from a structure function and "err" which designated the relative error of the value.
S_{pn}	pn junction area
S_{cup}	Reflector cup area
S_{cry}	Crystal chip area
N_{cup}	Average number of the photons bouncing off the cup walls
N_{cry}	Average number of crystal re-entries by photons
τ_z	Time
T_z	Temperature
ρ_z	Density
k_z	Thermal conductivity
C_z	Specific heat
	Spatial coordinate
q_z	Heat flux
H_z	Height
R_z	Radius
V_z	Volume
z	suffix that can be "a", "j", "d", "s", "sil", "pho", and "sap" denoting ambient, pn junction, LEDs' dome, pn junction substrate, silicone, phosphor, and sapphire materials, respectively
K_{uni_lat}	Unilateral heat propagation initial transient constant
K_{bi_lat}	Bilateral heat propagation initial transient constant
LEE	Total light extraction efficiency
LEE_{chip}	Chip to dome light extraction efficiency
LEE_{dome}	Dome to ambient light extraction efficiency

Appendix A

This appendix enables an analytical estimation of the reabsorption coefficient Λ dependence on the LEDs dome and cup geometries.

The optical losses from the portion of the light when it left the crystal for the first time P_{h_opt} are equal to:

$$P_{h\_opt} = IQE(I) \cdot LEE_{chip} \cdot (1 - LEE_{dome}) \tag{A1}$$

These losses are dissipated either on the reflector cup surface as P_{hC} or inside of the crystal during light re-entries as P_{h_re}:

$$P_{h\_opt} = P_{hC} + P_{h\_re} \tag{A2}$$

The equation can be rewritten with the coefficient Λ:

$$P_{hC} = P_{h\_opt} \cdot \Lambda \tag{A3}$$

$$P_{h\_re} = P_{h\_opt} \cdot (1 - \Lambda) \tag{A4}$$

The ratio between P_{hC} and P_{hJ_re} is determined by the number of the cup reflection events N_{cup}, the crystal transitions events N_{cry} and the correspondent probabilities of light absorption during these events. The absorption probabilities are determined by the cup walls reflection coefficient R_{cup} and the crystal transition coefficient T_{cry}:

$$\frac{P_{hC}}{P_{h\_re}} = \frac{N_{cup}\left(1 - R_{cup}\right)}{N_{cry}\left(1 - T_{cry}\right)} \tag{A5}$$

We assume that the average number of the photons bouncing off the cup walls and the number of crystal re-entries by the trapped light are proportional to their surface areas exposed to the dome. Thereby, the ratio between these quantities is:

$$\frac{N_{cup}}{N_{cry}} \cong \frac{S_{cup}}{S_{cry}} \tag{A6}$$

where S_{cup} is the area of the dome cup walls, S_{cry} is the area of the sapphire crystal in contact with the encapsulating dome. Now, assuming that the crystal transition probability is approximately equal to the light extraction efficiency of the chip: $T_{cry} \cong LEE_{chip}$, Λ coefficient can be determined based on Equations (A3)–(A6):

$$\Lambda \cong \left(1 + \frac{S_{cup}\left(1 - R_{cup}\right)}{S_{cry}\left(1 - LEE_{chip}\right)}\right)^{-1} \tag{A7}$$

Appendix B

In this appendix, we present an analytical solution for the transient temperature response of the interface between two semi-infinite bodies when a uniform heat source is located at the contact. One of the bodies represents the crystal substrate, the other one corresponds to the dome. Figure A1 illustrates the problem.

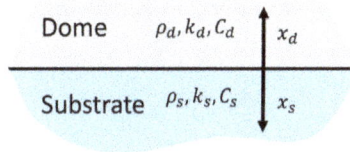

Figure A1. Transient heat conduction on the edge of the sapphire crystal and dome. x_d and x_s are the spatial coordinates.

The problem is one-dimensional. At the initial time $\tau = 0$ the bodies have uniform identical temperatures T_0. A constant and uniform heat source with power density q is activated at the interface at $\tau = 0$. The heat transfer governing equations are:

$$\rho_d C_d \frac{\partial T_d}{\partial \tau} = k_d \frac{\partial^2 T_d}{\partial x_d^2} \tag{A8}$$

$$\rho_s C_s \frac{\partial T_s}{\partial \tau} = k_s \frac{\partial^2 T_s}{\partial x_s^2} \tag{A9}$$

The initial and the interfacial conditions are:

$$\begin{aligned} T_d(x_d, 0) &= T_0 \\ T_s(x_s, 0) &= T_0 \end{aligned} \tag{A10}$$

$$T_d(0, \tau) = T_s(0, \tau) \tag{A11}$$

We define the heat fluxes q_d and q_s that leave the interface towards the dome and the sapphire, respectively. According to the law of conservation of energy, their sum is equal to q. Thus:

$$q_d = -k_d \frac{\partial T_d(0, \tau)}{\partial x_d}\bigg|_{x_d=0} \tag{A12}$$

$$q_s = -k_s \frac{\partial T_s(0, \tau)}{\partial x_s}\bigg|_{x_s=0} \tag{A13}$$

$$q = q_d + q_s \qquad (A14)$$

From Equations (A8), (A9), (A11)–(A13), it can be seen that q_d and q_s are constant and they depend only on the thermal properties of the sapphire crystal and the dome. The ratio between the heat fluxes is:

$$\frac{q_d}{q_s} = \sqrt{\frac{k_d \rho_d C_d}{k_s \rho_s C_s}} \qquad (A15)$$

Therefore, the bilateral heat propagation problem can be reduced to two unilateral problems with heat fluxes q_d and q_s. The heat flux through the sapphire crystal can be expressed as:

$$q_s = \frac{q}{1 + \sqrt{\frac{k_d \rho_d C_d}{k_s \rho_s C_s}}} \qquad (A16)$$

Thus, the surface temperature response is equal to:

$$T(t) = T_0 + \frac{2q}{\sqrt{\pi}\left(\sqrt{k_s \rho_s C_s} + \sqrt{k_d \rho_d C_d}\right)} \sqrt{\tau} \qquad (A17)$$

Appendix C

The calculations presented in this appendix show a way to estimate the parameters of our thermal resistance model shown in Figure 15c.

Using simple circuit analysis, we define the dependence of the heat flows $Q_{D2J}, Q_{J2T}, Q_{D2J}$, and Q_{T2C} on the corresponding thermal resistances and the average temperature of the junction $\overline{T}_j$, the average temperature of the upper surface of the thermal pad $\overline{T}_t$, the average case temperature $\overline{T}_c$ and the chosen measure of the dome temperature T_d:

$$Q_{D2J} = \frac{T_d - \overline{T}_j}{R_{th\_D2J}} \qquad (A18)$$

$$Q_{J2T} = \frac{\overline{T}_j - \overline{T}_t}{R_{th\_J2T}} \qquad (A19)$$

$$Q_{D2T} = \frac{T_d - \overline{T}_t}{R_{th\_D2T}} \qquad (A20)$$

$$Q_{T2C} = \frac{\overline{T}_t - \overline{T}_c}{R_{th\_T2C}} \qquad (A21)$$

The Kirchhoff's current law bounds the heat flows and the considered heat sources:

$$P_{hD} = Q_{D2J} + Q_{D2T} \qquad (A22)$$

$$P_{hJ} = Q_{J2T} - Q_{D2J} \qquad (A23)$$

$$P_{hC} = Q_{T2C} - Q_{J2T} - Q_{D2T} \qquad (A24)$$

The collection of Equations (A18)–(A24) creates an underdetermined system with 15 variables. However, $\overline{T}_j$ and $\overline{T}_c$ can be estimated by a thermal transient measurement and a dual interface material method. We have shown that R_{th_J2T} can be found from an SF if the junction thermal power P_{hJ} is used for thermal transient analysis instead of the total thermal power P_h. A method of estimation of P_{hJ} and the sum of the secondary heat sources P_{hD} and P_{hC} is also described in this paper.

Firstly, we consider an LED without phosphor light conversion layer. In this case, $P_{hD} = 0$ and the dome temperature is not a critical reliability parameter. Moreover, the parallel heat path through the dome formed by R_{th_D2J} and R_{th_D2T} has very high characteristic thermal resistance compared to the main thermal path due to low thermal conductivity of silicone [43]. Thus, Equations (A18), (A20), and (A22) associated with the dome node can be taken out of the consideration. Q_{D2J} and Q_{D2T} can be assumed to be equal to zero. Now, the system can be solved and the rest of the model parameters can be determined mathematically.

Secondly, we consider an LED with a phosphor light conversion layer. The phosphor particles encapsulated in the dome suppress multiple TIRs due to light re-emission and light scattering. Thus, we can assume that the trapped light reflection losses P_{hC} are negligible and all the secondary heat losses are primarily due to P_{hD}. The temperature of the phosphor dome is not uniform in a general case. More research is required to understand the temperature profile of it [70]. Yet, a number of methods, like infrared imaging, thermal couple measurements, or spectroscopic approaches, are known for phosphor layer temperature characterization [71–74]. Any of these methods can be chosen to provide a characteristic measure of the dome temperature T_d. Part of the heat produced by the phosphor dome dissipates through the pn junction and the sapphire, the other part goes directly to the thermal pad. Thus, the ratio of R_{th_D2J} and R_{th_D2T} can be approximated by the reciprocal of the ratio of the crystal and the cup surfaces S_{cry} and S_{cup}. Now the considered system of equations becomes determined. Therefore, all the model parameters can be calculated.

The accuracy of phosphor and junction temperature modeling by a thermal resistor network is validated in [22]. Our model can be mathematically transformed into the one used in the reference by network analysis techniques, thus, it can also be considered as validated. Yet, our model provides a physics-based explanation of the thermal behavior of an LED.

The presented method for parameter extraction enables analytical derivation of the models parameters. Nevertheless, such parameters as R_{th_D2J} and R_{th_D2T} can only be coarsely estimated by SF and LED package geometry analysis. As discussed earlier in the paper, such parameters can be extracted with a higher accuracy from calibrated thermal finite element models.

References

1. Schilders, W.H.; Van der Vorst, H.A.; Rommes, J. (Eds.) *Model Order Reduction: Theory, Research Aspects and Applications*; Mathematics in Industry; Springer: Berlin/Heidelberg, Germany, 2008; Volume 13, ISBN 978-3-540-78840-9.
2. Lasance, C.; Vinke, H.; Rosten, H.; Weiner, K.-L. A novel approach for the thermal characterization of electronic parts. In Proceedings of the 1995 IEEE/CPMT 11th Semiconductor Thermal Measurement and Management Symposium (SEMI-THERM), San Jose, CA, USA, 7–9 February 2011; Volume 36, pp. 1–9.
3. Sabry, M. Compact thermal models for electronic systems. *IEEE Trans. Components Packag. Technol.* **2003**, *26*, 179–185. [CrossRef]
4. Alexeev, A.; Bornoff, R.; Lungten, S.; Martin, G.; Onushkin, G.; Poppe, A.; Rencz, M.; Yu, J. Requirements specification for multi-domain LED compact model development in Delphi4LED. In Proceedings of the 2017 18th International Conference on Thermal, Mechanical and Multi-Physics Simulation and Experiments in Microelectronics and Microsystems (EuroSimE), Dresden, Germany, 3–5 April 2017; pp. 1–8.
5. Bornoff, R.; Hildenbrand, V.; Lugten, S.; Martin, G.; Marty, C.; Poppe, A.; Rencz, M.; Schilders, W.H.A.; Yu, J. Delphi4LED—From measurements to standardized multi-domain compact models of LED: A new European R & D project for predictive and efficient multi-domain modeling and simulation of LEDs at all integration levels along the SSL supply chain. In Proceedings of the 2016 22nd International Workshop on Thermal Investigations of ICs and Systems (THERMINIC), Budapest, Hungary, 21–23 September 2016; pp. 174–180.
6. Lungten, S.; Bornoff, R.; Dyson, J.; Maubach, J.M.L.; Schilders, W.H.A.; Warner, M. Dynamic compact thermal model extraction for LED packages using model order reduction techniques. In Proceedings of the 2017 23rd International Workshop on Thermal Investigations of ICs and Systems (THERMINIC), Amsterdam, the Netherlands, 27–29 September 2017; pp. 1–6.
7. Bornoff, R.; Farkas, G.; Gaal, L.; Rencz, M.; Poppe, A. LED 3D thermal model calibration against measurement. In Proceedings of the 2018 19th International Conference on Thermal, Mechanical and Multi-Physics Simulation and Experiments in Microelectronics and Microsystems (EuroSimE), Toulouse, France, 15–18 April 2018; pp. 1–7.
8. Meyer, J.; Thomas, C.; Tappe, F.; Ogbazghi, T. In Depth Analyses of LEDs by a Combination of X-ray Computed Tomography (CT) and Light Microscopy (LM) Correlated with Scanning Electron Microscopy (SEM). *J. Vis. Exp.* **2016**. [CrossRef] [PubMed]
9. Ma, B.; Lee, K.H. Evaluation of the internal quantum efficiency in light-emitting diodes. *J. Korean Phys. Soc.* **2015**, *67*, 658–662. [CrossRef]
10. Ryu, H.Y.; Ryu, G.H.; Lee, S.H.; Kim, H.J. Evaluation of internal quantum efficiency in blue and green light-emitting diodes using rate equation model. In Proceedings of the 2013 Conference on Lasers and Electro-Optics Pacific Rim (CLEOPR), Kyoto, Japan, 30 June–4 July 2013; pp. 1–2.
11. Lin, Y.; Tran, N.; Zhou, Y.; He, Y.; Shi, F. Materials Challenges and Solutions for the Packaging of High Power LEDs. In Proceedings of the 2006 International Microsystems, Package, Assembly Conference Taiwan, Taipei, Taiwan, 18–20 October 2006; pp. 1–4.
12. Diodes, R.P.L.; Ma, Y.; Yu, X.; Xie, B.; Hu, R.; Luo, X. Analysis of Phosphor Heat Generation and Temperature Distribution in Analysis of Phosphor Heat Generation and Temperature Distribution in Remote- plate

Phosphor-Converted Light-Emitting Diodes. In Proceedings of the Asian Conferenceon Thermal Sciences 2017, Jeju Island, Korea, 26–30 March 2017; pp. 1–6.

13. Juntunen, E.; Tapaninen, O.; Sitomaniemi, A.; Heikkinen, V. Effect of Phosphor Encapsulant on the Thermal Resistance of a High-Power COB LED Module. *IEEE Trans. Components Packag. Manuf. Technol.* **2013**, *3*, 1148–1154. [CrossRef]

14. Kim, J.K.; Luo, H.; Schubert, E.F.; Cho, J.; Sone, C.; Park, Y. Strongly Enhanced Phosphor Efficiency in GaInN White Light-Emitting Diodes Using Remote Phosphor Configuration and Diffuse Reflector Cup. *Jpn. J. Appl. Phys.* **2005**, *44*, L649–L651. [CrossRef]

15. Alexeev, A.; Martin, G.; Hildenbrand, V.; Bosschaart, K.J. Influence of dome phosphor particle concentration on mid-power LED thermal resistance. In Proceedings of the 2016 32nd Thermal Measurement, Modeling & Management Symposium (SEMI-THERM), San Jose, CA, USA, 14–17 March 2016; pp. 33–43.

16. Treurniet, T.; Lammens, V. Thermal management in color variable multi-chip led modules. In Proceedings of the Twenty-Second Annual IEEE Semiconductor Thermal Measurement And Management Symposium, Dallas, TX, USA, 14–16 March 2006; pp. 173–177.

17. Poppe, A.; Zhang, Y.; Wilson, J.; Farkas, G.; Szabo, P.; Parry, J.; Rencz, M.; Szekely, V. Thermal Measurement and Modeling of Multi-Die Packages. *IEEE Trans. Components Packag. Technol.* **2009**, *32*, 484–492. [CrossRef]

18. Mah, J.W.; Lee, B.K.; Devarajan, M. Thermal Impedance Measurement on Different Chip Arrangements for Various Multichip LEDs Application. *IEEE Trans. Electron Devices* **2015**, *62*, 2906–2912. [CrossRef]

19. Qian, X.; Zou, J.; Shi, M.; Yang, B.; Li, Y.; Wang, Z.; Liu, Y.; Liu, Z.; Zheng, F. Development of optical-thermal coupled model for phosphor-converted LEDs. *Front. Optoelectron.* **2019**. [CrossRef]

20. Luo, X.; Hu, R. Calculation of the phosphor heat generation in phosphor-converted light-emitting diodes. *Int. J. Heat Mass Transf.* **2014**, *75*, 213–217. [CrossRef]

21. Alexeev, A.; Cassarly, W.; Hildenbrand, V.D.; Tapaninen, O.; Sitomaniemi, A.; Wondergem, A. Simulating light conversion in mid-power LEDs. In Proceedings of the 2016 17th International Conference on Thermal, Mechanical and Multi-Physics Simulation and Experiments in Microelectronics and Microsystems (EuroSimE), Montpellier, France, 18–20 April 2016; pp. 1–7.

22. Ma, Y.; Hu, R.; Yu, X.; Shu, W.; Luo, X. A modified bidirectional thermal resistance model for junction and phosphor temperature estimation in phosphor-converted light-emitting diodes. *Int. J. Heat Mass Transf.* **2017**, *106*, 1–6. [CrossRef]

23. Lasance, C.J.M.S. *Thermal Management for LED Applications*; Lasance, C.J.M., Poppe, A., Eds.; Solid State Lighting Technology and Application Series; Springer: New York, NY, USA, 2014; Volume 2, ISBN 978-1-4614-5090-0.

24. Profumo, F.; Tenconi, A.; Faceili, S.; Passerini, B. Implementation and validation of a new thermal model for analysis, design, and characterization of multichip power electronics devices. *IEEE Trans. Ind. Appl.* **1999**, *35*, 663–669. [CrossRef]

25. Poppe, A.; Szalai, A. Practical aspects of implementation of a multi-domain LED model. In Proceedings of the 2014 Semiconductor Thermal Measurement and Management Symposium (SEMI-THERM), San Jose, CA, USA, 9–13 March 2014; pp. 153–158.

26. Mitterhuber, L.; Defregger, S.; Magnien, J.; Rosc, J.; Hammer, R.; Goullon, L.; Hutter, M.; Schrank, F.; Hörth, S.; Kraker, E. Thermal transient measurement and modelling of a power cycled flip-chip LED module. *Microelectron. Reliab.* **2018**, *81*, 373–380. [CrossRef]

27. Monavarian, M. Beyond Conventional c-Plane GaN-Based Light Emitting Diodes: A Systematic Exploration of LEDs on Semi-Polar Orientations. Ph.D. Disseration, Virginia Commonwealth University, Richmond, VA, USA, 2016.

28. Van Bien, T.; Szekely, V. Fine structure of heat flow path in semiconductor devices: A measurement and identification method. *Solid. State. Electron.* **1988**, *31*, 1363–1368.

29. Székely, V. A new evaluation method of thermal transient measurement results. *Microelectron. J.* **1997**, *28*, 277–292. [CrossRef]

30. Rencz, M.; Szekely, V. Structure function evaluation of stacked dies. In Proceedings of the Twentieth Annual IEEE Semiconductor Thermal Measurement and Management Symposium (IEEE Cat. No.04CH37545), San Jose, CA, USA, 11 March 2004; pp. 50–54.

31. Poppe, A. Testing of Power LEDs: The Latest Thermal Testing Standards from JEDEC. *Electron. Cool. Mag.* **2013**. Available online: https://www.electronics-cooling.com/2013/09/testing-of-power-leds-the-latest-thermal-testing-standards-from-jedec/ (accessed on 13 May 2019).

32. JESD51-50—Overview of Methodologies for the Thermal Measurement of Single- and Multi-Chip, Single- and Multi-PN-Junction Light-Emitting Diodes (LEDs). Available online: https://www.jedec.org/system/files/docs/JESD51-50.pdf (accessed on 13 May 2019).

33. JESD51-51—Implementation of the Electrical Test Method for the Measurement of Real Thermal Resistance and Impedance of Light-Emitting Diodes with Exposed Cooling. Available online: https://www.jedec.org/system/files/docs/JESD51-51.pdf (accessed on 13 May 2019).

34. JESD51-52—Guidelines for Combining CIE 127-2007 Total Flux Measurements with Thermal Measurements of LEDs with Exposed Cooling Surface. Available online: https://www.jedec.org/system/files/docs/JESD51-52.pdf (accessed on 13 May 2019).

35. JESD51-53—Terms, Definitions and Units Glossary for LED Thermal Testing. Available online: https://www.jedec.org/system/files/docs/JESD51-53.pdf (accessed on 13 May 2019).

36. Rencz, M.R.; Székely, V. Measuring partial thermal resistances in a heat-flow path. *IEEE Trans. Components Packag. Technol.* **2002**, *25*, 547–553. [CrossRef]

37. Rencz, M.; Szekely, V.; Morelli, A.; Villa, C. Determining partial thermal resistances with transient measurements, and using the method to detect die attach discontinuities. In Proceedings of the Eighteenth Annual IEEE Semiconductor Thermal Measurement and Management Symposium, San Jose, CA, USA, 12–14 March 2002; pp. 15–20.

38. He, P.; Zhang, J.; Zhang, J.; Yin, L. Effects of Die-Attach Quality on the Mechanical and Thermal Properties of High-Power Light-Emitting Diodes Packaging. *Adv. Mater. Sci. Eng.* **2017**, *2017*, 8658164. [CrossRef]

39. Chen, C.J.; Chen, C.M.; Horng, R.H.; Wuu, D.S.; Hong, J.S. Thermal management and interfacial properties in high-power GaN-based light-emitting diodes employing diamond-added Sn-3 wt.%Ag-0.5 wt.%Cu solder as a die-attach material. *J. Electron. Mater.* **2010**, *39*, 2618–2626. [CrossRef]

40. Bornoff, R.; Merelle, T.; Sari, J.; Di Bucchianico, A.; Farkas, G. Quantified Insights into LED Variability. In Proceedings of the 2018 24rd International Workshop on Thermal Investigations of ICs and Systems (THERMINIC), Stockholm, Sweden, 26–28 September 2018; pp. 1–6.

41. Alexeev, A.; Martin, G.; Onushkin, G.; Linnartz, J.-P. Accurate Thermal Transient Measurements Interpretation of Monochromatic LEDs. In Proceedings of the Semi-Therm 35, San-Jose, CA, USA, 18–22 March 2019, unpublished.

42. Tran, N.T.; Shi, F.G. LED package design for high optical efficiency and low viewing angle. In Proceedings of the 2007 International Microsystems, Packaging, Assembly and Circuits Technology, Taipei, Taiwan, 1–3 October 2007; pp. 10–13.

43. Alexeev, A.; Martin, G.; Onushkin, G. Multiple heat path dynamic thermal compact modeling for silicone encapsulated LEDs. *Microelectron. Reliab.* **2018**, *87*, 89–96. [CrossRef]

44. Arik, M.; Becker, C.A.; Weaver, S.E.; Petroski, J. Thermal management of LEDs: Package to system. In Proceedings of the Third International Conference on Solid State Lighting, Denver, CO, USA, 3–6 August 2004; pp. 64–76.

45. Hantos, G.; Hegedus, J.; Poppe, A. Different questions of today's LED thermal testing procedures. In Proceedings of the 2018 34th Thermal Measurement, Modeling & Management Symposium, San Jose, CA, USA, 19–23 March 2018; pp. 63–70.

46. Hantos, G.; Hegedus, J. K-factor calibration issues of high power LEDs. In Proceedings of the 2017 23rd International Workshop on Thermal Investigations of ICs and Systems (THERMINIC), Amsterdam, The Netherlands, 27–29 Septermber 2017; pp. 1–6.

47. Schubert, E.F. *Light-Emitting Diodes*; Cambridge University Press: Cambridge, UK, 2006; ISBN 9780511790546.

48. Zou, H.; Wang, J.; Feng, M.; Shieh, B.; Lee, S.W.R. A novel chip-on-board white light-emitting diode design for light extraction enhancement. In Proceedings of the 2016 13th China International Forum on Solid State Lighting (SSLChina), Beijing, China, 15–17 November 2016; Volume 2, pp. 24–27.

49. Wu, D.; Wang, K.; Liu, S. Enhancement of light extraction efficiency of multi-chips light-emitting diode array packaging with various microstructure arrays. In Proceedings of the 2011 IEEE 61st Electronic Components and Technology Conference, Lake Buena Vista, FL, USA, 31 May–3 June 2011; pp. 242–245.

50. Li, Z.T.; Wang, Q.H.; Tang, Y.; Li, C.; Ding, X.R.; He, Z.H. Light extraction improvement for LED COB devices by introducing a patterned leadframe substrate configuration. *IEEE Trans. Electron Devices* **2013**, *60*, 1397–1403. [CrossRef]

51. Chen, T.; Hsu, T.-C.; Luo, C.; Hsu, M.; Lee, T. Improvement in light extraction efficiency of high brightness In GaN-Based Light Emitting Diodes. In *Gallium Nitride Materials and Devices IV*; SPIE - International Society For Optics and Photonics: San Jose, CA, USA, 2009; Volume 7216, pp. 72161T1–72161T10.

52. Chung, S.-C.; Li, D.-R.; Lee, T.-X.; Yang, T.-H.; Ho, P.-C.; Sun, C.-C. Effect of chip spacing on light extraction for light-emitting diode array. *Opt. Express* **2015**, *23*, A640. [CrossRef] [PubMed]

53. Schweitzer, D.; Pape, H.; Kutscherauer, R.; Walder, M. How to evaluate transient dual interface measurements of the Rth-JC of power semiconductor packages. In Proceedings of the 2009 25th Annual IEEE Semiconductor Thermal Measurement and Management Symposium, San Jose, CA, USA, 15–19 March 2009; pp. 172–179.

54. Glavanovics, M.; Zitta, H. Thermal destruction testing: An indirect approach to a simple dynamic thermal model of smart power switches. In Proceedings of the 27th European Solid-State Circuits Conference, Villach, Austria, 18–20 September 2001; pp. 2–5.

55. Carslaw, H.S.; Jaeger, J.C. *Conduction of Heat in Solids*; Clarendon Press: Oxford, UK, 1959.

56. Pietrak, K.; Winiewski, T.S. A review of models for effective thermal conductivity of composite materials. *J. J. Power Technol.* **2015**, *95*, 14–24.

57. Hasselman, D.P.H.; Johnson, L.F. Effective Thermal Conductivity of Composites with Interfacial Thermal Barrier Resistance. *J. Compos. Mater.* **1987**, *21*, 508–515. [CrossRef]

58. Nan, C.W.; Birringer, R.; Clarke, D.R.; Gleiter, H. Effective thermal conductivity of particulate composites with interfacial thermal resistance. *J. Appl. Phys.* **1997**, *81*, 6692–6699. [CrossRef]

59. Every, A.G.; Tzou, Y.; Hasselman, D.P.H.; Raj, R. The effect of particle size on the thermal conductivity of ZnS/diamond composites. *Acta Metall. Mater.* **1992**, *40*, 123–129. [CrossRef]

60. Zhang, Q.; Pi, Z.; Chen, M.; Luo, X.; Xu, L.; Liu, S. Effective thermal conductivity of silicone/phosphor composites. *J. Compos. Mater.* **2011**, *45*, 2465–2473. [CrossRef]

61. Submitted, D.T. Thermal Properties of Yttrium Aluminum Garnett From Molecular Dynamics Simulations. In Proceedings of the ASME/JSME 2011 8th Thermal Engineering Joint Conference, Hawaii, HI, USA, 13–17 March 2011.

62. Dal Lago, M.; Meneghini, M.; Trivellin, N.; Mura, G.; Vanzi, M.; Meneghesso, G.; Zanoni, E. Phosphors for LED-based light sources: Thermal properties and reliability issues. *Microelectron. Reliab.* **2012**, *52*, 2164–2167. [CrossRef]

63. Fulmek, P.; Nicolics, J.; Nemitz, W.; Schweitzer, S.; Sommer, C.; Hartmann, P.; Schrank, F.; Wenzl, F.P. The impact of the thermal conductivities of the color conversion elements of phosphor converted LEDs under different current driving schemes. *J. Lumin.* **2016**, *169*, 559–568. [CrossRef]

64. Chung, T.Y.; Chiou, S.C.; Chang, Y.Y.; Sun, C.C.; Yang, T.H.; Chen, S.Y. Study of Temperature Distribution Within pc-WLEDs Using the Remote-Dome Phosphor Package. *IEEE Photonics J.* **2015**, *7*, 1–11. [CrossRef]

65. Hu, R.; Luo, X. A model for calculating the bidirectional scattering properties of phosphor layer in white light-emitting diodes. *J. Light. Technol.* **2012**, *30*, 3376–3380. [CrossRef]

66. Huang, M.; Yang, L. Heat generation by the phosphor layer of high-power white LED emitters. *IEEE Photonics Technol. Lett.* **2013**, *25*, 1317–1320. [CrossRef]

67. Schweitzer, D.; Pape, H.; Chen, L.; Kutscherauer, R.; Walder, M. Transient dual interface measurement A new JEDEC standard for the measurement of the junction-to-case thermal resistance. In Proceedings of the 2011 27th Annual IEEE Semiconductor Thermal Measurement and Management Symposium, San Jose, CA, USA, 20–24 March 2011; pp. 222–229.

68. Chen, H.T.; Tan, S.C.; Hui, S.Y.R. Analysis and modeling of high-power phosphor-coated white light-emitting diodes with a large surface area. *IEEE Trans. Power Electron.* **2015**, *30*, 3334–3344. [CrossRef]

69. Alexeev, A.; Martin, G.; Hildenbrand, V. Structure function analysis and thermal compact model development of a mid-power LED. In Proceedings of the 2017 33rd Thermal Measurement, Modeling & Management Symposium (SEMI-THERM), San Jose, CA, USA, 13–17 March 2017; pp. 283–289.

70. Yan, B.; Tran, N.T.; You, J.P.; Shi, F.G. Can junction temperature alone characterize thermal performance of white LED emitters? *IEEE Photonics Technol. Lett.* **2011**, *23*, 555–557. [CrossRef]

71. Kusama, H.; Sovers, O.J.; Yoshioka, T. Line shift method for phosphor temperature measurements. *Jpn. J. Appl. Phys.* **1976**, *15*, 2349–2358. [CrossRef]

72. Khalid, A.H.; Kontis, K. Thermographic phosphors for high temperature measurements: Principles, current state of the art and recent applications. *Sensors* **2008**, *8*, 5673–5744. [CrossRef] [PubMed]

73. Fuhrmann, N.; Baum, E.; Brübach, J.; Dreizler, A. High-speed phosphor thermometry. *Rev. Sci. Instrum.* **2011**, *82*. [CrossRef]

74. Yang, T.H.; Huang, H.Y.; Sun, C.C.; Glorieux, B.; Lee, X.H.; Yu, Y.W.; Chung, T.Y. Noncontact and instant detection of phosphor temperature in phosphor-converted white LEDs. *Sci. Rep.* **2018**, *8*, 296. [CrossRef]

energies

MDPI

Article

Extraction of Boundary Condition Independent Dynamic Compact Thermal Models of LEDs—A Delphi4LED Methodology

Robin Bornoff

Mentor–A Siemens Business, 81 Bridge Rd, Molesey, East Molesey KT8 9HH, UK; robin_bornoff@mentor.com

Received: 28 March 2019; Accepted: 23 April 2019; Published: 29 April 2019

Abstract: Multi-domain electro-thermal-optical models of LEDs are required so that their thermal and optical behavior may be predicted during a luminaire design process. Today, no standardized approach exists for the extraction of such models. Therefore, models are not readily provided by LED suppliers to end-users. This results in designers of LED-based luminaires wasting time on LED characterization and ad hoc model extraction themselves. The Delphi4LED project aims to address these deficiencies by identifying standardizable methodologies to extract both electro-optical and thermal compact models of LEDs that together can be used in a multi-domain simulation context. This article describes a methodology to extract compact thermal models of LEDs that are dynamic, in that they accommodate transient thermal effects, and are boundary condition-independent, in that their accuracy is independent of their thermal operating environment. Such models are achieved by first proposing an equivalent thermal nodal network topology. The thermal resistances and capacitances of that network are identified by means of optimization so that the transient thermal response of the network matches that of either an equivalent calibrated 3D thermal model or a transient thermal measurement of a physical sample. The accuracy of the thermal network is then verified by comparing the thermal compact model with a 3D detailed model, which predicts thermal responses within a 3D system-level model.

Keywords: LED; compact thermal model; boundary condition independent

1. Introduction

The DELPHI project (Development of Libraries of Physical models for an Integrated design environment) was a European community-funded 3-year research program that ran from 1993 to 1996. It was instrumental in both coining the term 'compact thermal model' (CTM) and elaborating a methodology to extract such models to accurately represent packaged integrated circuits thermally, regardless of the thermal operating environment in which they operated [1–3]. A follow-up project, SEED (Supplier Evaluation and Exploitation of DELPHI), saw an evaluation of the DELPHI methodology by component suppliers [4]. A final project, PROFIT (Prediction of Temperature Gradients Influencing the Quality of Electronic Products), extended the DELPHI methodology to the transient domain [5]. Compact thermal models need to satisfy thermal predictive accuracy requirements, as well as obfuscate any IP regarding their construction, materials and processes used in their manufacture. The abstracting of the complex conductive 3D heat flow paths within an IC package into an equivalent thermal network implicitly achieves the obfuscation of the IP but presents challenges in terms of retaining predictive accuracy. The DELPHI model extraction methodology addresses accuracy by seeking to guarantee that the CTM is accurate in a range of operating environments, specifically with differing peripheral heat transfer coefficients that the IC package might experience in operation. In this, the DELPHI project was successful, and it is an extension of this methodology to represent LEDs that forms the basis of this study. Standardized via JEDEC JESD15-4, the use of so-called DELPHI models has become commonplace when applying thermal simulation in the design of electronic products or products that contain electronics.

The rapid adoption of LEDs by the lighting industry has resulted in a coming-together of the longtime established lighting world and the relatively new semiconductor world. With this comes the need for the lighting industry to adopt processes and technologies that have evolved with the use of semiconductors, e.g., power and digital electronic applications. This adoption requires an understanding of the thermal behavior of semiconductors through the modeling of that behavior and thus the extraction of representative thermal models.

Whereas DELPHI models rely solely on the input of a power dissipation and provide operating junction temperature as an output, thermal models of LEDs cannot be considered from a purely thermal perspective. There is a coupling between the optical, electrical and thermal behavior of an LED to the extent where a model needs to represent all three 'physics.' The Delphi4LED project has proposed [4,5] that this multi-domain model be considered in two parts: an opto-electro model that is simulated in conjunction with a thermal model (Figure 1).

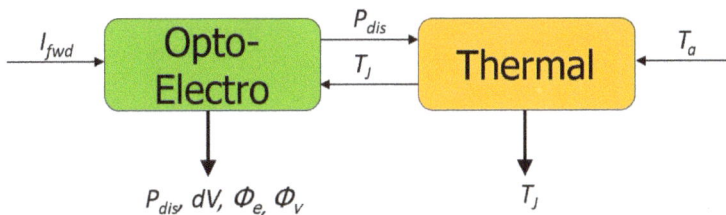

Figure 1. Co-simulation framework of a multi-domain LED compact model.

The opto-electro model requires forward current and operating chip junction temperature as input and provides power dissipation, voltage drop, emitted optical energy and luminous flux as output. The thermal model requires power dissipation and an ambient temperature as input and provides operating junction temperature (and potentially other temperature points, such as solder temperature) as output. These two models are to be solved in a co-simulation framework until such time as their results converge.

The extraction of the opto-electro model can be achieved via parameter fitting directly from the measurement data [6–8]. The method to extract the thermal model is described in this article.

2. Materials and Methods

The DELPHI methodology can be described by the following steps:

1. Perform thermal measurements of an IC package.
2. Calibrate a 'detailed' 3D thermal model of the IC package so that it replicates the measurements.
3. Simulate the detailed model in several different expected thermal operating environments (peripheral heat transfer coefficient boundary conditions) and note the resulting junction temperatures and surface heat fluxes.
4. Propose a CTM nodal network topology with assumed inter-node thermal resistance values.
5. Define an objective cost function that quantifies the difference between detailed model and compact model predicted junction temperatures and surface heat fluxes.
6. Optimize the CTM thermal resistance values until such time as the objective cost function has been minimized for all heat transfer coefficient conditions.
7. Quantify the accuracy of the CTM by validation.

The Delphi4LED methodology is based on the DELPHI methodology but extended in the following ways:

- To accommodate transient effects by the inclusion of thermal capacitances in the CTM network.
- To allow for the possibility of multiple heat sources due to additional phosphor conversion and optical reflection losses.
- To consider an objective cost function that quantifies the difference in transient Z_{th} thermal response curves (responses to a step change in power dissipation).
- To propose a nodal network topology that is tailored to and representative of LED devices (Figure 2).

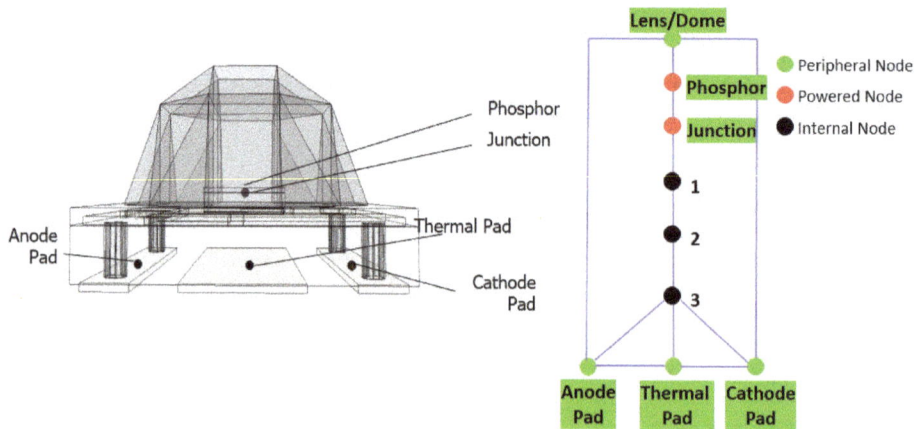

Figure 2. Delphi4LED-proposed compact thermal model (CTM) network topology (thermal resistances between all nodes, thermal capacitances at each node).

This article focuses on stage 3 and onward of the Delphi4LED methodology, applied to a mid-power Nichia NF2L757GRT-V1 warm white LED part. Stages 1 and 2 of the process, the optically corrected thermal transient measurement of the LED—measured in compliance with JESD51-51, JESD51-52 and CIE 225:2017—and the calibration of an equivalent 3D detailed thermal model, are described in [9].

As this part had no thermal pad, and prediction of the Phosphor temperature was not required by the intended design requirements, the nodal topology displayed in Figure 3 was used.

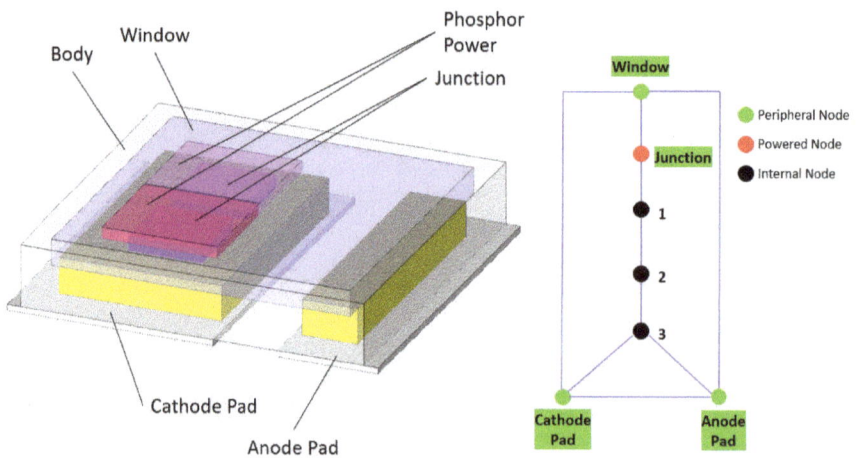

Figure 3. Nichia NF2L757GRT-V1 3D detailed model and equivalent CTM nodal topology.

The initial portion of the heat flow path, represented in Figure 3 by the Junction, Node 1, Node 2 and Node 3, was considered to be independent of the thermal operating environment into which the CTM was placed. As such, the thermal capacitances at these nodes, as well as the thermal resistances between, were extracted directly from the structure function, which was derived directly from the measured thermal response [10]. The remaining thermal resistances and capacitances were determined by a process of optimization. First, the thermal response of junction and cathode pad of the 3D detailed model were simulated in three differing heat transfer coefficient (HTC) environments. These were realized by applying differing HTCs at the cathode and anode pad peripheral faces of the 3D detailed model (Table 1). These were intended to cover a range of possible operating environments—in this case, differing substrate types and substrate circuitry onto which the LED part may be placed. Note that the high HTC values were due to the very small areas involved in the physical size of the pads.

Table 1. Three heat transfer coefficient sets.

W/m^{2}K	HTC Set 1	HTC Set 2	HTC Set 3
Anode	6000	5000	200
Cathode	30,000	50,000	2000

Due to the predominantly 1D nature of the heat flow, where most of the dissipated power was conducted down through the package and into the substrate, the relatively low value of HTC on the top and sides of the LED was ignored.

With assumed initial values of those thermal resistances and capacitances of nodes lying on the periphery of the package (i.e., Window, Cathode and Anode nodes) and the already-extracted values for all other nodes, the CTM was then simulated in the same three HTC environments, and the thermal response of the Junction and Cathode were recorded. The objective cost function was calculated as the sum of the areas between 3D detailed and CTM-predicted Z_{th} responses. With three environments and two temperature points, this cost function quantified the separation between six pairs of thermal responses. An optimization process then used multiple simulations of the CTM, with differing thermal resistance and capacitance values, until the objective cost function was minimized. This was achieved when the thermal resistance and capacitance values of the CTM were found to be the same thermal responses as the calibrated 3D detailed model.

To verify the boundary condition independence of the CTM, its thermal response was compared to that of the 3D detailed model when both were placed in a simulation model whose HTC environment differed from the three sets that were used in the CTM extraction process.

All thermal simulations and optimizations were carried out using Version 12.2 of the 'Simcenter Flotherm' software, commercially available from Mentor, a Siemens Business. Flotherm is a 3D CFD simulation software, tailored to thermal applications in electronics, that solves for all modes of heat transfer, i.e., convective, conductive, and radiative. It uses a structured Cartesian mesh to discretize a 3D model, as well as a flexible multi-grid approach for the solution of the linearized equations [11].

3. Results

3.1. Compact Thermal Model Extraction

Although the detailed model was calibrated against a single transient junction temperature thermal response, the 3D model could provide thermal responses at multiple points (control volumes). In addition to the Junction temperature, the maximum solder point temperature is of particular interest, which was the temperature at the Cathode solder point for the LED considered in this study. The method to extract the CTM was similar to that of the calibration of the detailed model against measurement [9]. The CTM was calibrated against the (calibrated) detailed model. Network thermal resistances and capacitances were optimized (calibrated) until the CTM thermal response matched the 3D detailed model.

The thermal resistances between and thermal capacitances at the Junction, Node 1, Node 2, and Node 3 were obtained directly from the thermal transient measurement. The remaining nine degrees of freedom of the rest of the network—six thermal resistances and three thermal capacitances—were identified by 200 computational experiments, followed by a gradient-based sequential optimization that was required to minimize the objective cost function (the sum of the areas between all six curve pairs). The initial comparison Z_{th} and final calibrated Z_{th}, as well as comparisons between the CTM and the detailed model, are shown in Figure 4a,b.

The thermal resistance values of the calibrated CTM are shown in Table 2 and the thermal capacitances are shown in Table 3.

Table 2. CTM thermal resistance values.

From Node	To Node	Thermal Resistance (K/W)
Junction	1	0.6593
1	2	1.8364
2	3	4.2061
3	Cathode	5.0598
3	Anode	351.1
Cathode	Anode	12549
Junction	Window	8799
Window	Cathode	5000
Window	Anode	9969

Table 3. CTM thermal capacitance values.

Node	Thermal Capacitance (J/K)
Junction	1.673×10^{-5}
1	0.0001639
2	0.000268
3	0.00075
Cathode	0.00588
Anode	0.0025
Window	0.0005

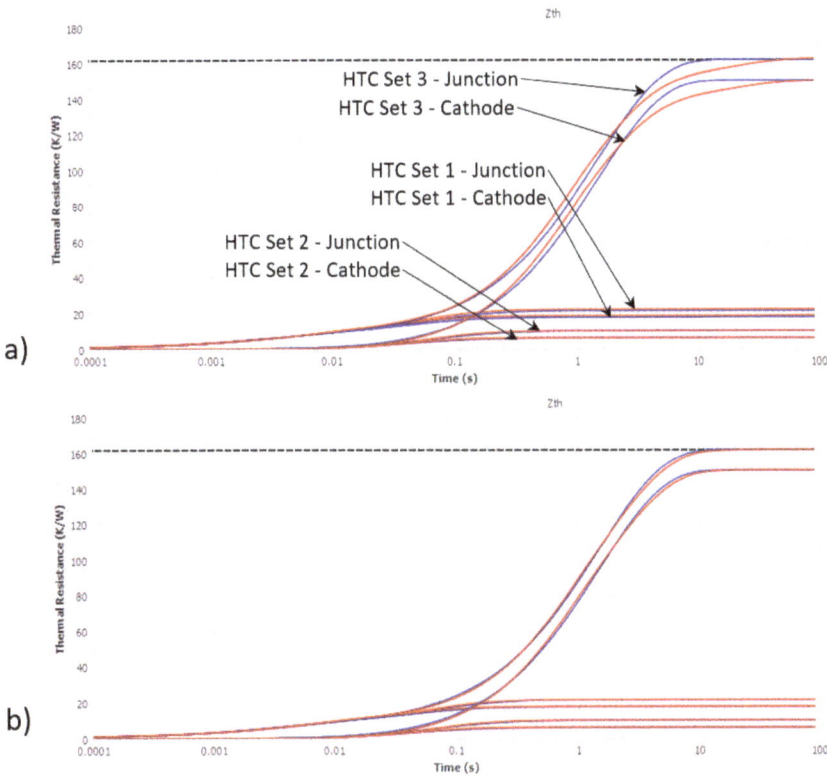

Figure 4. CTM (red curves) and detailed model (blue curves) Junction and Cathode thermal response comparisons for the three HTC sets: (**a**) for the initial assumed CTM resistance and capacitance values; (**b**) for the final calibrated CTM.

3.2. Validation of the Extracted CTM

The boundary condition independent accuracy of the CTM was implicitly assured for the three HTC sets that were considered during its extraction. To validate its accuracy for an intermediate HTC set, a final comparison was performed. Both CTM and detailed LED models were simulated independently in a 3D model of the environment that was originally used for the measurement (Figure 5). The LED was mounted on a ceramic substrate, which was itself mounted on a coldplate.

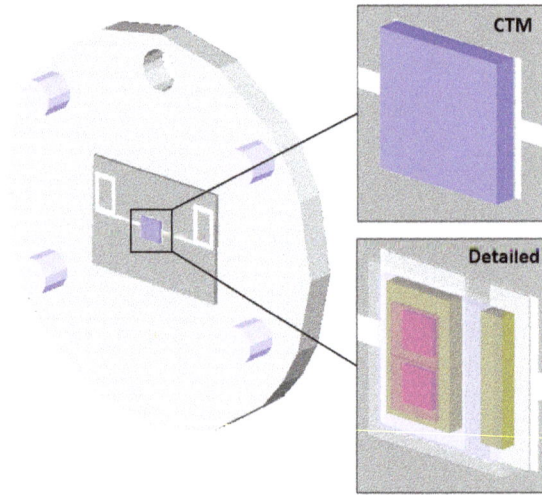

Figure 5. CTM validation configuration.

Compared to just imposing fixed HTCs on both the Anode and Cathode faces, as was done during the CTM extraction, this was a more complex and realistic operating environment. The HTCs experienced by the Cathode and Anode faces were a function of the effective thermal resistance of the rest of the heat flow path in the substrate, substrate circuitry and coldplate. The resulting Z_{th} thermal responses of both model types were then compared, including the error between detailed and CTM responses, normalized by the final steady-state temperature rise (Figure 6).

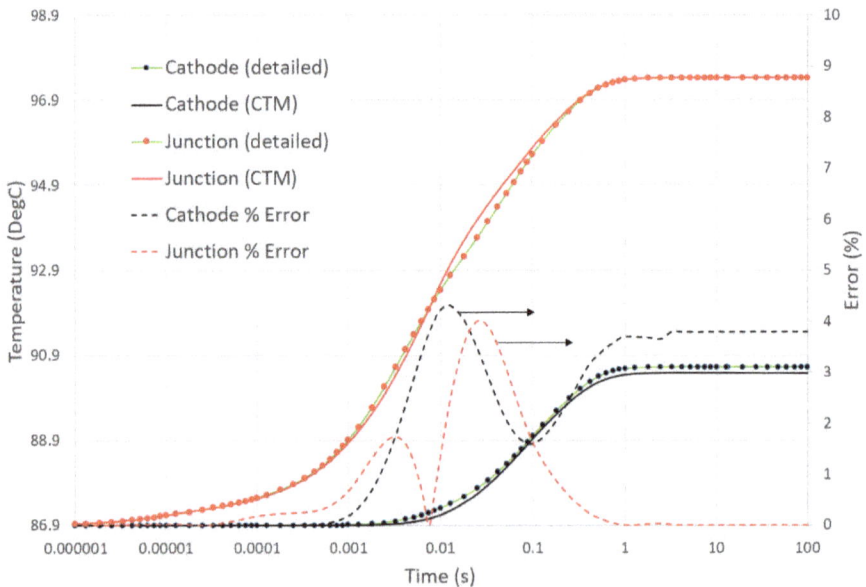

Figure 6. CTM vs. detailed error comparison for both Junction and Cathode thermal responses.

4. Discussion

The accuracy of the CTM is dependent on several factors:

1. Having sufficient nodes and inter-nodal resistance links to accommodate changes to the internal heat flow paths that occur due to changes in the peripheral HTCs.
2. Having sufficient nodes to accurately discretize the thermal capacitance of those heat flow paths to ensure the correct transient thermal responses.
3. Extraction of the CTM using a sufficient number of external HTC boundary condition sets to reflect the potential variation of end-user operating environments.

Even with a simple network topology, containing only seven nodes, and extracting the CTM using a training set of only three HTC combinations, the maximum error in the resulting transient thermal response of the CTM was only 4.3%. The original DELPHI methodology proposed up to 44 HTC sets, though they had to cover a wider range of expected operating environment, such as spray cooling and the mounting of heatsinks on the IC package top. Due to the light-emitting nature of an LED, the range of thermal operating conditions is lower, though a set of HTCs representing those environments is yet to be proposed. Discussion of such a set is outside the scope of this paper but will be addressed subsequently.

There can be multiple distinct heat sources within an LED, e.g., at the chip junction(s), where Phosphor conversion occurs (white LEDs only) and at any surface within the package where optical reflection occurs. The amount of power of the latter two can be derived from that at the chip Junction via a Phosphor conversion efficiency and a light extraction efficiency [12]. In the example presented in this article, despite there being two chips with four heat sources represented in the detailed model (two Junctions and two Phosphor powers), the CTM considered only a single Junction node with a single power. This was to simplify the extraction of the CTM (three fewer nodes), to reflect the fact that, for the intended usage of the CTM, Phosphor temperature prediction was not required and to recognize the fact that both chip junctions exhibited very similar thermal responses. The combined two Junctions and two Phosphor powers were dissipated at the single powered Junction node. The CTM extraction process then sought to guarantee that the CTM was capable of predicting the measured thermal response of the junction, which was itself an average of the two chip junctions.

A more complex nodal topology that contained two Junction nodes and two Phosphor power nodes could be considered. The applied CTM extraction methodology would be the same, though more computational experiments would be required to optimize the reduction of the objective cost function, due to there being more nodes and thus more thermal resistance and thermal capacitance values to identify.

The CTM, solved together with an opto-electro model of the LED, is capable of predicting operating chip junction temperature, optical performance in terms of luminous flux and emitted optical energy, as well as voltage drop and thus total consumed power. All such parameters are considered as design goals or constraints which, when predicted during early-stage architecting of a proposed luminaire design, will minimize the risk of identifying that a first physical prototype fails to achieve its designed aim. The use of a luminaire design tool providing the ability to solve such a multi-domain compact model of an LED, including a validation of its accuracy, was described by [13].

Thermal modeling of electronics applications falls into two groups; those that rely on a discretized 3D geometric definition of all objects on the heat flow path (a so-called 'detailed' modeling approach) and those that abstract the thermal behavior into a form that obfuscates the application geometry and results in much faster simulation than the detailed approach (a so-called 'compact' model). For compact modeling, the use of equivalent thermal networks is well established and supported by a range of 3D thermal simulation tools. However, model order reduction techniques have emerged over the last decade and now form the state of the art in compact modeling [14,15]. Unlike thermal networks, whose accuracy is predicated on the choice of network topology and the effectiveness of an optimizer to identify appropriate thermal resistances and capacitances, a reduced order model may be extracted directly from a detailed model to a defined accuracy and is suited for both multi-heat source and transient applications. Although such reduced order models may be solved as a 'standalone' with prescribed peripheral HTCs, their solution in conjunction with a 3D detailed representation of their operating environment has not yet been demonstrated but will form the state of the art once done so.

Funding: This research received funding from the European Union's Horizon 2020 research and innovation program through the H2020 ECSEL project Delphi4LED (grant agreement 692465).

Acknowledgments: Support from Delphi4LED project partners, especially from Genevieve Martin (Signify), Andras Poppe (Mentor), Lajos Gaál (Mentor), Marta Rencz (Mentor) and Gabor Farkas (Mentor) is acknowledged.

Conflicts of Interest: The author declares no conflict of interest.

Nomenclature

CFD	computational fluid dynamics
CIE	Commission internationale de l'éclairage
CTM	compact thermal model
HTC	heat transfer coefficient
IC	integrated circuit
I_{fwd}	forward current
IP	intellectual property
JEDEC	Joint Electron Device Engineering Council
LED	light emitting diode
P_{dis}	power dissipation
T_j	junction temperature
T_a	ambient temperature
dV	voltage drop
Z_{th}	transient thermal impedance curve
Φ_e	emitted optical power
Φ_v	luminous flux

References

1. Lasance, C.J.M.; Rosten, H.I.; Vinke, H.; Weiner, K.-L. A Novel Approach for the Thermal Characterization of Electronic Parts. In Proceedings of the Eleventh IEEE SEMI-THERM Symposium, San Jose, CA, USA, 7–9 February 1995.
2. Rosten, H.; Parry, J.; Lasance, C.J.M.; Vinke, H.; Temmerman, W.; Nelemans, W.; Assouad, Y.; Gautier, T.; Slattery, O.; Cahill, C.; et al. Final Report to SEMI-THERM XIII on the European-Funded Project DELPHI—The Development of Libraries and Physical Models for an Integrated Design Environment. In Proceedings of the Thirteenth IEEE SEMI-THERM Symposium, Austin, TX, USA, 28–30 January 1997.
3. Rosten, H.; Lasance, C.J.M.; Parry, J. The World of Thermal Characterization According to DELPHI—Part I: Background to DELPHI. *IEEE Trans. CHMT* **1997**, *20*, 384–391. [CrossRef]
4. Bornoff, R.; Hildenbrand, V.; Lugten, S.; Martin, G.; Marty, C.; Poppe, A.; Rencz, M.; Schilders, W.; Yu, J. Delphi4LED—From Measurements to Standardized Multi-Domain Compact Models of LEDs: A New European R&D Project for Predictive and Efficient Multidomain Modeling and Simulation of LEDs at all Integration Levels Along the SSL Supply Chain. In Proceedings of the Therminic 2016, Budapest, Hungary, 21–23 September 2016.
5. Poppe, A. Simulation of LED Based Luminaires by Using Multi-Domain Compact Models of LEDs and Compact Thermal Models of their Thermal Environment. *Microelectron. Reliab.* **2017**, *72*, 65–74. [CrossRef]
6. Farkas, G.; Gaál, L.; Bein, M.; Poppe, A.; Ress, S.; Rencz, M. LED Characterization within the Delphi4LED Project. In Proceedings of the ITHERM'18, San Diego, CA, USA, 29 May–1 June 2018; p. 393.
7. Farkas, G.; Gaál, L.; Poppe, A.; Rencz, M.; Bornoff, R. LED multiphysics modeling for "Industry 4.0", an approach proposed by the Delphi4LED European project. In Proceedings of the 2018 IEEE Electronics Packaging Technology Conference (EPTC), Singapore, 4–7 December 2018; p. 268.

8.   Farkas, G.; Poppe, A.; Gaál, L.; Hantos, G.; Berényi, C.; Rencz, M. Structural analysis and modelling of packaged light emitting devices by thermal transient measurements at multiple boundaries. In Proceedings of the 24th International Workshop on THERMal INvestigation of ICs and Systems (THERMINIC 2018), Stockholm, Sweden, 26–28 September 2018; p. 152.

9.   Bornoff, R.; Farkas, G.; Gaal, L.; Rencz, M.; Poppe, A. LED 3D thermal model calibration against measurement. In Proceedings of the 2018 19th International Conference on Thermal Mechanical and Multi-Physics Simulation and Experiments in Microelectronics and Microsystems (EuroSimE) Toulouse, Toulouse, France, 15–18 April 2018; pp. 1–7.

10.   Van Bien, T.; Szekely, V. Fine structure of heat flow path in semiconductor devices: A measurement and identification method. *Solid State Electron.* **1988**, *31*, 1363–1368.

11.   Technical Specifications. Available online: https://www.mentor.com/products/mechanical/flotherm/flotherm/technical-specifications (accessed on 21 April 2019).

12.   Alexeev, A.; Martin, G.; Onushkin, G.; Linnartz, J. Accurate Thermal Transient Measurements Interpretation of Monochromatic LEDs. In Proceedings of the Thirty Fifth IEEE SEMI-THERM Symposium, San Jose, CA, USA, 18–22 March 2019.

13.   Marty, C.; Poppe, A.; Martin, G. LED Luminaire Design Using Virtual Twin: Development of a Multidomain Compact Model in the Delphi4LED European Project. In Proceedings of the Strategies in Light, Las Vegas, NV, USA, 27 February–1 March 2019.

14.   Codecasa, L.; d'Alessandro, V.; Magnani, A.; Rinaldi, N. Matrix Reduction Tool for Creating Boundary Condition Independent Dynamic Compact Thermal Models. In Proceedings of the 21st International Workshop on Thermal Investigations of ICs and Systems (THERMINIC 2015), Paris, France, 30 September–2 October 2015.

15.   Sabry, M.N. Dynamic Compact Thermal Models Used for Electronic Design: A Review of Recent Progress. In Proceedings of the Interpack, Maui, HI, USA, 6–11 July 2003.

energies

MDPI

Article
Parametric Compact Thermal Models of Power LEDs

Marcin Janicki [1,*], **Tomasz Torzewicz** [1], **Przemysław Ptak** [2], **Tomasz Raszkowski** [1], **Agnieszka Samson** [1] and **Krzysztof Górecki** [2]

[1] Department Microelectronics and Computer Science, Lodz University of Technology, Łódź, Wólczańska 221/223, 90-924 Łódź, Poland; torzewicz@dmcs.pl (T.T.); traszk@dmcs.pl (T.R.); asamson@dmcs.pl (A.S.)

[2] Department of Marine Electronics, Gdynia Maritime University, Gdynia, Morska 81-87, 81-225 Gdynia, Poland; p.ptak@we.umg.edu.pl (P.P.); p.gorecki@we.umg.edu.pl (K.G.)

* Correspondence: janicki@dmcs.pl; Tel.: +48-42-6312654

Received: 1 April 2019; Accepted: 30 April 2019; Published: 7 May 2019

Abstract: Light-emitting diodes are nowadays the most dynamically developing type of light sources. Considering that temperature is the main factor affecting the electrical and lighting parameters of these devices, thermal models are essential subcomponents of the multidomain models commonly used for simulation of their operation. The authors investigated white power light-emitting diodes soldered to Metal Core Printed Circuit Boards (MCPCBs). The tested devices were placed in a light-tight box on a cold plate and their cooling curves were registered for different diode heating current values and various preset cold plate temperatures. These data allowed the computation of optical and real heating power values and consequently the generation of compact thermal models in the form of Foster and Cauer RC ladders. This also rendered possible the analysis of the influence of the considered factors on the compact model element values and their parametrization. The resulting models yield accurate values of diode junction temperature in most realistic operating conditions and they can be easily included in multidomain compact models of power light emitting diodes.

Keywords: LED compact thermal models; heating and optical power; Cauer RC ladder

1. Introduction

Power light emitting diodes are nowadays rapidly replacing traditional incandescent bulbs and they are ever more and more frequently being used in numerous lighting applications [1]. From the consumer point of view, the most important parameter describing the performance of LEDs is the emitted optical power. However, its value is a function of many different factors, such as the device operating point or the cooling conditions [2–4]. Thus, the analysis of LED operation involves various coupled phenomena of different nature: electrical, optical and thermal. Especially important for the operation of these devices is temperature, which affects all other physical domains. Consequently, the thermal management of LED applications is currently the objective of extensive research [5].

For these reasons, thermal simulations of either individual devices or their luminaries together with cooling assemblies are always carried out during the design phase. These simulations could be performed using software based on some numerical method, such as the finite element one [6–8]. Unfortunately, numerical methods require the knowledge of exact structure geometry and material thermal properties, which might not be readily available. Additional problems poses the complexity of structure discretization mesh which could render the simulation time unacceptably long [9].

Thus, when the knowledge of detailed temperature field distribution in space is not required and the main goal of simulations is the determination of device junction temperature, an attractive alternative to detailed 3D thermal models are Compact Thermal Models (CTMs), which typically consist of a very limited number of thermal resistors and capacitors, yet they provide satisfactory simulation accuracy. An additional advantage of CTMs is that, using the electro-thermal analogy, they render possible the execution of multidomain simulations in some ordinary SPICE-like circuit simulators [10–12]. Obviously, a standardized form of CTMs are the DELPHI style ones, which are boundary condition independent and do not disclose any proprietary information, but again the data necessary to generate them can be provided only by device manufacturers [13,14].

Another form of CTMs are the RC ladder models which can be obtained directly from system dynamic thermal responses using different model order reduction methods [15,16]. Unfortunately, compared to the DELPHI style models, the ladder ones are not boundary condition independent and they can be employed only in a specific LED operating point and cooling conditions. Hence, the main goal of the research presented in this paper was to develop a methodology for the generation of parametrized ladder models, which could be applied in a wide range of LED operating and cooling conditions. Such thermal networks might be included then in more complex compact multidomain models taking into account also mutual dependencies between multiple LEDs contained in a single luminaire or those capable of determining the heating and optical power for a given operating point and cooling conditions [5,17–19]. However, the research on the development of such models for LEDs is still ongoing, e.g. in the frames of the Delphi4LED European project [20].

The following section presents the experimental setup as well as the results of device thermal and optical measurements which allowed the determination of the optical and real heating power. Then, the computed spectra of thermal time constants for the recorded cooling curves are analyzed in order to generate device CTMs in the form of Cauer ladders. Next, parametric expressions for the computation of model element values are proposed. Finally, the cooling curves simulated with these models are compared with the measured ones.

2. Measurement Results

2.1. Measurement Equipment and Experimental Setup

The devices investigated in this paper were commercially available white power LEDs from the XLamp® XP-L family manufactured by the Cree Company (Durham, North Carolina, USA). The diodes were soldered to the MCPCBs made of an aluminum alloy having the dimensions of 2.5 cm × 2.5 cm × 0.2 cm and pictured in Figure 1a. During all measurements the substrates were attached with the application of thermal grease to the 10.0 cm × 8.0 cm × 1.8 cm copper cold plate through which continuous water flow was forced as shown in Figure 1b. The temperature values of the cold plate and the MCPCB were measured by thermocouples whose locations are indicated in the figure. Based on the temperature values provided by the thermocouples the water temperature was automatically adjusted to maintain the preset cold plate temperature. The LED, which is lit-on in the figure, was positioned on the cold plate at the bottom of the box exactly in its middle.

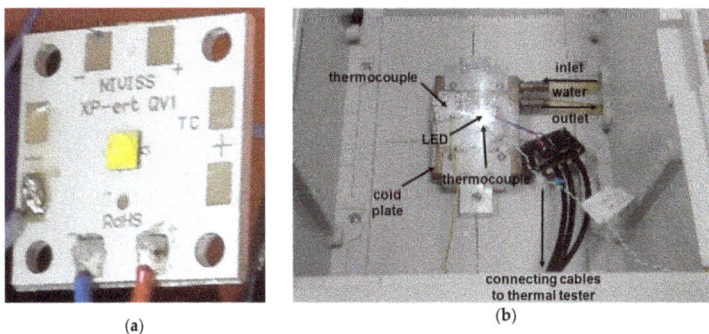

(a)

(b)

Figure 1. Photos of the experimental setup: (**a**) the investigated LED soldered to the MCPCB; (**b**) the MCPCB placed on a cold plate inside the light tight box.

Initially, the investigated LEDs were calibrated on a cold plate for the forward current of 10 mA. The measured voltage drop across the diode junction decreased linearly with temperature at the rate of −1.34 mV/K. During the actual measurements, first the diode was heated with a preset current value until the thermal steady state was reached. Then, the power was switched off and the device cooling curves were measured by forcing the constant current value for which the diodes were previously calibrated. Each time the heating and cooling phases lasted 30 s. All measurements were taken with the thermal transient tester T3Ster®, manufactured by Mentor Graphics (Wilsonville, OR USA), which renders possible the recording of thermal responses with a microsecond resolution. Besides, this equipment provides also an entire range of software thermal analysis tools used extensively later on in this paper.

The electrical power supplied to LED terminals is the sum of the emitted optical radiant power and the real heating power. The ratio of the optical power to the electrical one, expressed in percent, is known as the radiant efficiency. Consequently, the measurement of the optical power is required in order to carry out any thermal analysis because it allows also the determination of the real heating power, which should be used as an input quantity for an LED thermal model [21].

The JEDEC standard recommends that the value of this power should be found by performing the measurements of the total emitted radiant flux in an integrating sphere [22]. However, in this paper the optical power was computed based on the irradiance value measured with the Delta OHM HD2302 photoradiometer (Caselle di Selvazzano, Italy), which was placed in the lid of the light-tight box directly over the diode at the distance of 17 cm. This distance, which is 50 times greater than the size of the LED, is large enough so as to consider the diode as a point light source for geometrical analyses. Then, the optical power emitted by the diode was evaluated, as outlined in [23], using the radiation spatial distribution chart provided in the datasheet by the LED manufacturer.

2.2. Results of Electrical Measurements

The diode thermal responses were recorded for 8 different forward current values ranging from 0.1 A to 2.0 A and 9 preset cold plate temperature varying between 10°C and 90°C with the step of 10°C. Each time the measurements were repeated with the LED thermal pad not soldered and then again with the pad properly soldered. This procedure is similar to the dual interface measurement method suggested for the identification of the junction-to-case thermal resistance in the JEDEC standard [24]. Altogether, 144 diode cooling curves were measured. Selected measurement results, shown in Figure 2, are represented as the diode heating curves, which were obtained by subtracting the measured values from the respective steady state values.

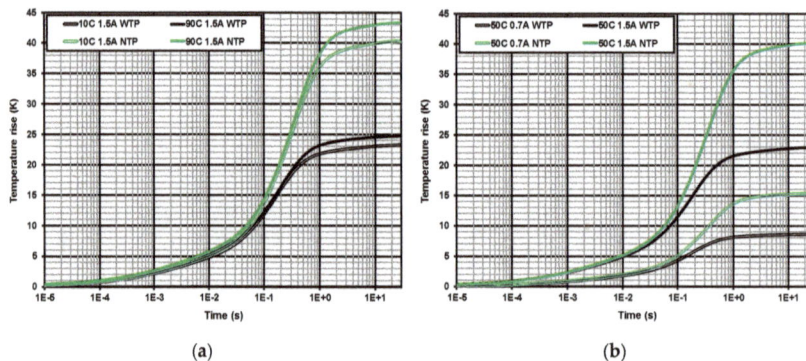

(a)   (b)

Figure 2. Recorded LED heating curves: (**a**) results for heating current of 1500 mA and cold plate temperature values 10 °C and 90 °C; (**b**) results for heating currents of 700 mA and 1500 mA for cold plate temperature 50 °C.

This approach is correct if material thermal properties can be considered temperature independent, what was justified here taking into account that the junction temperature rise values were relatively small because of the forced water cooling. Figure 2 presents the curves recorded for the LED heating current of 1.5 A and the cold plate temperature maintained at 10 °C and 90 °C, whereas Figure 2b shows the curves registered for the current values of 0.7 A and 1.5 A when the cold plate temperature was maintained at 50 °C. The black lines represent the measurement results obtained with the thermal pad soldered (WTP) and the lighter ones when the thermal pad was not soldered (NTP). The double lines refer either to lower current or temperature values. The curves, as can be seen in the charts, are only moderately affected by the cold plate temperature and the junction steady state temperature rise value increases with temperature by less than 10%. On the other hand, the soldering of the thermal pad reduces effectively the temperature rise value by approximately 40%, but any effect is visible only after some 0.5 s.

2.3. Optical Measurements

The optical (P_o) and real heating (P_h) power values as well as the radiant efficiency computed for the two limit temperature values using the earlier described methodology are shown in Figure 3. The optical power is plotted in Figure 3a using the solid lines, whereas the dashed ones represent the heating power. Similarly as previously, the cases when the thermal pad is not soldered are denoted by the double lines. The heating power, as can be seen, is fairly independent from temperature, but its value increases rapidly for high current values, reaching even 2 W. On the other hand, the increase of the optical power is less important for high diode currents and its value is visibly affected both by the cold plate temperature and the device soldering manner. For the heating current of 2.0 A, the optical power is reduced by 15% because of the higher cold temperature and another 10% when the thermal pad is not soldered. Therefore, the radiant efficiency curves plotted in Figure 3b have their distinct maxima, which correspond to the LED forward current values ranging from 40 mA to 120 mA. Both the proper soldering of the thermal pad and the increase of cold plate temperature shift the maxima towards higher current values and reduce the efficiency.

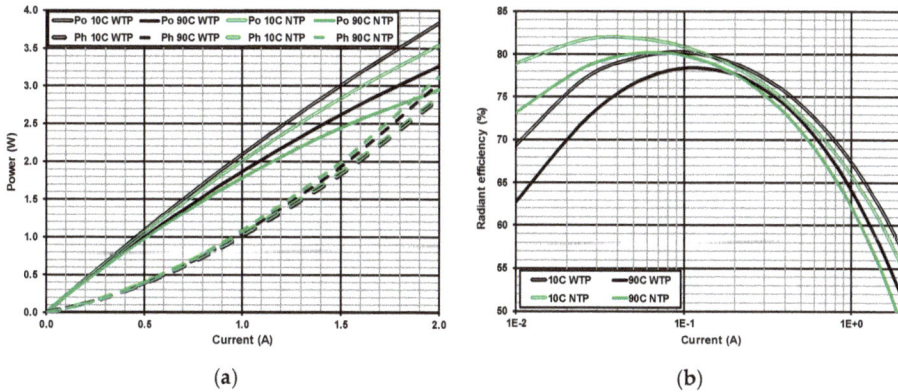

(a)   (b)

Figure 3. Measured LED power and radiant efficiency for cold plate temperature values of 10 °C and 90 °C: (a) heating and optical power; (b) radiant efficiency.

3. Thermal Modeling and Simulations

3.1. Time Constant Spectra

The measurements of the optical power emitted by the investigated device, as stated previously, allowed the computation of the heating power and the application of the Network Identification by Deconvolution (NID) thermal analysis method. According to this method, the device responses to power step excitations, sampled at time instants equidistant on the logarithmic time scale, are numerically differentiated and then the frequency spectra of thermal time constants contained in the responses can be computed by performing the numerical deconvolution [25]. The spectra obtained for the previously presented heating curves using the software implementing the NID method and provided together with the thermal tester are shown in Figure 4.

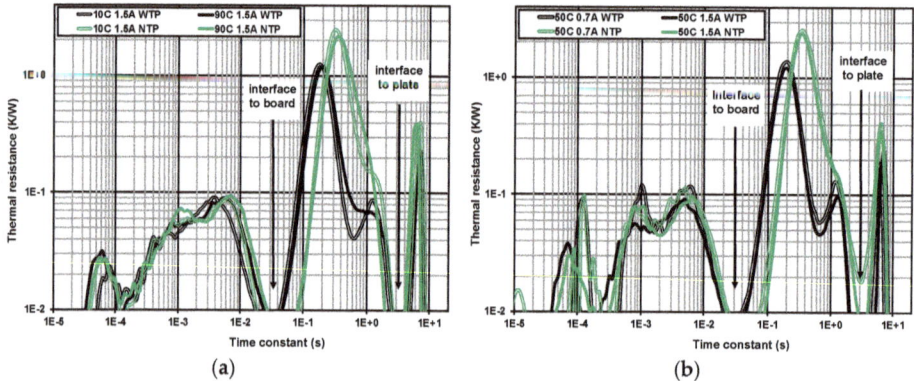

Figure 4. Computed thermal time constant spectra: (**a**) results for heating current of 1500 mA and cold plate temperature values 10 °C and 90 °C; (**b**) results for heating currents of 700 mA and 1500 mA for cold plate temperature 50 °C.

According to the theory, the minima in the spectra indicate the locations where heat diffuses into another material. Here, there are two distinct minima located around 30 ms and 3 s, which are marked in the charts by arrows. These minima, as discussed in [26], correspond to the time instants when heat diffuses from the LED package to the MCPCB and then to the cold plate. Consequently, the thermal resistance of the leftmost section characterizes the heat flow from the junction through the package. The middle part of the spectra corresponds to the interface with the MCPCB. Therefore, leaving the thermal pad unsoldered visibly increases the magnitude of these peaks and shifts them to the right, hence delaying thermal responses. Finally, the rightmost tiny peaks reflect the heat exchange with the cold plate.

3.2. Compact Thermal Models

The division of the time constant spectra into individual sections in the locations of the minima indicated in Figure 4 leads directly to the three-stage compact models in the form of the Foster RC ladder, pictured in Figure 5a. In this circuit the real heating power dissipated in the junction (the current source) flows through three consecutive RC stages towards the ambient (the thermal ground). The time constants of these three stages are equal to the products of the respective thermal resistances and capacitances. The currents represent the heat flow and the voltages correspond to the temperature rise over the ambient.

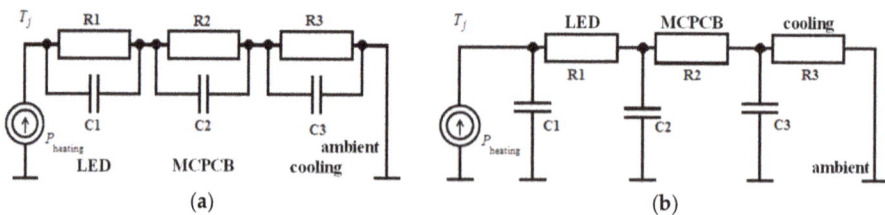

Figure 5. RC ladder compact thermal models: (**a**) Foster ladder; (**b**) Cauer ladder.

The division of the spectra directly determined the resistor values of the Foster RC ladder. Then, the thermal time constant values were found employing the built-in Matlab® Levenberg-Marquardt constrained optimization routine [27]. The optimal time constant values within each of the allowable ranges, delimited by the earlier discussed minima located at some 30 ms and 3 s, were computed by minimizing the error between simulations and measurements. Finally, the thermal capacitor values were found by dividing the respective time constant and resistor values.

Unfortunately, the element values of CTMs in the form of Foster RC ladders cannot have any physical interpretation since capacitors form a direct path from the junction to the ambient. Hence, heat dissipated in the source can be instantaneously sensed as a non-zero heat flux (capacitor current) virtually at any location in this thermal network, thus implying the infinite speed of heat diffusion. A physically correct CTM can be obtained by converting the Foster RC ladder into the mathematically equivalent Cauer RC network shown in Figure 5b. The element values of this model could already have physical interpretation since all the capacitors are connected to the thermal ground. Consequently, heat does not reach remote locations of the network until the thermal capacitors are charged, so the speed of heat diffusion is limited. This conversion can be carried out by executing the continued fraction algorithm, which for a low number of RC components is numerically stable [28].

3.3. Parametric Model

The Cauer RC ladder CTMs generated for different heating current and cold plate temperature values were analyzed so as to identify the influence of these factors on the CTM element values. The results showed that only the resistances R1, R2 and the capacitance C2 vary noticeably, but unlike in the case discussed in [29], they depend only on one of the considered factors at a time. Thus, it was possible to use the following simple parametric formulas to compute model element values:

$$Y(X) = a \times X^b + c, \tag{1}$$

where Y is either a resistor or capacitor value and X is a heating current or cold plate temperature value. The values of coefficients a, b, c were determined again using the built-in Matlab® optimization routines (The MathWorks, Natick, Massachusetts USA).

The final expressions obtained for the computation of the parametrized RC Cauer ladder CTM element values are provided in Tables 1 and 2, separately for each manner of diode soldering considered here. The current should be expressed here in amperes and the temperature in degrees Celsius. Moreover, the variable resistor values computed with the models are plotted in Figure 6 and the variable capacitance in Figure 7a. As can be seen in the tables, it was assumed that the value of resistor R1 depends on temperature and the values of the elements in the middle RC stage depend on the current. In all the considered cases very high values of the determination coefficient R^2 were attained what proves a reasonable goodness of fit.

Table 1. Element values of the parametrized Cauer CTM for LED with the thermal pad soldered.

Thermal Resistances R (K/W)	Thermal Capacitances C (J/K)
$R1 = 1.757 \times 10^{-15} \cdot T^{7.321} + 3.018$	$C1 = 0.00144$
$R2 = 23.3277 \cdot I^{-0.0976} - 12.9997$	$C2 = 0.005074 \cdot I^{0.6914} + 0.01149$
$R3 - 0.48$	$C3 = 14.71$

Analyzing the thermal resistances, the resistor R1, which corresponds to the standard junction-to-case resistance, has a fairly constant value up to the cold plate temperature of 50 °C. Then this value gradually increases, by approximately 10% at 90 °C. Referring to the value of 2.2 K/W given in the datasheet as the typical junction to the solder point resistance, the value of slightly more than 3.0 K/W seems very realistic, since the manufacturer did not specify any measurement conditions or the type of thermal model applied. The resistance values in the case when the thermal pad is not soldered are always by some 0.2 K/W higher, what is justified by the fact that in this case the heat diffusion to the MCPB is impeded and the resistance increases.

The value of the resistor R2, which represents the resistance of the interface between the package and the MCPCB, decreases significantly with the current and falls in the range from 8.8 to 14.2 K/W in the case when the thermal pad is soldered. In the other case this value is much higher, and ranges from 16.5 to 26.5 K/W, what demonstrates the importance of the proper use of the thermal pad. Owing to the forced water cooling, the value of the resistor R3 is the smallest one and it is equal to 0.48 K/W in the first case or 0.95 K/W in the latter one. This discrepancy will be discussed later on during the analysis of capacitance values.

Table 2. Element values of the parametrized Cauer CTM for LED with the thermal not connected.

Thermal Resistances R (K/W)	Thermal Capacitances C (J/K)
$R1 = 1.252 \times 10^{-13} \cdot T^{6.361} + 3.239$	$C1 = 0.002$
$R2 = 15.12 \cdot I^{-0.2534} + 3.786$	$C2 = 0.2071 \cdot I^{-0.01749} + 0.223$
$R3 = 0.95$	$C3 = 7.07$

Looking now at the capacitances, the capacitor C1 values, though constant, differ significantly for the two considered soldering cases and amount to 1.4 mJ/K and 2.0 mJ/K. This time again it can be explained by the fact that generated heat, when it is not properly sunk into the substrate spreads throughout a relatively larger volume of the package, hence the capacitance value of the heated volume is higher. The modeled value of capacitance C2, plotted in Figure 7a, slightly increases with current and at the value of 2.0 A is equal to some 20 mJ/K and 40 mJ/K. This time again the higher capacitance value for the case when the thermal pad is not soldered can be explained by the existence of a poor thermal contact, which is characterized also by a higher capacitance. Finally, the values of the capacitance C3 differ significantly, similarly as the ones for the resistor R3, however it should be pointed that they both correspond to virtually the same thermal time constant value of around 7 s. Looking at Figure 4, one can notice that the black peaks, which correspond to the thermal pad soldered, are indeed visibly lower, so probably this difference results from the identification errors introduced by the NID method which has a limited resolution.

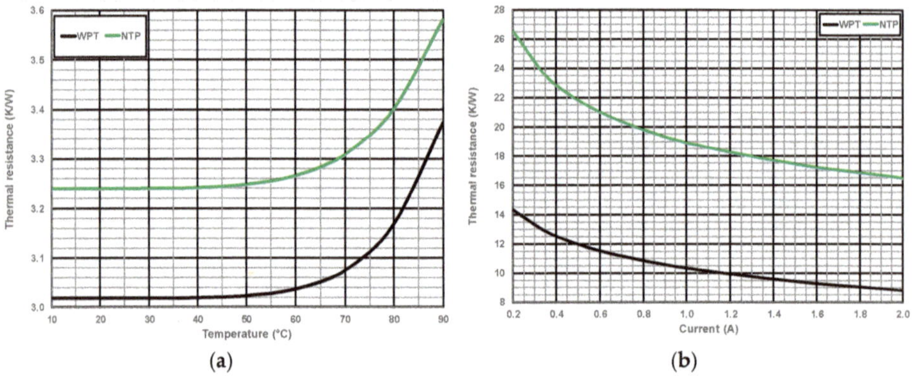

Figure 6. Modeling and simulation with Cauer RC ladder: (**a**) modeled values of the junction to solder point thermal resistance R1; (**b**) modeled values of the solder point to cold plate thermal resistance R2.

3.4. Thermal Simulations

Considering that the detailed thermal model of the LED was not available, the parametric CTMs were validated by simulating the diode heating curves and comparing them with the measured ones. The resulting curves are plotted in Figure 7b for the two extreme cold plate temperature values and the heating currents of 0.5 A and 2.0 A. This time the measured curves are represented with black lines the simulated ones with lighter lines. Unlike in Figure 2, the double logarithmic scale was used in order to assess better the simulation errors for short heating times. As can be seen, the simulated curves follow accurately the measured ones at all the time instants over 1 ms. This relatively simple CTM provided in all cases an excellent accuracy with simulation errors remaining under 3% of the steady state temperature rise.

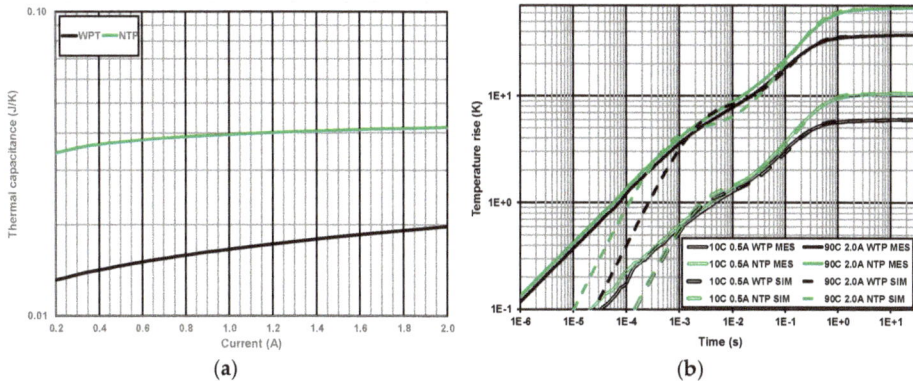

Figure 7. Modeling and simulation with Cauer RC ladder: (**a**) modeled values of the capacitance C2; (**b**) comparison of measured and simulated LED heating curves.

Undoubtedly, the simulation accuracy for short time instants could be improved by the addition of the fourth RC stage inserted next to the LED junction and corresponding to the peaks located to the left of the tiny minima at around 0.1±0.2 ms visible in Figure 4, which probably could be attributed to the die attach. However, taking into account that the details of the internal package structure are unknown, this stage was not included in the present model because it would increase greatly the computational effort of parametrization.

4. Conclusions

This paper demonstrated, based on a practical example, that with the proper determination of the real heating power and the division of thermal time constant spectra in the locations of their minima it is possible to obtain compact thermal models in the form of Cauer ladders with physically meaningful element values. Furthermore, it was shown that in the case of LEDs repeating the thermal measurements with the thermal pad soldered and left unsoldered might constitute an interesting alternative to the dual interface measurement method suggested in the JEDEC standard [24], where the thermal contact resistance with the heat sink is changed by applying thermal grease or loosening the attaching screw.

Moreover, extending the research presented by the authors in [26], the parametric relations describing the dependence of thermal resistances and capacitances on device operating current and temperature were proposed for the computation of CTM element values. The diode heating curves simulated with these parametric models showed very good agreement with the experimental results. Such models could become then thermal components of larger multidomain LED compact models coupling electrical, thermal and optical phenomena.

The parametrization of ladder CTMs rendered such models more versatile because now they could be used in different ambient conditions, similarly as the Delphi-style boundary condition independent models. Furthermore, the generation of such models does not require any proprietary information from a manufacturer because it is carried out based only on the knowledge of measured dynamic thermal responses. Finally, considering that the ladder thermal model components have physical

Energies **2019**, *12*, 1724

meaning, i.e., they correspond to the semiconductor die, the package or the heat sink, the change of a certain component requires the replacement of only one RC stage in the model without the necessity to regenerate the entire model.

Author Contributions: The measurements presented in this manuscript and their evaluation was carried out by T.T. and P.P. The research on the generation of compact thermal models and their parametrization was done by A.S. and T.R. Finally, M.J. and K.G. supervised the research and prepared the manuscript.

Funding: This research was founded from the Polish Ministry of Science and Higher Education programme "Regional Excellence Initiative" 2019-2022 project No. 006/RID/2018/19, the sum of financing 11 870 000 PLN.

Conflicts of Interest: The authors declare no conflict of interest.

References

1. Weir, B. Driving the 21st century's lights. *IEEE Spectr.* **2012**, *49*, 42–47. [CrossRef]
2. Schubert, E.F. *Light Emitting Diodes*, 3rd ed.; Rensselaer Polytechnic Institute: Troy, NY, USA, 2018.
3. Biber, C. LED Light Emission as A Function of Thermal Conditions. In Proceedings of the 24th IEEE Semiconductor Thermal Measurement and Management Symposium, San Jose, CA, USA, 16–20 March 2008; pp. 180–184. [CrossRef]
4. Górecki, K.; Ptak, P.; Janicki, M.; Torzewicz, T. Influence of Cooling Conditions of Power LEDs on Their Electrical, Thermal and Optical Parameters. In Proceedings of the 25th International Conference Mixed Design of Integrated Circuits and Systems, Gdynia, Poland, 21–23 June 2018; pp. 237–242.
5. Lasance, C.J.M.; Poppe, A. (Eds.) *Thermal Management for LED Applications*; Springer: Dordrecht, Holland, 2014.
6. Bender, V.C.; Iaronka, O.; Marchesan, T.B. Study on The Thermal Performance of LED Luminaire using Finite Element Method. In Proceedings of the 39th Annual Conference of the IEEE Industrial Electronics Society IECON, Vienna, Austria, 10–13 November 2013; pp. 6099–6104. [CrossRef]
7. Leng, L.S.; Retnasamy, V.; Sauli, Z.; Vairavan, R.; Shahimin, M.M.; Ong, N.R.; Kirtsaeng, S. Finite Element Modelling of Thermal Performance of LED Lamp using Open Source Software - Salome and Elmer. In Proceedings of the IEEE Symposium on Computer Applications & Industrial Electronics ISCAIE, Langkawi, Malaysia, 24–25 April 2017; pp. 65–68. [CrossRef]
8. Lee, H.E.; Lee, D.; Lee, T.I.; Shin, J.H.; Choi, G.M.; Kim, C.; Lee, S.H.; Lee, J.H.; Kim, Y.H.; Kang, S.M.; et al. Wireless powered wearable micro light-emitting diodes. *Nano Energy* **2019**, *55*, 454–462. [CrossRef]
9. Tandon, S.; Liu, E.; Zahner, T.; Besold, S.; Kalb, W.; Elger, G. Transient Thermal Simulation of High Power LEDs and Its Challenges. In Proceedings of the 18th International Conference on Thermal, Mechanical and Multi-Physics Simulation and Experiments in Microelectronics and Microsystems, Dresden, Germany, 3–5 April 2017; pp. 45–52. [CrossRef]
10. Sabry, M.N. Compact thermal models for electronic systems. *IEEE Trans. Compon. Packag.* **2003**, *26*, 179–185. [CrossRef]
11. Mawby, P.A.; Igic, P.M.; Towers, M.S. Physically based compact device models for circuit modelling applications. *Microelectr. J.* **2001**, *32*, 433–447. [CrossRef]
12. Menozzi, R.; Cova, P.; Delmonte, N.; Giuliani, F.; Sozzi, G. Thermal and electrothermal modeling of components and systems: Review of the research at the University of Parma. *Facta Univ. Ser.: Electron. Energetics* **2015**, *28*, 325–344. [CrossRef]
13. Lasance, C.J.M. Ten years of BCI compact thermal modeling of electronic parts: Review. *Heat Transf. Eng.* **2008**, *29*, 149–168. [CrossRef]
14. Standard JESD15-4. *DELPHI Compact Thermal Model Guideline*; JEDEC: Arlington, VA, USA, 2008.
15. Sofia, J.W. Analysis of thermal transient data with synthesized dynamic models for semiconductor devices. *IEEE Trans. Compon. Packag. A* **1995**, *18*, 39–47. [CrossRef]
16. Masana, F.N. A new approach to the dynamic thermal modelling of semiconductor packages. *Microelectron. Reliab.* **2001**, *41*, 901–912. [CrossRef]
17. Górecki, K. Modelling mutual thermal interactions between power LEDs in SPICE. *Microelectron. Reliab.* **2015**, *55*, 389–395. [CrossRef]

18. Górecki, K.; Ptak, P. Modelling mutual thermal coupling in LED modules. *Microelectron. Int.* **2015**, *32*, 152–157. [CrossRef]
19. Górecki, K.; Ptak, P. Modeling LED lamps in SPICE with thermal phenomena taken into account. *Microelectron. Reliab.* **2017**, *79*, 440–447. [CrossRef]
20. Poppe, A. Simulation of LED based luminaires by using multi-domain compact models of LEDs and compact thermal models of their thermal environment. *Microelectron. Reliab.* **2017**, *72*, 65–74. [CrossRef]
21. Standard JESD51-51. *Implementation of the Electrical Test Method for the Measurement of Real Thermal Resistance and Impedance of Light-Emitting Diodes with Exposed Cooling Surface*; JEDEC: Arlington, VA, USA, 2012.
22. Standard JESD51-52. *Guidelines for Combining CIE 127-2007 Total Flux Measurement with Thermal Measurement of LED with Exposed Cooling Surface*; JEDEC: Arlington, VA, USA, 2012.
23. Górecki, K.; Ptak, P. Simple Method of Measuring Photometric and Radiometry Parameters of Power LEDs. In Proceedings of the 2018 Baltic URSI Symposium (URSI), Poznan, Poland, 15–17 May 2018; pp. 121–124. [CrossRef]
24. Standard JESD51-14. *Transient Dual Interface Test Method for The Measurement of The Thermal Resistance Junction-to-Case of Semiconductor Devices with Heat Flow through A Single Path*; JEDEC: Arlington, VA, USA, 2010.
25. Szekely, V. A new evaluation method of thermal transient measurement results. *Microelectr. J.* **1997**, *28*, 277–292. [CrossRef]
26. Janicki, M.; Torzewicz, T.; Samson, A.; Raszkowski, T.; Napieralski, A. Experimental identification of LED compact thermal model element values. *Microelectron. Reliab.* **2018**, *86*, 20–26. [CrossRef]
27. Marquardt, D. Algorithm for least-squares estimation of non-linear parameters. *SIAM J. Appl. Math.* **1963**, *11*, 431–441. [CrossRef]
28. Gerstenmaier, Y.C.; Kiffe, W.; Wachutka, G. Combination of Thermal Subsystem Modeled by Rapid Circuit Transformation. In Proceedings of the 13th International Workshop on Thermal Investigation of ICs and Systems, Budapest, Hungary, 17–19 September 2007; pp. 115–120. [CrossRef]
29. Torzewicz, T.; Ptak, P.; Samson, A.; Raszkowski, T.; Janicki, M.; Górecki, K. Parametric Compact Thermal Modeling of Power LEDs. In Proceedings of the 18th Intersociety Conference on Thermal and Thermo-mechanical Phenomena in Electronic Systems, Las Vegas, NV, USA, 28–31 May 2019. in press.

energies

MDPI

Article

Modelling a Switching Process of IGBTs with Influence of Temperature Taken into Account [†]

Paweł Górecki and Krzysztof Górecki *

Department of Marine Electronics, Gdynia Maritime University, 81-225 Gdynia, Poland;
p.gorecki@we.umg.edu.pl
* Correspondence: k.gorecki@we.umg.edu.pl
† This paper is an extended version of our paper published in 24th International Workshop on Thermal
  Investigations of ICs and Systems Therminic 2018, Kista, 164 40 Stockholm, Sweden, 26–28 September 2018,
  doi:10.1109/THERMINIC.2018.8592873.

Received: 1 April 2019; Accepted: 14 May 2019; Published: 18 May 2019

Abstract: In this article the problem of modelling a switching process of Insulated Gate Bipolar Transistors (IGBTs) in the SPICE software is considered. The new form of the considered transistor model is presented. The model includes controlled voltage and current sources, resistors and voltage sources. In the model, influence of temperature on dc and dynamic characteristics of the IGBT is taken into account. A detailed description of the dynamic part of this model is included in the article and some results of experimental verification are shown. Verification is performed for a transistor IRG4PC40UD by International Rectifier. The presented results of computations and measurements show clearly influence of temperature on on-time and off-time, and additionally switching energy losses are observed. Moreover, the results of investigations performed with the use of the new model are compared to the results of computations performed with classical models of the considered device given in the literature. It is proved that the new model makes it possible to obtain a better match to the results of measurements than the considered models described in the literature.

Keywords: power semiconductor devices; IGBT; modelling; transient analysis; SPICE; switching; thermal phenomena

1. Introduction

Insulated Gate Bipolar Transistors (IGBTs) belong to the group of power semiconductor devices [1]. They are commonly used as electronic switches in power electronic applications [2–4]. In a process of designing such circuits computer simulations are typically realized.

In order to perform reliable simulations of an electronic circuit, proper software and accurate models of all electronic components included in this circuit are indispensable. One of the most popular software for a computer analysis of electronic and power electronic circuits is SPICE [2,4–10]. In this software, models of many semiconductor devices, including a model of the IGBT, are built-in. Additionally, in many papers, e.g., [11–18], different forms of IGBT models are described. Of course, these models differ from one another by accuracy and the form. In the mentioned models, physical phenomena occurring in the IGBT could be described using equations of different forms.

In the authors' previous papers [16,17,19] accuracy of selected models of the considered transistor given in the literature are analysed and new forms of dc model of the considered transistor correctly describing influence of temperature on its dc characteristics are presented. The results of computations and measurements shown in the cited papers prove that the considered literature models correctly describe dc characteristics of the considered transistor at room temperature only, whereas visible differences between the results of computations and measurements can be obtained at a high temperature value (over 100 °C) and in the sub-threshold range of operation.

Typically, IGBTs are used as electronic switches operating in switch-mode power converters [2–4,20–22]. In such converters dynamic properties of the considered transistor are very important. These properties limit the value of switching frequency and they are characterised by such parameters as, e.g., by turn-off and turn-on times [3]. Values of these parameters change when temperature changes [23,24].

In this paper, which is an extended version of the conference paper [25], the problem of modelling dynamic properties of the IGBT is considered. Properties of selected literature models of the IGBT and their usefulness for proper computations of waveforms of terminal voltages and currents of this transistor during a switching process in a wide range of temperature are investigated. A new dynamic model of the IGBT for SPICE is proposed and its correctness is verified experimentally for different values of temperature.

In Section 2 the elaborated model is described, whereas in Section 3 some results of computations performed with the use of the new model and literature models as well as the results of measurements are shown and discussed.

2. Model Description

On the basis of the previously formulated by the authors dc model of the IGBT dedicated for the SPICE software [16], a new dynamic model of this device is formulated. In the new dynamic model controlled current sources describing currents flowing through parasitic capacitances of the considered transistor are added. In some papers, e.g., [1,11,26] the IGBT is typically modelled using the network representation of this device shown in Figure 1. As it is visible in the structure of the IGBT, three components can be distinguished: The MOSFET, the BJT and the diode.

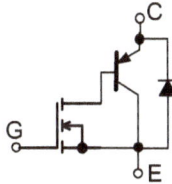

Figure 1. Network representation of the typical structure of the Insulated Gate Bipolar Transistor (IGBT).

On the basis of this classical IGBT structure presented in Figure 1 a new dynamic model taking into account parasitic capacitances is elaborated. The network representation of the new model, based on the conception presented in Figure 1, is shown in Figure 2.

In Figure 2, electronic components describing dc characteristics of the IGBT are marked with yellow colour and labelled as a dc model. In turn, controlled current sources G_{CGE}, G_{CCE} and G_{CGC} model currents flowing through internal capacitances C_{GE}, C_{CE} and C_{GC}, respectively, of the transistor. In analytical formulas describing the considered controlled current sources influence of temperature on the value of the mentioned capacitances is taken into account.

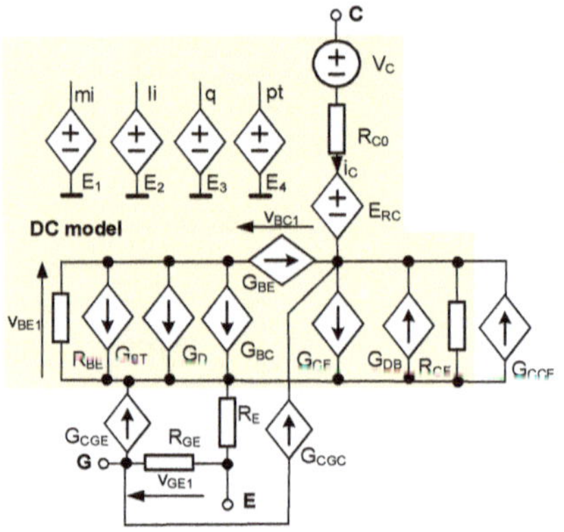

Figure 2. Network representation of the elaborated dynamic model of the IGBT.

In the dc model of the IGBT, drain current of the internal MOS structure is represented by two controlled current sources: G_D describing channel current of the MOS transistor and G_{ST} modelling the sub-threshold component of this current. To compute the values of currents and voltages in the model of the internal MOS structure, auxiliary controlled voltage sources (E_1, E_2, E_3 and E_4) are indispensable.

The controlled current source G_{BE} represents current flowing between the base and the emitter of the bipolar transistor contained in the structure of the modelled transistor. In turn, the controlled current source G_{BC} describes current flowing between the base and the collector of the bipolar transistor contained in the modelled IGBT.

The controlled current source G_{CE} models the main current of the IGBT. The controlled current source G_{DB} models dc characteristic of the diode. All the equations used to describe all the mentioned controlled current sources are described and discussed in paper [16].

In order to model electric inertia of the IGBT controlled current sources G_{CGE}, G_{CCE} and G_{CGC} are used. The mentioned current sources model current flowing through parasitic capacitances of the transistor: C_{GE}, C_{CE} and C_{GC}, respectively. These currents are described with formulas presented below. Capacitance between the collector and the emitter is described by formula of the form analogous to formula describing p-n junction capacitance. Because this junction does not rapidly change into the forward mode, in the equation describing this capacitance the diffusive component is omitted, but influence of temperature on this capacitance is described by following formula:

$$C_{CE} = C_{CE0} \cdot \left(1 + \frac{V_{CE}}{V_j}\right)^{-M_j \cdot (1 + r \cdot (T_j - T_0))} \tag{1}$$

where V_{CE} marks the collector emitter voltage, C_{CE0}—output capacitance of the IGBT at zero voltage on the diode, V_j—built-in potential of reverse diode, M_j—parameter describing the doping profile of the junction, r—the temperature coefficient of parameter M_j, T_j—internal temperature of the IGBT, T_0—reference temperature.

Capacitance between the gate and the emitter of the transistor is described by the following spline:

$$C_{GE} = C_{GE0} \cdot w + x \cdot (T_j - T_0) +$$
$$+ \begin{cases} 0 \; if \; V_{GE} < V_{GEmin} + V_{C1E} \\ C_{0x} \cdot k_1 \cdot (C_{G1} \cdot V_{GC1} + C_{G2}) \; if \; V_{GEmin} + V_{C1E} < V_{GE} \leq V_{GEmax} + V_{C1E} \\ C_{0x} \; if \; V_{GE} > V_{GEmax} + V_{C1E} \; and \; V_{C1E} < 0 \\ C_{0x} \cdot \left[u \cdot (T_j - T_0) + 0.75 \cdot \left(\frac{V_{GC1}}{2 \cdot V_{GE} + V_{C1E}} \right)^2 \right] \; if \; V_{GE} > V_{GEmax} + V_{C1E} \; and \; V_{C1E} \geq 0 \end{cases} \tag{2}$$

where C_{ox} is capacitance dependent on thickness of the layer of oxide under the gate t_{OX}, w—width of the channel of the MOS structure, C_{GE0}—capacitance per unit of channel widths between the gate and the emitter, V_{GEmax}, V_{GEmin}, C_{G1}, C_{G2}, x, u and k_1—other parameters of the model, whereas V_{GE} and V_{C1E} are voltages marked in Figure 2.

Capacitance C_{ox} occurring in Equation (2) is given by formula of the form [8,27]:

$$C_{ox} = \frac{\varepsilon_0 \cdot \varepsilon_{ox} \cdot L \cdot w}{t_{OX}} \tag{3}$$

where L denotes the length of the channel of the MOS structure, ε_0 is dielectric permeability of free air, and ε_{ox} relative dielectric permeability of silicon oxide.

Capacitance between the gate and the collector is described by the dependence of the form:

$$C_{GC} = C_2 \cdot \frac{i_{C} \cdot q}{k \cdot T_j} + \begin{cases} C_1 \cdot w \ if \ V_{GC1} < V_{GCmin} + e \cdot \left(T_j - T_0\right) \\ \left[C_{GD0} + y \cdot \left(T_j - T_0\right)\right] \cdot w \cdot \left(1 + \frac{V_{C1E}}{V_{jC}}\right)^{-M_{j2} + n \cdot \left(T_j - T_0\right)} \ if \ V_{GC1} \geq V_{GCmin} + e \cdot \left(T_j - T_0\right) \end{cases} \tag{4}$$

where C_{GD0} is capacitance per the unit of channel width between the gate and the drain on the unit of the channel width, C_1, C_2, y, e, n, V_{GCmin}—other parameters of the model, M_{j2}—parameter describing the doping profile of this junction, V_{jC}—potential of the base-emitter junction, q—electron charge k—the Boltzmann constant, i_C—collector current.

In practice, in order to model each of these capacitances, the use of a subcircuit shown in Figure 3 is needed.

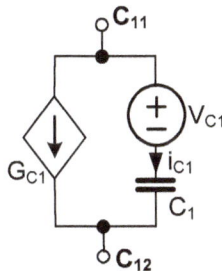

Figure 3. Manner of modelling internal capacitances of the transistor.

This subcircuit consists of linear capacitor C_1 of the fixed value of capacitance, voltage source V_{C1} of the zero value and the controlled current source G_{C1}. The voltage source V_{C1} is used to monitor the value of current i_{C1}, which is proportional to time derivatives of voltage on the current source G_{C1} and on the capacitor C_1. The current flowing through the controlled current source G_{C1} modelling parasitic capacitance C_X is given by the following formula:

$$I_{GC1} = \frac{C_X - C_1}{C_1} \cdot i_{C1} \tag{5}$$

In the presented dynamic model of the IGBT output currents of the considered controlled current source (G_{CCE}, G_{CGE} and G_{CGC}) describing current flowing through parasitic capacitances of the IGBT are described by Equation (5), in which capacitance C_X is equal to the appropriate capacitances described with Equations (1)–(4).

3. Results of Computations and Measurements

In order to verify usefulness of the worked out dynamic IGBT model many computations and measurements of waveforms of collector current while switching-on and switching-off this device were performed. Measurements were realised in the measurement set-up shown in Figure 4.

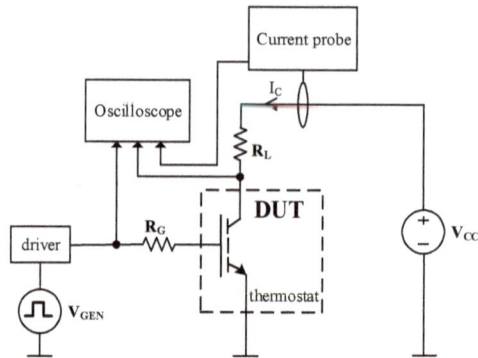

Figure 4. Diagram of a set-up to measure waveforms of voltage and currents of the IGBT while switching.

In the presented measurement set-up, voltage source V_{CC} feeds the collector of the investigated device. The control signal is generated by the function generator and is amplified by the driver MCP1305. Resistor R_L limits collector current and resistor R_G limits the maximum current of the gate of the tested IGBT. In the measurement set-up the oscilloscope Rigol DS1052E, the current probe Tektronix TCPA 300, the feeder NDN of the type DF1760SL10A and the functional generator of the type NDN JC5603P are used. The case temperature of the investigated device is measured by the pyrometer Optex PT-3S. The band of the applied oscilloscope is 50 MHz, and the current probe −100 MHz. The applied driver is characterised with the maximum output current of 4.5 A and with the maximum output voltage amounting to 18 V. During measurements the investigated transistor is situated in the thermostat.

Investigations were performed for type the IRG4PC40UD transistor [28] manufactured by International Rectifier company. During measurements this transistor was screwed to a large aluminium heat-sink situated in the thermostat. The gate of this transistor was controlled by a rectangular pulses train. The amplitude of this signal was equal to 15 V, whereas its frequency was equal to 20 kHz. Resistor R_G of resistance equal to 210 Ω was connected in series with the gate of the transistor. In order to minimise influence of a self-heating phenomenon on the investigated waveforms, all the measurements presented in this section were obtained just after switching-on or switching-off the control signal.

In Figure 5 influence of temperature on the measured waveforms of collector current obtained at selected temperature values in the range from 21 to 114 °C is shown. Such a range of investigations was selected arbitrarily taking into account three reasons. Firstly, the lowest value of temperature should be equal to typical room temperature. Secondly, the highest value of temperature must be lower than the maximum allowable temperature of operation of the considered transistor. Thirdly, differences between the highest and lowest temperature should be possibly big to visibly illustrate influence of temperature on dynamic properties of the tested transistor.

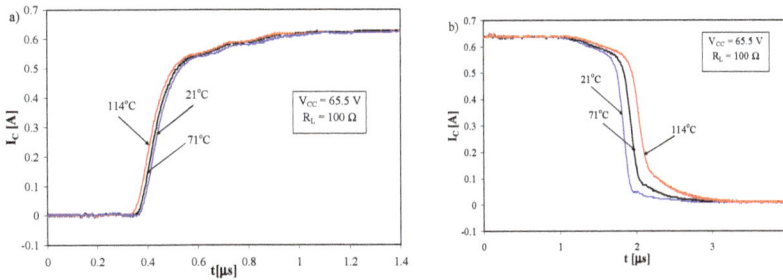

Figure 5. Measured waveforms of collector current of the considered transistor in (**a**) switching-on and (**b**) switching-off the IGBT at different values of temperature.

During measurements, the collector circuit of the tested IGBT was fed from the source V_{CC} of constant voltage equal to 65.5 V Resistor R_L of resistance 100 Ω which was connected between this source and the collector of the tested transistor. As it is visible, temperature very weakly influences waveforms $I_C(t)$ while switching-on, whereas influenced of temperature on these waveforms was strong while switching off. In particular, considerable extension of the so-called current tail could be observed at a temperature rise. The temperature rise caused acceleration of the switching-on process of the investigated transistor by 50 ns and a slowdown of the switching-off process of this transistor by almost 500 ns.

Using the presented in the previous section dynamic model of the IGBT, waveforms of voltages and currents of the considered device during the switching process at fixed values of temperature were computed. The results of computations obtained using the authors' new model are presented in Figures 6–8 as red lines. They are compared with the results of measurements (points) and the results of computations obtained by means of two popular literature models—the Hefner model [27,29] built-in in the SPICE (black lines) and the model available at the website of International Rectifier [15] (blue lines).

The considered literature models were described in detail in paper [16]. Figure 6 shows waveforms of collector current during the switching-off process of the considered device supplied in the identical manner as in the case considered in Figure 5. The results shown in Figure 6a correspond to temperature equal to 22 °C, whereas the results shown in Figure 6b correspond to temperature equal to 114 °C.

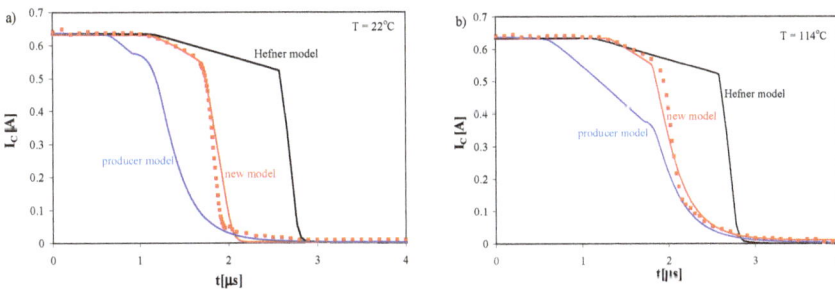

Figure 6. Computed and measured waveforms of collector current during the switching-off process at temperature equal to (**a**) 22 °C and (**b**) 114 °C.

As it is visible, good agreement was obtained between the results of computations and measurements at both the considered values of temperatures only for the authors' model. For the producer model [15], the switching-off process ran too quickly, especially for temperature T = 22 °C. In temperature equal to 114 °C, essential improvement in accuracy of computations performed with the described model within the range of current values below 300 mA could be observed. In turn, the results of computations obtained with the use of the Hefner model were characterised by the overlong switching-off process. Similarly, as it was presented for dc characteristics in paper [16], this model does not take into account influence of temperature on the shape of characteristics of the considered device.

Figure 7 shows waveforms of collector current of the investigated IGBT during the switching-on process for temperature equal to 22 °C (Figure 7a) and 114 °C (Figure 7b).

At temperature equal to 22 °C good agreement between the results of measurements and computations was obtained for all the tested models. In contrast, for temperature T = 114 °C good agreement between the results of computations and measurements was obtained with the use of the authors' model, and acceptable agreement between these results was assured also by the literature models. The producer model is characterised by good accuracy only within for values of collector current below 300 mA.

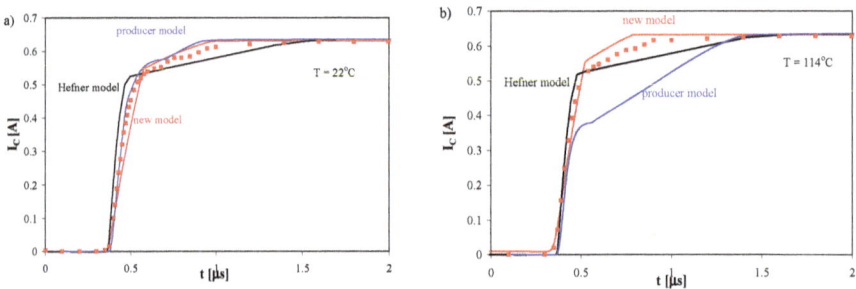

Figure 7. Measured and computed waveforms of collector current while switching-on at temperature equal to (**a**) 22 °C and (**b**) 114 °C.

Figure 8 illustrates waveforms of collector current during the switching-on process (Figure 8a) and the switching-off process (Figure 8b) of the tested IGBT operating at room temperature and load resistance R_L equal to 14.7 Ω.

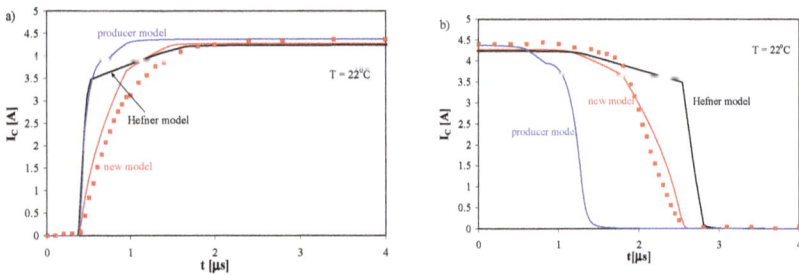

Figure 8. Measured and computed waveforms of collector current while (**a**) switching-on and (**b**) switching-off at R_L = 14.7 Ω.

Only for a new model good agreement between the results of measurements and computations was obtained. Computations performed with the use of both the literature models showed incorrectly too short duration time of the switching-on process. In turn, the computed with the use of these models waveform of collector current during the switching-off process showed too short duration time of this process for the producer model and too long—for the Hefner model.

Comparing the results of measurements and computations obtained for different values of load resistance (Figures 6a, 7a and 8), it is visible that at a higher value of this resistance both the considered switching times are shorter than for a lower value of resistance R_L.

Table 1 shows the values of times t_{on} and t_{off} computed with the use of three considered models and measured values which correspond to different values of resistance R_L and temperature. As one can notice, in five out of six considered cases the value of the mentioned times was computed with the use of the author's model. In the case of time t_{off} computed at temperature 114 °C and resistance $R_L = 100\ \Omega$ the value of this time was obtained with the use of the producer model. However, as it is visible in Figure 6b, this model does not describe exactly the whole course of collector current during the switching-off process, so the accurately computed value of time t_{off} results from good modelling of the coordinates of only two points in the waveform $I_C(t)$ used in computing values of parameters shown in this table.

Table 1. Computed and measured values of times t_{on} and t_{off}.

$R_L[\Omega]$	$T_a[°C]$	t_{on}[ns]				t_{off}[ns]			
		Producer Model	Hefner Model	Authors' Model	Measured	Producer Model	Hefner Model	Authors' Model	Measured
100	22	525	794	560	618	1736	2722	1882	1880
100	114	1033	808	460	566	2312	2725	2243	2370
14.7	22	542	870	1072	1336	1295	2592	2460	2236

As it visible in the presented comparison of the results of computations and measurements (Table 1), only the authors' model assured good accuracy. The computed by means of both literature models waveforms of collector current while switching significantly differ from the results of measurements. The proposed model is correct over a wide range of changes of temperature and load resistance.

As it can be observed, the literature models assured good agreement between the results of computations and measurements only in the case of the process of switching-on at room temperature. In other operating conditions visible differences between the results of computations and measurements were observed. In turn, the authors' model correctly described influence of temperature on waveforms of collector current both while switching-on and switching-off the transistor.

Using the authors' model, waveforms of the collector current while switching-off (Figure 9) and switching-on (Figure 10) the considered transistor are computed and the obtained results of computations are compared to the results of measurements.

In Figure 9 it is visible that the time of switching-off increases together with an increase of the switched current. In the investigated range of current i_C values, time of switching-off increases from 1.22 µs to 2.34 µs. This increase is linear and time of switching-off increases with an increase of current with the slope equal to about 94 ns/A.

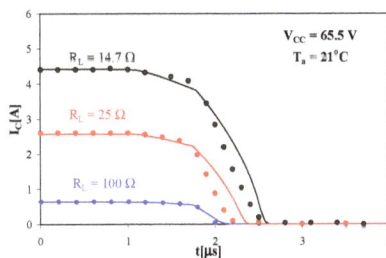

Figure 9. Computed and measured waveforms of current i_C while switching-off the transistor.

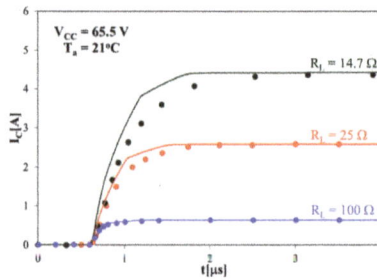

Figure 10. Computed and measured waveforms of current i_C while switching-on the transistor.

In Figure 10 it is visible that the on-time increases together with an increase of the switched current. In the investigated range of values of currents i_C, the on-time increases from 620 ns to 1.69 µs. This increase is linear, and the on-time increases with an increase of current of 186 ns/A. This increase is double compared to the case of switching-off.

From the point of view of switched-mode electronic circuits energy losses during the switching process are essential. In Figures 11 and 12 the computed and measured waveforms of power losses in the investigated transistor at two different values of temperature are presented. Figure 11 shows a shift between the maxima of courses p(t) obtained while switching-off in the considered temperatures, whereas the corresponding waveforms p(t) while switching-on shown in Figure 12 differs between one another imperceptibly.

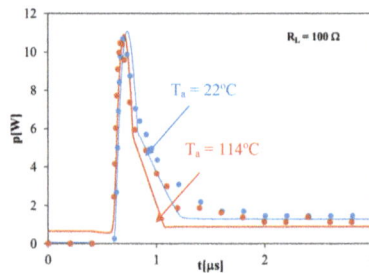

Figure 11. Measured and computed waveforms of power dissipated in the transistor while switching-on.

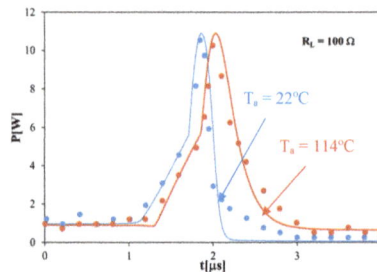

Figure 12. Measured and computed waveforms of power dissipated in the transistor while switching-off.

In Figure 13 the computed and measured waveforms of gate-emitter voltage while charging the gate of the investigated transistor are presented. These waveforms were measured and computed for collector current $I_C = 20$ A and voltage feeding the collector $V_{CC} = 400$ V. Current of the gate during the test amounted to 100 mA. As it is visible, the results of computations and measurements are convergent and satisfactory; the differences between them do not exceed 10%. This confirms correctness of the description of input capacitance of the considered transistor.

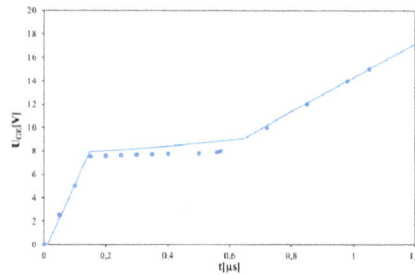

Figure 13. Computed and measured waveforms of gate-emitter voltage while charging the gate of the transistor.

4. Conclusions

In this article a new form of the IGBT model dedicated for the SPICE software is proposed. This model takes into account influence of temperature on dynamic properties of the considered device. In the description of the proposed model new dependences of device parasitic capacitances on temperature and on the co-ordinates of the device operating point are included in the form of analytical formulas.

The usefulness of the elaborated model is verified for the transistor operating as a switch. Waveforms of collector current are computed with the use of the new model and two models given in the literature. Computations are performed at different values of temperature and load resistance. The results of computations are compared to the results of measurements. The waveforms of collector current while switching-on and switching-off obtained using the new model fit well the results of measurements in the whole considered range of changes of temperature and load resistance. In contrast, the results of computations obtained using the considered literature models visibly differ from results of measurements. The difference increase while load resistance decreases.

The performed computations and measurements prove correctness of the new model. It is also visible that temperature strongly influences the value of off-time, but its influence on on-time could be omitted. Moreover, it is also shown that the new model properly describes waveform of power dissipated in the investigated transistor while it is being switched. Correctness of the used description of the transistor input capacitance prove good agreement between the computed and measured waveforms of gate-emitter voltage while charging the gate.

The disadvantage of the proposed model is a complicated description of internal capacitances making use of splines. It could result in some problems with convergence of the analysis performed for a long time—equal to hundreds or thousands of periods of the signal controlling the gate of the considered transistor.

The presented dynamic model of the IGBT could be useful for designers of power electronics circuits, making possible realisation of more accurate computer simulation of such circuits. This model could be also useful in didactics to present influence of temperature changes on switching-on and switching-off processes in power converters including IGBTs.

As it is commonly known, temperature of the IGBT could change as a result of self-heating phenomena [17]. Therefore, in further investigations the authors will elaborate an electrothermal model of the IGBT on the basis of the dynamic model presented in this article.

Author Contributions: The measurements presented in this manuscript and their evaluation was carried out by P.G. Manuscript was prepared and the equations were formulated by P.G. and K.G. Finally, K.G. supervised the research.

Funding: The scientific work was financed with the Polish science budget resources in the years 2017–2021, as the investigation project within the framework of the program "Diamentowy Grant". The project was financed within the program of the Ministry of Science and Higher Education called "Regionalna Inicjatywa Doskonałości" in the years 2019–2022, the project number is 006/RID/2018/19, the sum of financing is 11 870 000 PLN.

Conflicts of Interest: The authors declare no conflict of interest.

References

1. Baliga, B.J.; Adler, M.S.; Love, R.P.; Gray, P.V.; Zommer, N.D. The insulated gate transistor—A new 3-terminal MOS-controlled bipolar power device. *IEEE Trans. Electron Devices* **1984**, *31*, 821–828. [CrossRef]
2. Ericson, R.; Maksimovic, D. *Fundamentals of Power Electronics*, 2nd ed.; Kluwer Academic Publisher: Norwell, MA, USA, 2001.
3. Rashid, M.H. *Power Electronic Handbook*, 2nd ed.; Academic Press: Oxford, UK, 2007.
4. Kazimierczuk, M.K. *Pulse-width Modulated DC-DC Power Converters*, 2nd ed.; John Wiley & Sons: Hoboken, NJ, USA, 2008.
5. Maksimovic, D.; Stankovic, A.M.; Thottuvelil, V.J.; Verghese, G.C. Modelling and simulation of power electronic converters. *Proc. IEEE* **2001**, *89*, 898–912. [CrossRef]
6. Mohan, N.; Robbins, W.P.; Undeland, T.M.; Nilssen, R.; Mo, O. Simulation of Power Electronic and Motion Control Systems—An Overview. *Proc. IEEE* **1994**, *82*, 1287–1302. [CrossRef]
7. Rashid, M. *SPICE for Power Electronics and Electronic Power*, 3rd ed.; Taylor and Francis Group: Oxfordshire, UK, 2016.
8. Wilamowski, B.M.; Jaeger, R.C. *Computerized Circuit Analysis Using SPICE Programs*, 1st ed.; McGraw-Hill: New York, NY, USA, 1997.
9. Górecki, K.; Zarębski, J. Influence of MOSFET model form on characteristics of the boost converter. *Inf. MIDEM* **2011**, *41*, 1–7.
10. Bargieł, K.; Zarębski, J.; Bisewski, D. SPICE-aided modeling of high-voltage silicon carbide JFETs. In *IOP Conference Series: Materials Science and Engineering*; IOP Publishing Ltd.: Bristol, UK, 2016; Volume 104. [CrossRef]
11. Hefner, A.R.; Diebolt, D.M. An experimentaly verified IGBT model implemented in the Saber circuit simulator. *IEEE Trans. Power Electron.* **1994**, *9*, 532–542. [CrossRef]
12. Baliga, B.J. Analytical modelling of IGBTs: Challenges and solutions. *IEEE Trans. Electron Devices* **2013**, *60*, 535–543. [CrossRef]
13. Apeldoorn, O.; Schmitt, S.; De Doncker, R.W. An electrical model of a NPT-IGBT including transient temperature effects realized with PSpice device equations modeling. In Proceedings of the IEEE International Symposium on Industrial Electronics, Guimaraes, Portugal, 7–11 July 1997; pp. 223–228.
14. Wu, R.; Wang, H.; Pedersen, K.B.; Ma, K.; Ghimire, P.; Iannuzzo, F.; Blaabjerg, F. A temperature-dependent thermal model of IGBT modules suitable for circuit-level simulations. *IEEE Trans. Ind. Appl.* **2016**, *52*, 3306–3314. [CrossRef]
15. Spice Models and Saber Models. Web-Site of International Rectifier. Available online: http://www.irf.com/product-info/models/saber/ (accessed on 4 February 2019).
16. Górecki, P.; Górecki, K.; Zarębski, J. Modelling the temperature influence on dc characteristics of the IGBT. *Microelectron. Reliab.* **2017**, *79*, 96–103. [CrossRef]
17. Górecki, K.; Górecki, P. Modelling the influence of self-heating on characteristics of IGBTs. In Proceedings of the 21st International Conference Mixed Design of Integrated Circuits and Systems MIXDES, Lublin, Poland, 19–21 June 2014; pp. 298–302.
18. Sheng, K.; Williams, B.W.; Finney, S.J. A review of IGBT models. *IEEE Trans. Power Electron.* **2000**, *15*, 1250–1266. [CrossRef]
19. Górecki, K.; Górecki, P. Modelling dynamic characteristics of the IGBT with thermal phenomena taken into account. *Microelectron. Int.* **2017**, *34*, 160–164. [CrossRef]
20. Liu, Y.T.; Liang, S.Q.; Jin, D.H.; Peng, J.C. Performance comparison of Si IGBT and SiC MOSFET power devices based LCL three-phase inverter with double closed-loop control. *IET Power Electron.* **2019**, *12*, 322–329. [CrossRef]

21. Seo, J.H.; Choi, C.H.; Hyun, D.S. A new simplified space-vector PWM method for three-level inverters. *IEEE Trans. Power Electron.* **2001**, *16*, 545–550.
22. Blaabjerg, F.; Jaeger, U.; Munknielsen, S.; Pedersen, J.K. Power Losses in PWM-VSI Inverter Using NPT or PT IGBT Devices. *IEEE Trans. Power Electron.* **1995**, *10*, 358–367. [CrossRef]
23. Górecki, P. Investigation of the influence of thermal phenomena on dynamic parameters of the IGBT. In Proceedings of the 25th International Conference on Mixed Design of Integrated Circuits and Systems MIXDES, Gdynia, Poland, 21–23 June 2018; pp. 243–247.
24. Avenas, Y.; Dupont, L.; Khatir, Z. Temperature Measurement of Power Semiconductor Devices by Thermo-Sensitive Electrical Parameters—A Review. *IEEE Trans. Power Electron.* **2012**, *27*, 3081–3092. [CrossRef]
25. Górecki, P.; Górecki, K. Modelling influence of temperature on the switching process of IGBTs. In Proceedings of the 24th International Workshop on Thermal Investigations of ICs and Systems Therminic 2018, Stokholm, Sweden, 26–28 September 2018. [CrossRef]
26. Napieralski, A.; Napieralska, M. *Polowe Półprzewodnikowe Przyrządy Dużej Mocy*, 1st ed.; Wydawnictwa Naukowo-Techniczne: Warsaw, Poland, 1995.
27. Hefner, A.R. An investigation of the drive circuit requirements for the power insulated gate bipolar transistor (IGBT). *IEEE Trans. Power Electron.* **1991**, *6*, 208–219. [CrossRef]
28. IRG4PC40UD Insulated Gate Bipolar Transistor with Ultrafast Soft Recovery Diode, Data Sheet, International Rectifier. 1997. Available online: https://www.infineon.com/dgdl/irg4pc40udpbf.pdf?fileId=5546d462533600a4015356442ac722dc (accessed on 15 May 2019).
29. PSpice A/D Reference Guide. Product Version 15.7. 2006. Available online: siyh.byethost11.com/PDF/PSpiceGuide.pdf?i=1 (accessed on 15 May 2019).

energies

MDPI

Article

Influence of Power Losses in the Inductor Core on Characteristics of Selected DC–DC Converters

Krzysztof Górecki * and Kalina Detka

Department of Marine Electronics, Gdynia Maritime University, Morska 83, 81-225 Gdynia, Poland;
k.detka@we.umg.edu.pl
* Correspondence: k.gorecki@we.umg.edu.pl

Received: 1 April 2019; Accepted: 20 May 2019; Published: 24 May 2019

Abstract: The paper presents the results of a computer simulation illustrating the influence of power losses in the core of an inductor based on the characteristics of buck and boost converters. In the computations, the authors' model of power losses in the core is used. Correctness of this model is verified experimentally for three different magnetic materials. Computations are performed with the use of this model and the Excel software for inductors including cores made of ferrite, powdered iron, and nanocrystalline material in a wide range of load resistance, as well as input voltage of both the considered converters operating at different values of switching frequency. The obtained computation results show that power losses in the inductor core and watt-hour efficiency of converters strongly depend on the material used to make this core, in addition to the input voltage and parameters of the control signal and load resistance of the considered converters. The obtained results of watt-hour efficiency of the considered direct current (DC)–DC converters show that it changes up to 30 times in the considered ranges of the mentioned factors. In turn, in the same operating conditions, values of power losses in the considered cores change from a fraction of a watt to tens of watts. The paper also considers the issue of which material should be used to construct the inductor core in order to obtain the highest value of watt-hour efficiency at selected operation conditions of the considered converters.

Keywords: DC–DC converters; ferromagnetic cores; modeling; power losses

1. Introduction

Direct current (DC)–DC converters, which are used, e.g., in switch mode power supplies and semiconductor devices, also contain magnetic elements used to store electrical energy [1]. In recent years, more and more papers focused on modeling power losses in magnetic elements and on examining the influence of power losses in these elements on the characteristics of electronic equipment [1–12].

The aspiration to minimize the sizes of power supplies and to increase the frequency of the control signal causes an increase in power loss density in ferromagnetic cores and, consequently, limits the watt-hour efficiency of DC–DCconverters [2,8–10,12]. Losses in magnetic material can be caused, e.g., by hysteresis of the magnetizing curve or by eddy currents [13]. Values of power loss components depend on many factors, e.g., the type of magnetic material used, the amplitude and frequency of the waveforms of magnetic flux density, and their shape [5,13–20].

In previous studies [13,16–19], the methods for computing power losses in magnetic materials based on modifications of the Steinmetz model, defined as improved Steinmetz equation (ISE) or modified Steinmetz equation (MSE), were presented. In these models, the shape of the waveform of the excitation signal is taken into account; however, the influence of temperature on the properties of magnetic materials and on the parameters of the considered model is not taken into account. Yet, in the presented research results, among others, ferrite materials such as N87 were used, whose properties strongly depend on temperature. In addition, as stated in the cited papers, these models are valid only for a low value of frequency up to 20 kHz.

In turn, in Reference [15], a new model called I2SE was proposed. This model is an extended version of the ISE model. In this model, five additional parameters related to the relaxation process are added. However, the mentioned model also does not take into account the influence of temperature on the properties of magnetic materials and their power loss. Additionally, the presented description does not allow its implementation in the SPICE software. In Reference [5], three models of a ferromagnetic core were proposed. The first was a linear model, the second was a non-linear isothermal model that uses the classic Jiles–Atherton model to describe the magnetization curve, and the third was an electrothermal model of the ferromagnetic core. The electrothermal model took into account influence of temperature on core properties, but unfortunately contained a formal error, which did not allow computing power losses based on the magnetization curve obtained from this model.

Reference [12] presented a method of modeling power losses in the core of the inductor, which is a component of the boost converter. The results of computations using the various models described, among others [13,16,17,19], were compared to the results obtained using the authors' model given in Reference [3], which took into account the influence of temperature, frequency, and the shape of magnetic flux density on the properties of magnetic materials and power losses in the ferromagnetic core of an inductor.

In References [2,14,21,22], much emphasis was given to the analysis of the influence of losses in semiconductor devices on the characteristics of DC–DC converters. Yet, losses in the cores of magnetic elements were omitted in these analyses. The problem of the influence of models of power losses in ferromagnetic cores on the characteristics of buck and boost converters was also considered in References [13,20]. In the cited papers, it was shown that the use of classical models described References [14,23] can cause significant underestimation of power losses in the core. The observed changes in the values of power losses can increase three-fold. In addition, most of the presented models are based on a linear approximation of the real characteristics of inductors or transformers. Such an approach may cause overstated or understated watt-hour efficiency of electronic circuits [19,24,25].

On the other hand, in Reference [26], the average electrothermal model of an inductor was described; it took into account the dependence of power losses in ferromagnetic materials on frequency and temperature. The investigations were carried out for the boost converter. Reference [27] presented the results of computations for the boost converter using the electrothermal model of an inductor and hybrid models of diodes and transistors.

The aim of this paper was to analyze the influence of power losses in the inductor core on the characteristics of selected DC–DC converters. Investigations are performed for buck and boost converters including inductors with cores made of different ferromagnetic materials. The authors' model of losses in the ferromagnetic core is used in computations performed with the use of the Excel software. The dependences of power losses in the inductor core and watt-hour efficiency of the considered DC–DC converters on load resistance, input voltage, and switching frequency are computed. The influence of selection of the material used to make the inductor core on the properties of the considered converters is discussed.

In Section 2, the Steinmetz, the natural Steinmetz, and the authors' models of power losses occurring in magnetic materials are described. Section 3 presents the investigated converters, while Section 4 outlines the results of computations obtained using the chosen models.

2. Model of Power Losses in the Core of an Inductor

In the literature, many papers focusing on power losses per unit of volume (P_V) in magnetic materials proposed power loss models of the inductor core. Unfortunately, some models did not have the analytical dependence describing losses in the ferromagnetic core. This makes it impossible to implement these models in an easy way using computer programs. Many models were described in References [12,20]; however, they concerned models of power losses in magnetic materials which took into account the triangular shape of amplitude of magnetic flux density in the core. The Steinmetz model [23] was also considered in these papers as the first model of power losses in magnetic materials, shown below.

$$P_{loss} = k \cdot f^{\alpha} \cdot B_m^{\beta}, \tag{1}$$

where k, α, and β are material parameters, B_m is the amplitude of magnetic flux density, and f is the frequency.

The results given in Reference [20] showed that differences between values of power dissipated in the inductor core computed with the use of the models described in References [13,23] and the catalog data reached as high as 60%. In contrast, the results of computations performed with the use of the core model described in Reference [3] adequately fit the catalog data. In practice, this means that the use of classical models to describe power losses in the core, given in References [13,23] makes it impossible to obtain correctly computed values of power losses. Differences between the values obtained with the authors' model given in Reference [3] and the considered literature models are largest at high values of temperature, frequency of the control signal, and input voltage [20]. In turn, the results given in Reference [12] showed that, in order to compute power losses of the inductor core included in the boost converter, it is essential to take into account the shape of the waveform of magnetic flux density and the influence of temperature and frequency on the properties of magnetic materials. Omitting the influence of the mentioned factors could cause overestimation of power losses in the core of an inductor, exceeding even 75% [12].

In addition, the use of the classical Steinmetz model [23] to compute power losses in magnetic materials is justified in the range of low frequency not exceeding 25 kHz [12]. For frequency values higher than 25 kHz, power losses in the magnetic material per unit volume increase by up to 15% [3,12].

In Reference [28], it was shown that the total power losses in the toroidal core of a transformer are the sum of hysteresis losses and eddy current losses. The eddy current losses are computed on the basis of Equation (2).

$$P_w = \frac{\pi}{6} \cdot \sigma_F \cdot dd^2 \cdot B_m{}^2 \cdot f^2, \tag{2}$$

where σ_F is the conductivity of the material, B_m is the amplitude of sinusoidal magnetic flux density at frequency f, and dd is the thickness of the sheet.

Reference [29] presented the procedure for formulating a general model of power losses frequently appearing in previous models; however, the drawback of this model was the fact that it is dedicated to laminated sheets and was verified only for frequencies up to 2 kHz.

In Reference [30], the results of investigations on how to compute power losses in soft magnetic materials as a result of the movement of single Bloch walls during magnetization of magnetic material were presented. The well-defined domain structures of the equation with limited accuracy were additionally formulated. The conclusions presented in the cited paper allowed determining trends of power losses in different ferromagnetic materials [3].

For example, Reference [13] showed a method of modeling power losses in magnetic materials, called the natural Steinmetz extension (NSE). This method can be used for sinusoidal and triangular waveforms of magnetic flux density. In the cited paper, it was shown that the coefficients α, β, and k_w of the proposed model significantly depend on frequency and temperature. Yet, no analytical dependence describing such influence was given.

In Reference [3], dependences describing power losses in magnetic materials which took into account the influence of temperature and frequency of B(t) on the power losses and on the parameter β were proposed. The formula describing power loss density has the following form [3]:

$$P_v = P_{v0} \cdot f^{\alpha} B_m^{\beta} \cdot (2 \cdot \pi)^{\alpha} \cdot \left(1 + \alpha_p \cdot (T_R - T_m)^2\right) \cdot (0.6336 - 0.1892 \cdot ln(\alpha)) \tag{3}$$

where α_p is the temperature coefficient of losses in ferromagnetic material, T_R is the core temperature, T_m is the temperature at which the material has the smallest loss, and parameter P_{v0} is described by Equation (4) [3].

$$P_{v0} = a \cdot \exp\left(-\frac{f + f_0}{f_3}\right) + a_1 \cdot (T_R - T_m) + a_2 \cdot \exp\left(\frac{f - f_2}{f_1}\right), \tag{4}$$

where $a, a_1, a_2, f_0, f_1, f_2,$ and f_3 are material parameters.

In turn, parameter β is described by Equation (5) [3].

$$\beta = \begin{cases} 2 \cdot (1 - \exp(-T_R/\alpha_T)) + 1.5 \ if \ 1 - \exp(-T_R/\alpha_T) > 0 \\ 1.5 \ if \ 1 - \exp(-T_R/\alpha_T) < 0 \end{cases}, \tag{5}$$

where α_T is a material parameter.

For the triangular waveform of magnetic flux density, power losses in magnetic material can be computed from the dependence shown in Equation (6) [3].

$$P_V = P_{v0} \cdot f^\alpha B_m^\beta \cdot 2^\alpha \cdot \left(1 + \alpha_p \cdot (T_R - T_m)^2\right) \cdot \left(d^{1-\alpha} + (1-d)^{1-\alpha}\right), \tag{6}$$

where d denotes the duty factor of the triangular waveform.

In Table 1, model parameter values for the core model from Reference [3] for selected magnetic materials are shown. The investigations were carried out for inductors containing cores made of powdered iron (RTP), ferrite material (RTF), and nanocrystaline material (RTN).

Table 1. Model parameters values of selected inductor core RTP—powdered iron; RTF—ferrite material; RTN—nanocrystalline material.

Parameter	RTP	RTF	RTN
α	1.15	1.29	1.88
α_p	0	0.00011	0
T_m (K)	368	343	358
a	10	1.1	100
f_0 (Hz)	2000	25,000	200
f_1 (kHz)	90	1,000,000	300
f_2 (Hz)	10	10	100
f_3 (Hz)	15,000	50,000	10
a_1	10	0.013	0.01
a_2	12	0.7	0.00019

The method of estimating the parameter values of the authors' model given in Reference [3] was described in detail in Reference [31], while material parameters (α) of the Steinmetz model described in Reference [23] had the same values as for the model parameters presented in Reference [3], and parameters k and β were computed using the transformed Equation (1) with the catalog data of power losses per unit volume (P_v) in magnetic material at a specific value of frequency (f) and amplitude of magnetic flux density (B_m).

Figure 1 shows the dependence of power loss density in ferrite material (RTF) for two core temperatures equal to 25 °C (a) and 100 °C (b) in a wide range of frequency changes. In Figures 1–3, points denote the catalog data [32], the solid lines denote the results of computations obtained using the model from Reference [3] described by Equations (3)–(5), and the dashed lines denote the results of computations using the model from Reference [23] described by Equation (1).

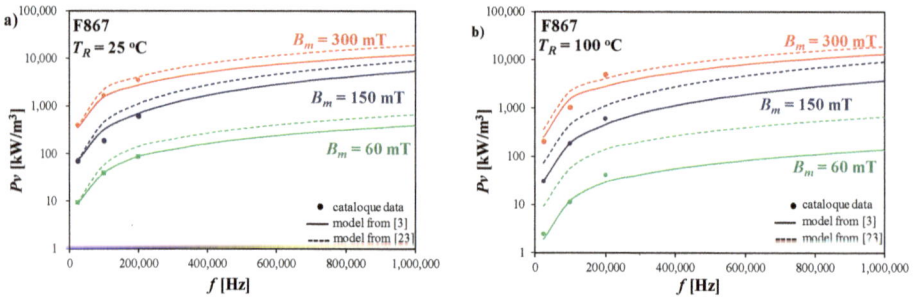

Figure 1. Dependence of power losses per unit volume (P_v) in the core made of ferrite material F867 on the frequency at a core temperature equal to 25 °C (**a**) and 100 °C (**b**).

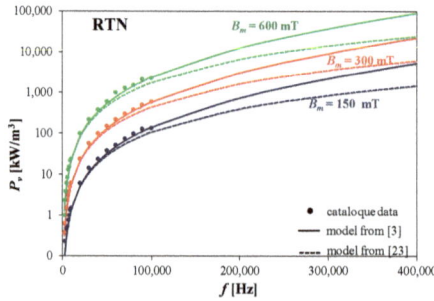

Figure 2. Dependence of power losses per unit volume (P_v) in the core made of nanocrystalline material (RTN) on the frequency.

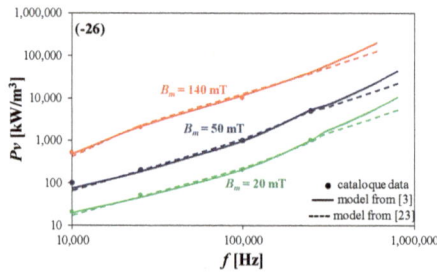

Figure 3. Dependence of power losses per unit volume (P_v) in the inductor core made of powdered iron material (RTP) on the frequency.

As can be seen in Figure 1, with an increase in frequency, power losses in the magnetic material increase at a different value of amplitude of magnetic flux density. It is also worth noting that good agreement between the catalog data and computations made with the use of the model given in Reference [3] was obtained for both considered values of temperature.

Figure 2 shows the dependence of power loss density in nanocrystalline material (RTN) on the frequency at three values of amplitude of magnetic flux density equal to 150, 300, and 600 mT. As can be seen, power losses in the nanocrystalline material increase with an increase in frequency. Also, good agreement between the catalog data [33] and the results of computations performed with the use of the model from Reference [3] was obtained.

Figure 3 presents the dependence of power losses per unit volume (P_v) in the core made of powdered iron (RTP) material on the frequency. As can be seen, as for cores made of the other considered materials (RTN, RTF), power losses per unit of volume increase with an increase in frequency, also leading to good agreement between the catalog data and the results of computations performed with the use of the model from Reference [3].

The results presented in Figures 1–3 show that the authors' model ensures good agreement between the computed and catalog dependences of power losses on frequency, amplitude of magnetic flux density, and temperature. In turn, the Steinmetz model gives the value of these losses even several times higher. Differences between the results of computations made using both the considered models are particularly high for the ferrite core (F867). For the core made of powdered iron (−26), the results obtained with both the models are practically identical. For the nanocrystalline core (RTN), the computed values of power loss using the Steinmetz model are lower than the values of power loss using the authors' model and the catalog data in the range of high frequency.

In order to compute power losses in the inductor core, it is necessary to multiply losses expressed with Equations (1), (3), and (6) by the equivalent volume of the core (V_e).

As mentioned earlier, omission of the dependence of power losses in the magnetic material on the frequency, temperature, and the shape of magnetic flux density can cause a significant overstatement of power losses in the inductor core. As shown in Reference [20], a significant effect on power losses in magnetic materials results from the shape of the magnetic flux density B(t) waveform. The data provided by the manufacturers usually refer to the case when the waveform of magnetic flux density B(t) is sinusoidal. The ferromagnetic core, being a component of such magnetic elements as an inductor, is generally used in systems in which the waveform of magnetic flux density is triangular. Therefore, in order to determine the influence of power losses on the characteristics of the buck and boost converter, computations were carried out taking into account the triangular waveform of magnetic flux density, and power losses in the inductor core were computed from the dependence in Equation (6). As can be seen from the dependence in Equation (6), power losses in the core of an inductor operating in a DC–DC converter depend on the frequency, amplitude of magnetic flux density, and duty factor of the signal controlling the transistor.

3. Investigated Converters

In order to investigate the influence of power losses in the inductor core on the properties of selected DC–DC converters, computations were performed for the boost converter and the buck converters, whose electrical diagrams are shown in Figure 4.

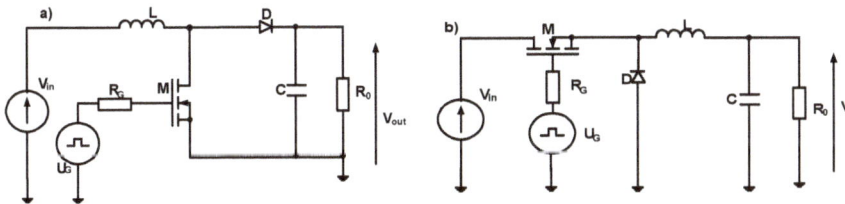

Figure 4. Diagrams of the boost (**a**) and the buck (**b**) converters.

In the tested converters, the diode and the transistor are modeled as ideal switches. In this case, power losses are dissipated in inductor L only. In order to compute such losses in magnetic materials, the value of amplitude of magnetic flux density should be computed using the following formula [12,22]:

$$B_m = \frac{L \cdot \Delta I_L}{z \cdot S_{Fe}},\qquad(7)$$

where S_{Fe} denotes the cross-section area of the core, z is the number of turns, and ΔI_L is the peak-to-peak value of inductor current. In the boost converter, the value of ΔI_L is computed with the following formula [12,22]:

$$\Delta I_L = \frac{V_{in} \cdot d \cdot T}{L},\qquad(8)$$

where V_{in} is the value of input voltage of the boost converter.

In turn, the peak-to-peak value of inductor current ΔI_L for the buck converter is computed with the following formula [12,22]:

$$\Delta I_L = \frac{(V_{in} - V_{out}) \cdot d \cdot T}{L}.$$
(9)

As resulting from Equation (7), the value of amplitude of magnetic flux density is a function of the number of turns, inductance, cross-section area of the core, input voltage of the converter, the duty cycle, and the period of switching. It is important that the converter operation mode does not influence the formula describing power losses of the inductor core.

For the boost converter operating in continuous conducting mode (CCM), the value of the output voltage depends only on input voltage and the control signal duty cycle, and it is described as follows:

$$V_{out} = V_{in} \cdot \frac{1}{1-d}.$$
(10)

On the other hand, for the boost converter operating in discontinuous conducting mode (DCM), the converter output voltage is described as follows [21]:

$$V_{out} = V_{in} \cdot \left[1 + \frac{V_{in} \cdot d \cdot T \cdot R_0}{2 \cdot L \cdot V_{out}}\right].$$
(11)

The boost converter operates in CCM mode if the following formula is fulfilled:

$$R_0 \geq \frac{2 \cdot L \cdot V_{out}}{V_{in} \cdot (1-d) \cdot d \cdot T}.$$
(12)

In turn, output voltage of the buck converter operating in CCM mode for a given input voltage (V_{in}) and duty cycle of the control signal (d) is described in Equation (13) [22].

$$V_{out} = d \cdot V_{in}.$$
(13)

Output voltage of the buck converter operating in DCM mode is given in Equation (14) [22],

$$V_{out} = \frac{V_{in}}{\frac{2 \cdot I_0 \cdot L}{d \cdot T \cdot V_{in}} + 1},$$
(14)

and depends on inductance of inductor L, output current I_0, and the period of the control signal T. The buck converter operates in DCM mode when the following dependence is fulfilled:

$$R_0 \geq \frac{2 \cdot L}{(1-d) \cdot T}.$$
(15)

4. Results of Computations

In order to verify the influence of power losses in the inductor core on selected DC–DC converters, computations using the model presented in Reference [3] and described by Equations (3)–(6) were performed with the Excel software. The accuracy of the calculations is limited by the accuracy of the used models of power losses in the inductor core and by the model DC–DC converter characteristics.

As a further consideration, the model of inductor core losses from Reference [3] was chosen because, as can be seen in Figures 1–3, it correctly models the losses in magnetic materials. In addition, the selected model takes into account the triangular waveform of magnetic flux density. In the considerations, it is assumed that inductance of the inductor is constant for all the considered materials as a result of the air gap occurring in the core in a given range of changes of the magnetic field, as shown in Figure 5 [34]. Investigations were carried out for ring cores made of ferrite (RTF), powdered iron

(RTP), and nanocrystalline material (RTN) [31–33,35]. In order to emphasize the influence of power losses on the characteristics of the considered circuits, inductor cores of large size were chosen for testing. The company Ferryster [35] offers toroidal cores with the outside diameter of the considered cores equal to 102 mm, the internal diameter equal to 57.3 mm, and the height equal to 33 mm. In the computations, each of the considered cores had identical size, and, due to differences in the length of the air gap, each inductor had the same inductance ($L = 100$ µH), while 27 winding turns and a cross-section area (S_{Fe}) of 737.55 mm^2 were adopted.

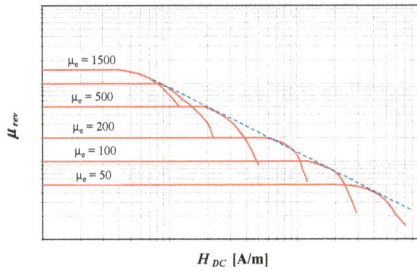

Figure 5. Typical dependence of magnetic permeability on magnetic force [34].

In order to compute the value of magnetic flux density, the dependence in Equation (7) was used assuming inductance L as constant. The peak-to-peak current value ΔI_L was computed using the dependence in Equation (8) for the boost converter and the dependence in Equation (9) for the buck converter.

Figure 6 presents the dependence of the amplitude of magnetic flux density on the input voltage of the boost converter (Figure 6a) and the buck converter (Figure 6b). In order to compute the value of amplitude of magnetic flux density, Equation (7) was used. All the presented characteristics in this section were determined on the assumption that the duty cycle was equal to 0.5.

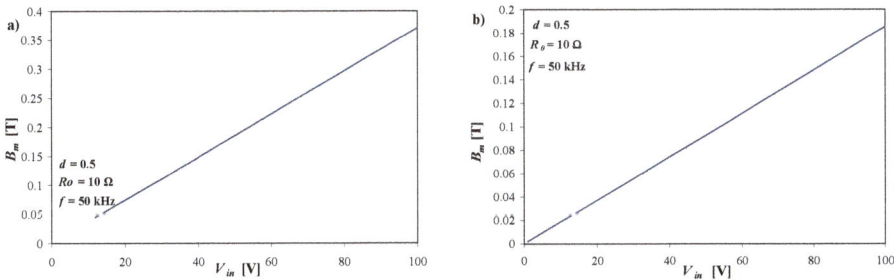

Figure 6. Dependence of the amplitude of magnetic flux density on the input voltage of the boost converter (**a**) and the buck converter (**b**).

As shown in Figure 6, the dependence of amplitude of magnetic flux increases linearly with an increase in input voltage, obtaining values up to 350 mT for the boost converter and 180 mT for the buck converter at an input voltage equal to 100 V.

Later in this section, ferromagnetic cores used in the inductor contained in the considered converters are denoted with as follows: powder iron core (−26, RTP), nanoctrystalline core (M070, RTN), and ferrite core (F867, RTF).

Figures 7–9 present the influence of input voltage V_{in}, control signal frequency, and load resistance of the boost converter on power losses in the inductor core and the watt-hour efficiency of the converter for three considered materials used to build the inductor core (i.e., RTP, RTF, and RTN). Values of watt-hour efficiency were computed as follows:

$$\eta = 1 - \frac{P_R \cdot R_0}{V_{out}^2}$$

(16)

where P_R denotes power losses in the inductor core computed using Equations (4)–(6).

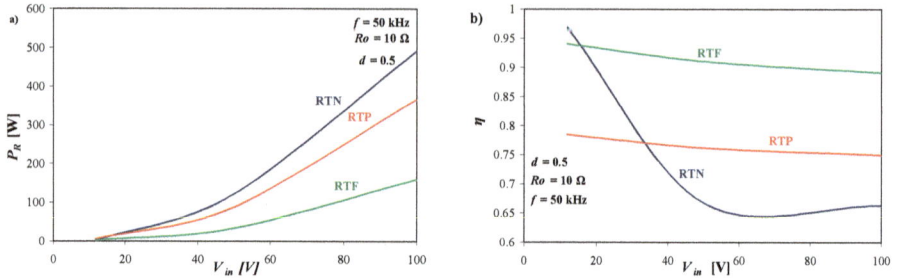

Figure 7. Dependence of power losses in the inductor core (**a**) and watt-hour efficiency of the boost converter (**b**) on input voltage.

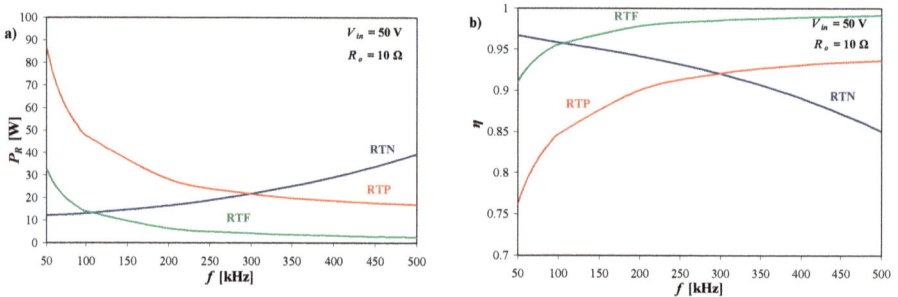

Figure 8. Dependence of power losses in the inductor core (**a**) and watt-hour efficiency of the boost converter (**b**) on frequency of the control signal.

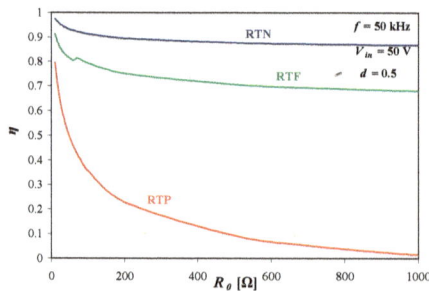

Figure 9. Dependence of the watt-hour efficiency of the boost converter on the load resistance.

Figure 7 presents the influence of input voltage on power losses in the inductor core and the watt-hour efficiency of the boost converter.

Dependences of power losses in the inductor core and the watt-hour efficiency as a function of input voltage of the considered converters presented in Figures 7 and 8, and in Figures 10 and 11 were computed using Equations (6) and (16), respectively.

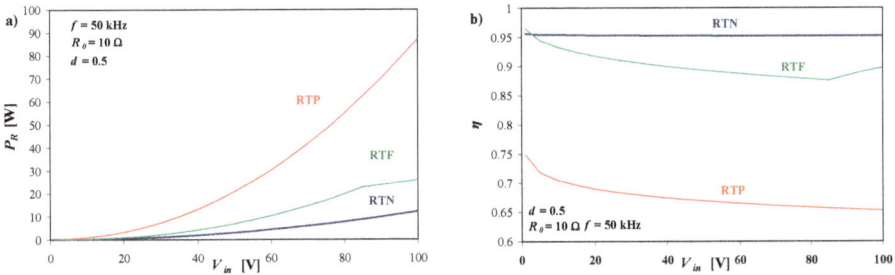

Figure 10. Dependence of power losses in the inductor core (**a**) and watt-hour efficiency of the buck converter (**b**) on input voltage.

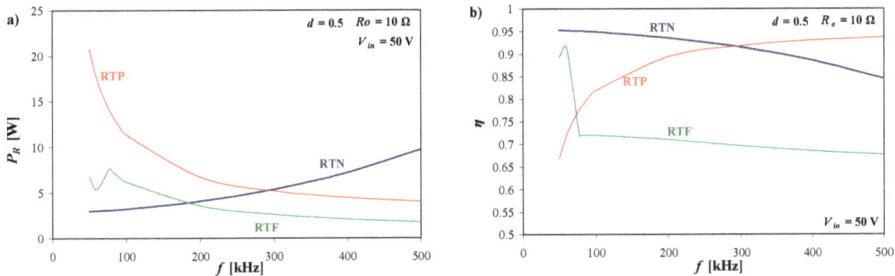

Figure 11. Dependence of power losses in the inductor core (**a**) and watt-hour efficiency of the buck converter (**b**) on frequency.

In order to compute watt-hour efficiency, it is necessary to compute power losses in the inductor core using Equation (6); the resistance R_0 value was deemed constant and equal to 10 Ω. At this value of resistance R_0, converters are in CCM mode. By changing the value of input voltage V_{in}, the output voltage V_{out} of the boost converter can be computed from the dependence in Equation (10) or from Equation (13) for the buck converter. Computations were performed at a constant value of the control signal duty cycle $d = 0.5$.

In turn, in order to compute the dependence of the watt-hour efficiency on load resistance in a wide range of R_0 changes, Equation (16) was used, and a change in the operation mode of the converter from CCM to DCM was taken into account. Limit values of resistance R_0, corresponding to transition between the range of CCM and DCM, were computed by means of the inequalities in Equations (12) and (15). Output voltage values were computed using the dependences in Equations (11) and (14) for CCM and DCM, respectively. The value of power losses (P_R) was computed using Equation (6).

As shown in the figures, an increase in input voltage of the boost converter causes an increase in power losses in the inductor core for all the considered materials. However, the highest value of these power losses was obtained for the inductor containing the core made of nanocrystaline material (about 500 W at V_{in} = 100 V), while the lowest value of these power losses was obtained for the ferrite core (120 W at V_{in} = 100 V).

In turn, watt-hour efficiency was highest for the inductor containing the core made of ferrite material in the whole considered range of input voltage ($\eta \geq 0.9$), and the lowest value of watt-hour efficiency was obtained for the inductor containing the nanocrystalline core. It is worth noting that, for the range of input voltage from 10 to 40 V, the watt-hour efficiency of the converter containing an inductor with the nanocrystalline core decreased from 0.9 to 0.65.

Figure 8 presents the influence of the control signal frequency on power losses in the inductor core and the watt-hour efficiency of the boost converter.

The dependence of power losses in the inductor core made of ferrite (RTF) and powdered iron (RTP) decreased with an increase in frequency. In the considered range of frequency for the RTP core, power losses decreased from 86 W to 16 W, whereas, for the RTF core, they decreased from 32 W to 3 W. In turn, for the core made of nanocrystalline material, the considered dependence was a monotonically increasing function, and power losses increased from 12 W to 39 W. The highest power losses in the core were observed at f = 50 kHz (87 W for RTP), whereas, at frequency f = 500 kHz, this power was equal to 41 W for RTN. In turn, the watt-hour efficiency of the boost converter was an increasing function of frequency for the inductor containing RTF or RTP cores, whereas this efficiency was a decreasing function of frequency for the inductor containing the RTN core. The highest watt-hour efficiency equal to 0.98 was obtained for the converter containing an inductor with the RTF core at frequency f > 250 kHz, whereas, for this converter containing the inductor with the RTP core, the highest value of watt-hour efficiency was equal to 0.92. In turn, the watt-hour efficiency of the boost converter with the inductor including the RTN core decreased in the considered range of frequency from 0.97 to 0.84.

Figure 9 presents the influence of load resistance on the watt-hour efficiency of the boost converter. Due to the constant value of amplitude of magnetic flux density and frequency, values of power losses for the boost converter in the load resistance function were constant and amounted to 73 W for the RTP core, 24 W for the RTF core, and 10 W for the RTN core. Therefore, only the dependence of watt-hour efficiency as a function of load resistance is presented.

The watt-hour efficiency was a decreasing function of load resistance for all the considered materials. While load resistance increased, watt hour efficiency decreased from 0.98 to 0.89 for the RTN core, from 0.91 to 0.73 for the RTF core, and from 0.79 to 0.03 for the RTP core. As shown in the figures, the lowest value of watt-hour efficiency equal to 0.03 was obtained for the RTP core.

Figures 10–12 present the influence of input voltage V_{in}, the control signal frequency f, and load resistance R_0 of the buck converter on power losses in the inductor core and the watt-hour efficiency of the buck converter for the three considered materials used to build the inductor core (i.e., RTP, RTF, and RTN).

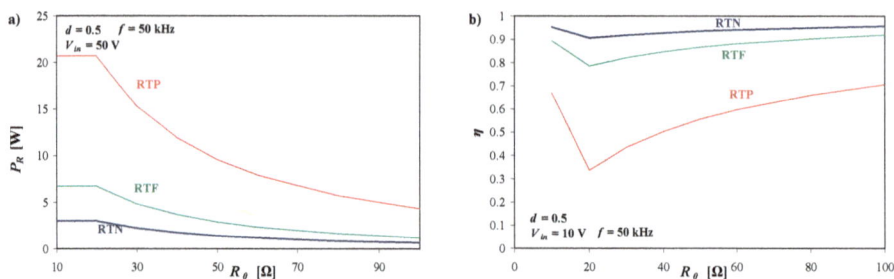

Figure 12. Dependence of power losses in the inductor core (**a**) and watt-hour efficiency of the buck converter (**b**) on load resistance.

As shown in Figure 10, an increase in input voltage of the buck converter caused an increase in power losses in the inductor core for all the considered materials. However, the highest value of power losses in the inductor core was obtained for the inductor containing the RTP core (up to 87 W at V_{in} = 100 V), while the lowest value of power losses in the inductor core was obtained for the RTN core (about 10 W at V_{in} = 100V). For the inductor core, the value of power obtained for the RTF core was equal to 24 W at V_{in} = 100 V.

In turn, the watt-hour efficiency for the inductor containing the RTN core in the whole considered range of input voltage had the same value equal to 0.95. The lowest value of watt-hour efficiency was observed for the inductor containing the RTP core. For this core, P_R decreased from 0.75 to about 0.65 only. The dependence $\eta(V_{in})$ for the RTF core possessed the minimum equal to 0.88 at V_{in} = 83 V, whereas the highest value was equal to 0.97 at V_{in} tending to 0.

Figure 11 presents the influence of the control signal frequency on power losses in the inductor core and the watt-hour efficiency of the buck converter.

Dependences of power losses and watt-hour efficiency had similar character as the dependences describing these parameters for the boost converter. The highest power losses in the core were obtained at f = 50 kHz for the RTP core, where losses were equal to about 20 W. These losses decreased with an increase in frequency to about 4 W at f = 500 kHz. In the considered frequency range, power losses decreased from 7 W to 2 W for the RTF core, whereas they increased from 3 W to 11 W for the RTN core.

In turn, the highest value of watt-hour efficiency was observed for the converter containing the inductor with the RTP core at f = 500 kHz. For the buck converter with the inductor including the RTP core, the watt-hour efficiency increased from 0.67 at f = 50 kHz to 0.94 at f = 500 kHz. In turn, this efficiency decreased from 0.95 to 0.84 for the inductor with the RTN core, and from 0.92 to 0.68 for the inductor with the RTF core.

Figure 12 presents the influence of load resistance on power losses in the inductor core (Figure 12a) and the watt-hour efficiency (Figure 12b) of the buck converter.

As shown in Figure 12, power losses in the core for R_0 < 20 Ω were constant. Above this value of load resistance, power losses decreased monotonically in the whole range of the considered load resistance changes. The highest values were obtained for the inductor containing the RTP core (decreasing from 20.7 W to 4.5 W), while intermediate values were obtained for the RTF core (decreasing from 6.5 W to 1.5 W), and the lowest values were obtained for the RTN core (decreasing from 3 W to 0.8 W).

On the other hand, the dependence of watt-hour efficiency had the local minimum equal to 0.33 for the RTP core, 0.78 for the RTF core, and 0.9 for the RTN core at a load resistance equal to 20 Ω. Above this value of load resistance, the watt-hour efficiency of the buck converter increased with an increase of load resistance. The highest value of watt-hour efficiency was obtained for the converter containing the inductor with the RTN core. The curves presented in Figure 12 had the characteristics of spline functions due to the fact that, for R_0 = 20 Ω, the considered converter operated in the critical mode, for R_0 < 20 Ω, it was characterized by the continuous conducting mode (CCM), and, for R_0 > 20 Ω, it operated in the discontinuous conducting mode (DCM).

5. Conclusions

The paper presented the results of investigations of buck and boost converters. It was shown that the authors' model of power losses in the ferromagnetic core ensures a better match of the computed values of losses in the inductor with the catalog data for cores made of different materials. Using this model, the dependence of power losses in the inductor core on input voltage, load resistance, and switching frequency of the converters was computed. Taking into account only losses in the core of the inductor, the watt-hour efficiency of converters was computed.

It was observed that the choice of core material could significantly influence the watt-hour efficiency of both considered DC–DC converters. In the cases considered in this paper, by changing the core material, it was possible to achieve up to a 40% change in the value of watt-hour efficiency for the buck converter and the boost converter. It is worth noting that, for each of the considered ferromagnetic materials, there exists a range of converter input voltage, load resistance, and frequency, in which the watt-hour efficiency of the converter containing the inductor with the core made of this material was higher than for inductors with cores made of other materials.

The results of investigations presented in this paper showed that the watt-hour efficiency of buck and boost converters can strongly depend on losses in the inductor cores. The observed influence of core material on the watt-hour efficiency of DC–DC converters can be stronger than the influence of power losses in semiconductor devices. The results of computations presented in this paper were based on very simple models of phenomena occurring in the considered converters, but they allowed isolating only one mechanism of losses in the investigated converters, as well as the examination of its influence on parameters of the considered circuits. The simple model of DC–DC converter characteristics used during computations showed that losses in the inductor core can visibly limit the watt-hour efficiency of buck and boost converters. Of course, the obtained results of investigations showed only the direction of changes in the watt-hour efficiency of these converters while changing selected factors. In order to perform an accurate analysis of these converters, the accurate models of all the converter components should be used.

The results of the performed analyses can be useful for designers of switch mode power supplies allowing a more accurate estimation of the watt-hour efficiency of these circuits and proper selection of the material used for the inductor core. They can also be used to illustrate the influence of selected factors on the watt hour efficiency of DC–DC converters.

Author Contributions: Conception of the paper (K.G.); Methodology (K.G. and K.D.); Computer analyses (K.D.); Writing and editing paper (K.G. and K.D.)

Funding: The scientific work was financed with the National Science Center Poland budget as the investigation project (single scientific study) within the framework of the program "Miniatura 1", project number 2017/01/X/SP7/00441. The project was also financed within the program of the Ministry of Science and Higher Education called "Regionalna Inicjatywa Doskonałości" in the years 2019–2022, project number 006/RID/2018/19, with the sum of financing equivalent to 11,870,000 Polish złoty (PLN).

Conflicts of Interest: The authors declare no conflict of interest.

Nomenclature

B	magnetic flux density
B_m	amplitude of magnetic flux density
d	duty cycle of the signal controlling DC–DC converters
f	frequency
I_0	average value of the converter output current
L	inductance of the inductor
P_R	power losses in the inductor core
P_V	power losses density per unit of volume
P_W	power losses in the winding
R_0	converter load resistance
S_{Fe}	cross-section of the core
T	period of the signal controlling DC–DC converters
T_R	core temperature
V_e	volume of the core
V_{in}	converter input voltage
V_{out}	converter output voltage
z	number of turns of the inductor
ΔI_L	peak-to-peak value of inductor current
η	watt-hour efficiency of the tested DC–DC converter

References

1. Hilal, A.; Raulet, M.A.; Martin, C.; Sixdenier, F. Power loss prediction and precise modeling of megnetic power components in dc–dc power converter application. *IEEE Trans. Power Electron.* **2015**, *30*, 2232–2238. [CrossRef]
2. Vorperian, V. *Fast Analitycal Techniques for Electrical and Electronic Circuits*; Cambridge University Press: Cambridge, UK, 2002.
3. Detka, K.; Górecki, K. Modelling the power losses in the ferromagnetic materials. *Mater. Sci. Pol.* **2017**, *35*, 398–404. [CrossRef]
4. Górecki, K.; Rogalska, M. The compact thermal model of the pulse transformer. *Microelectron. J.* **2014**, *45*, 1795–1799. [CrossRef]
5. Wilson, P.R.; Ross, J.N.; Brown, A.D. Simulation of magnetics components models in electrics circuits including dynamic thermal effects. *IEEE Trans. Power Electron.* **2002**, *17*, 55–65. [CrossRef]

6.   Jin, P.; Li, Y.; Li, G.; Chen, Z.; Zhai, X. Optimized hierarchical power oscillations control for distributed generation under unbalanced conditions. *Appl. Energy* **2017**, *194*, 343–352. [CrossRef]

7.   Duong, M.Q.; Nguyen, H.H.; Nguyen, T.H.D.; Nguyen, T.T.; Sava, G.N. Effect of component design on the dc/dc power converters dynamics. In Proceedings of the 10th International Symposium on Advanced Topics in Electrical Engineering, Bucharest, Romania, 23–25 March 2017.

8.   Gao, Y.; Liu, H.; Ai, J. Novel high step-up DC–DC converter with three-winding-coupled-inductors and its derivatives for a distributed generation system. *Energies* **2018**, *11*, 3428. [CrossRef]

9.   Knott, A.; Andersen, T.M.; Kamby, P.; Pedersen, J.A.; Madsen, M.P.; Kovacevic, M.; Andersen, M.A.E. Evolution of very high frequency power supplies. *IEEE J. Emerg. Sel. Top. Power Electron.* **2013**, *2*, 386–394. [CrossRef]

10.   Dou, Y.; Ouyang, Z.; Thummala, P.; Andersen, M.A.E. PCB embedded inductor for high-frequency ZVS SEPIC converter. In Proceedings of the 2018 IEEE Applied Power Electronics Conference and Exposition (APEC), San Antonio, TX, USA, 4–8 March 2018; pp. 98–104.

11.   Magambo, J.S.N.T.; Bakri, R.; Margueron, X.; Le Moigne, P.; Mahe, A.; Guguen, S.; Bensalah, T. Planar magnetic components in more electric aircraft: Review of technology and key parameters for DC–DC power electronic converter. *IEEE Trans. Transp. Electrif.* **2017**, *3*, 831–842. [CrossRef]

12.   Detka, K.; Górecki, K. Modelling power losses in an inductor contained in the boost converter. In Proceedings of the 2018 IEEE 12th International Conference on Compatibility, Power Electronics and Power Engineering (CPE-POWERENG 2018), Doha, Quatar, 10–12 April 2018. [CrossRef]

13.   Van den Bossche, A.; Valchev, V.C. *Inductors and Transformers for Power Electronics*; CRC Press: Boca Raton, FL, USA, 2005.

14.   Barlik, R.; Nowak, M. *Energoelektronika—Elementy, Podzespoły, Układy*; Oficyna Wydawnicza Politechniki Warszawskiej: Warszawa, Poland, 2014.

15.   Muhlethaler, J.; Biela, J.; Walter Kolar, J.; Ecklebe, A. Improved core—Loss calculation for magnetic components employed in power electronic systems. *IEEE Trans. Power Electron.* **2012**, *27*, 964–973. [CrossRef]

16.   Nakahara, M.; Wada, K. Loss analysis of magnetic components for a solid-state-transformer. *IEEJ J. Ind. Appl.* **2015**, *4*, 387–394. [CrossRef]

17.   Yue, S.; Yang, Q.; Li, Y.; Zhang, C. Core loss calculation for magnetic materials employed in SMPS under rectangular voltage excitations. *AIP Adv.* **2018**, *8*, 056121. [CrossRef]

18.   Reinert, J.; Brockmeyer, A.; De Doncker, R. Calculation of losses in ferro- and ferrimagnetic materials based on the modified Steinmetz equation. *IEEE Trans. Ind. Appl.* **2001**, *37*, 1055–1061. [CrossRef]

19.   Muhlethaler, J.; Biela, J.; Kolar, J.; Ecklebe, A. Core losses under the DC bias condition based on steinmetz parameters. *IEEE Trans. Power Electron.* **2012**, *27*, 953–963. [CrossRef]

20.   Górecki, K.; Detka, K. Analysis of influence of losses in the core of the inductor on parameters of the buck converter. In Proceedings of the 2018 Baltic URSI Symposium (URSI), Poznań, Poland, 14–17 May 2018; pp. 129–132. [CrossRef]

21.   Basso, C. *Switch-Mode Power Supply SPICE Cookbook*; McGraw-Hill: New York, NY, USA, 2001.

22.   Ericson, R.; Maksimovic, D. *Fundamentals of Power Electronics*; Kluwer Academic Publisher: Norwell, MA, USA, 2001.

23.   Steinmetz, C.P. On the law of hysteresis. *AIEE Trans.* **1892**, *9*, 3–64. [CrossRef]

24.   Rashid, M.H. *Power Electronic Handbook*; Elsevier Academic Press: Amsterdam, The Netherland, 2007.

25.   Kazimierczuk, M.K. *Pulse-Width Modulated DC-DC Power Converters*; John Wiley & Sons: Hoboken, NJ, USA, 2008.

26.   Górecki, K.; Detka, K. Application of average electrothermal models in the SPICE-aided analysis of boost converters. *IEEE Trans. Ind. Electron.* **2019**, *66*, 2746–2755. [CrossRef]

27.   Detka, K.; Górecki, K.; Zarębski, J. Modeling single inductor DC-DC converters with thermal phenomena in the inductor taken into account. *IEEE Trans. Power Electron.* **2017**, *32*, 7025–7033. [CrossRef]

28.   Parchomiuk, M. Badanie przekształtnika DC-DC z izolacją transformatorową przeznaczonego do zasilania potrzeb własnych pojazdów trakcyjnych. *IAPGOŚ* **2014**, *4*, 113–116. [CrossRef]

29.   Popescu, M.; Miller, T.; Ionel, D.M.; Dellinger, S.; Heidemann, R. On the physical basis of power losses in laminated steel and minimum-effort modeling in an industrial design environment. In Proceedings of the 42nd IAS Annual Meeting Industry Applications Conference, New Orleans, LA, USA, 23–27 September 2007. [CrossRef]

30. Bertotti, G.; Fiorillo, F.; Soardo, G. The prediction of power losses in soft magnetic materials. *J. Phys. Colloq.* **1988**, *49*, 1914–1919. [CrossRef]

31. Górecki, K.; Detka, K. The parameter estimation of the electrothermal model of inductors. *Inf. MIDEM* **2015**, *45*, 29–38.

32. F867—Material Characteristics. Available online: http://sklep.remagas.pl/images/W%C5%82a%C5%9Bciwo% C5%9Bci_materia%C5%82u_F-867.pdf (accessed on 16 May 2019).

33. Typical Material Properties of Nanoperm. Available online: https://www.magnetec.de/fileadmin/pdf/ catalogue_e.pdf (accessed on 16 May 2019).

34. Epcos Ferrites and Accessories, Data Book 2013. Available online: https://www.tdk-electronics.tdk.com/ download/519704/069c210d0363d7b4682d9ff22c2ba503/ferrites-and-accessories-db-130501.pdf (accessed on 16 May 2019).

35. Rdzenie Toroidalne Typu RTP. Available online: https://feryster.pl/rdzenie-proszkowe-rtp (accessed on 16 May 2019).

energies

MDPI

Article

Structural Analysis of Power Devices and Assemblies by Thermal Transient Measurements

Gabor Farkas [1], Zoltan Sarkany [1,2] and Marta Rencz [1,2,*]

1   Mentor, a Siemens business, Mechanical Analysis Division, Gábor Dénes utca 2, 1117 Budapest, Hungary
2   Department of Electron Devices, Budapest University of Technology and Economics, Magyar tudósok körútja 2, bldg. Q, 1117 Budapest, Hungary
*   Correspondence: rencz@eet.bme.hu

Received: 11 March 2019; Accepted: 3 July 2019; Published: 15 July 2019

Abstract: Power modules composed of semiconductor dice, thermal interface layers, and cooling mounts can be characterized by thermal transient testing at their actual position (in situ). This paper demonstrates that transient testing enables tracking of changes in material quality and structural details on the raw heating or cooling curves. Higher precision can be achieved with the structure function technique where absolute and partial thermal resistance and capacitance values can be used for unambiguous identification of structural elements in a heat conducting path. Measurement techniques are presented to characterize the self-heating of a die and heat transfer between dice. Change of the thermal interface material layers in assembly during test sequences is also highlighted by the structure function concept. The power distribution between dice and wiring is analyzed by the newly introduced "accordion" principle.

Keywords: thermal transient testing; non-destructive testing; thermal testability; in-situ characterization

1. Introduction

With the growing level of power density in up-to-date power devices and assemblies, it has become essential to support their manufacturing with testing tools enabling one to check the integrity of the realized appliances. There are several existing traditional non-destructive analysis techniques, such as x-ray or scanning acoustic microscopy (SAM), but their use is troublesome and may require the assembly to be dismounted, at least partly. There is a growing need to provide an appropriate testing methodology that can be used for in situ analysis of power devices and which is not restricted to checking just a certain part within the assembly. It is also often expected that after verifying the correctness of the design or the integrity of a sample by thermal testing, the product remains functional in its normal operation mode.

In literature we can find ways to use thermal transients induced by power change to characterize a part of an electronic appliance. Recognizing that the thermal transient testing always provides information about the whole heat conducting path from a heat source towards the measurement environment, in this paper we systematically follow the possibilities for analyzing complex structures built from power electronics devices, thermal interface materials, and cooling mounts.

The theoretical principles of structural transient testing are summarized in previous studies, such as [1–3]. In this paper we examine a series of examples of how this technique can be used for analyzing the health of the different structural elements at different locations in the assembly.

In many cases the structural analysis is simplified to condensing the detailed composition of the device into a single characteristic number. Such often used descriptors are the R_{thJC} junction to case thermal resistance for devices with a dedicated cooling surface ("case") or the R_{thJA} junction to ambient thermal resistance for assemblies. These data are useful for narrowing the selection of devices for an engineering task and for "back of the envelope" calculations but do not replace trial measurements and simulations for an actual design.

The well-established framework of thermal measurement standards [4–6], however, helps in planning the structural analysis process as well. These standards represent the condensed knowledge of many decades regarding the selection of assemblies, thermal boundary, electrical excitation, data acquisition, and the interpretation of the measured results.

The traditional standards and methods are highly based on the characteristic features of the silicon material used in power switching and amplifying appliances. Testing of silicon devices in regular packages more or less goes through a checklist. However, when these methods were used for large area power modules, containing several functional dice, some of the results proved to be inappropriate [7]. With the advent of the use of wide band-gap materials, many of the established testing methods have become impractical or even misleading [8].

In this paper, we will introduce cases where a sort of "Design of Experiment" is needed, although not in the broadest statistical sense. Instead, a proper re-interpretation of the standard framework has to be applied, involving trials on powering and sensing currents. In addition, standards give only detailed support for cases when the analyzed part of the assembly is the heat conducting path from the semiconductor to a physical package surface. If the part of interest in the assembly is another structural element, such as a thermal interface material (TIM) layer or a cooling mount, a proper re-definition of the methodology is needed.

It has to be noted that although thermal testing has become more important in achieving reliable operation over a long lifetime, the construction of complete appliances often overlooks thermal testability aspects. Consequently, these tests often need a workaround for accessing devices that are relevant for their power consumption or can be used as sensing points.

In the following sections we follow the steps of how thermal transient testing can be used for structure analysis. First, in Section 2, we present how the thermal transient testing can be accomplished in practice. In Section 3, we discuss how the measured results can be interpreted in structure analysis. Section 4 presents various case studies, with each demonstrating a special interpretation of the structure functions, depending on our knowledge of the actual structure. These examples also allow us to demonstrate how the difficulties in the thermal transient evaluation of these devices can be overcome. Finally, in Section 5 we summarize the findings and give recommendations about using thermal transient testing for structure analysis.

2. Thermal Transient Measurements

Generally, to carry out thermal transient measurements we need one or more heater elements and one or several temperature sensors. In most cases the heat source is a semiconductor, typically called the "chip" in the literature on system design and the "die" (plural dice, sometimes dies) in works on semiconductor technology and packaging. As defined in measurement standards [4,5], in discrete devices that have a solitary die, the powering and sensing can occur on the same device; more precisely, on its dissipating surface. In this way we can record the temperature change of the hottest point in an electronic circuitry.

In thermal measurements the traditional term "junction" is often used for the powered thin material layer of the semiconductor. In many device categories (diodes, insulated gate transistors, thyristors) the heat source that is also serving as the sensor is in fact a pn junction driven into forward operation.

The required sudden power change on a pn junction can be created by switching down from a high I_{HEAT} heating current to a low I_{SENSE} measurement current level (Figure 1).

The sensing current maintains a forward voltage on the device at a low power level. In a calibration step, this voltage can be mapped to the junction temperature. This is achieved by recording the voltage in a thermostat at different temperatures, using the same low current.

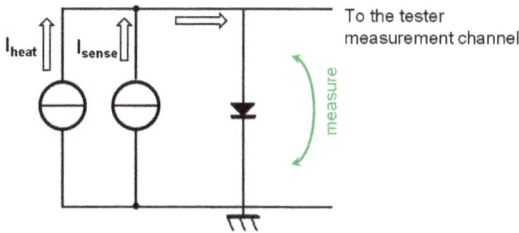

Figure 1. Scheme for the thermal transient measurement of a diode.

In the case of modules, the heating current(s) may flow through several devices, and more dice can be used for sensing, applying a sensor current on each of them. Many details of the powering and temperature sensing principles are given in a previous study [9].

2.1. Test Definition

Structural analysis by thermal transient testing is a robust technique, covering a very wide range of devices. The same principles and practical methods are valid for microelectro-mechanical systems (MEMS), locomotive engine drives, or processors. The range of applied heating currents is milliamperes for MEMS and kiloamperes for engine drives; the time for reaching thermal equilibrium is milliseconds for MEMS and hours for street luminaires, although the same methodology still applies.

As stated above, measurement standards offer appropriate guidelines for a test framework.

The JEDEC JESD 51-14 thermal transient measurement standard prescribes the careful definition of the *device type, environmental conditions, powering,* and *data acquisition* conditions, and *data evaluation* (Table 1 in the reference [6]).

Table 1. Power on the module with steady I_{HEAT} heating current, at cold plate temperature of 25 °C.

	"dry" Coldplate	"wet" Coldplate
I_{HEAT} (A)	P (W)	P (W)
10	9.3	9.3
20	19.7	20.0
30	31.2	31.7
40	43.5	44.5

Thermal testing and the evaluation of the results could be treated in a purely theoretical way, but it is easier to understand the concepts through a practical example. In this work, we present measurements on six quite different devices, each of which could be used as device type to expose the general principles. We present now the test of a typical high current module, a solid state relay. We have already used this device as a workhorse to present the thermal management concepts because of its ruggedness and flexibility in other works [10,11], but with a different focus. Now, after using it as an example for introducing the basic measurements in Section 2, we concentrate on its special features for the uses of a high sense current (Section 3.3), low heating current (Sections 3.3 and 4.6), and the roughness of its base plate, which necessitates a curing step when used with thermal interface pastes (Section 4.5).

The test definition is as follows.

Device type: standard Siemens AC solid state relay module, type 3RF2190. The module contains two thyristors (Silicon Controlled Rectifier (SCR)) and some control circuitry on a ceramic direct bonded copper (DBC) substrate. A simplified schematic of the device is shown in Figure 2a.

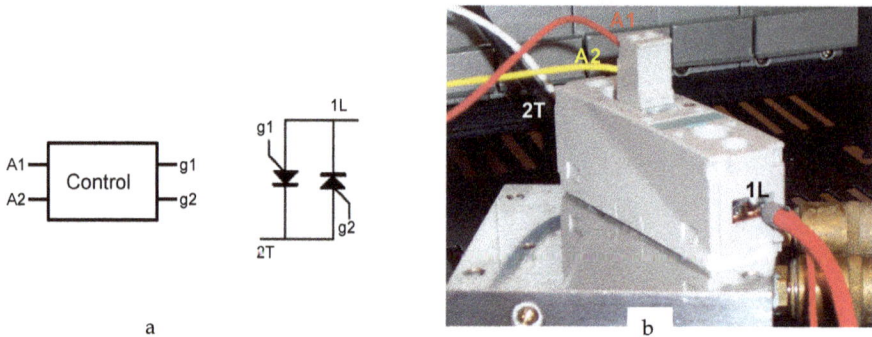

Figure 2. (**a**) Schematic of the solid state relay module; (**b**) sample on cold plate, with leads attached.

Assembly and environment: The samples were fixed on a water cooled cold-plate, with different thermal interfaces underneath. In order to distinguish between the portions of the heat conducting path belonging to the device, the TIM, and the cooling mount, we followed the JEDEC JESD 51-14 standard [6], fixing the samples first on a dry surface and then on the plate wetted with standard thermal grease (Figure 2b).

2.2. Determining the Current Levels and the Timing for the Test

In the daily practice of many laboratories, the applied heating and sensing currents are determined by experience with formerly tested low-power, discrete devices, by the limited range available with older thermal testers with poor specification, and also by some misconceptions.

Thermal testers can produce a power step on a semiconductor structure by a sudden change of voltage or current at some pins. In the case of thyristors, the regular way of heating is to apply a steady current between the anode and cathode for a prolonged time. More sophisticated concepts of power programming are detailed in a previous study [9].

In this test, the actual power that develops on the device depends on the current applied, and also on the actual thermal boundary. Better cooling lowers the die temperature, resulting in a higher forward voltage at the same heating current, and therefore in higher power on the device.

Applying different I_{HEAT} currents, we measured the heating power (I_F forward current times V_F forward voltage) listed in Table 1.

An important step in the design of this test is selecting the proper sense current based on trial measurements.

In the scheme of Figure 1, we recorded cooling transients on one thyristor in the module. The cooling occurred at different sense currents after switching off several amperes of heating current.

Figure 3 shows the recorded transient change of the V_F forward voltage at $I_{SENSE} = 0.2$ A and $I_{SENSE} = 2$ A sense current levels.

We can observe that as 0.2 A is still near to the hold current threshold of the thyristor, it "attempts" to turn off and recovers around 1 ms only. With $I_{SENSE} = 2$ A we can expect pure thermally induced transients after 100 µs. The actual "thermally induced" nature of a transient can be verified by comparing a set of measurements at different heating and sensing currents. One can claim that the root cause of the voltage change was the cooling of the device in that time interval, where the curves normalized by the power change coincide (see later Figure 13).

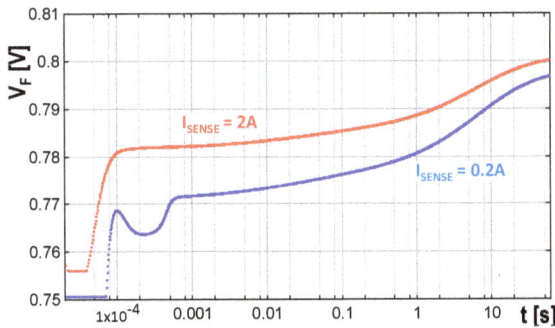

Figure 3. Transient change of the V_F forward voltage on a thyristor in the module of Figure 2 during cooling at different sense current levels.

2.3. Transient Test of the Module

With trial measurements similar to the one shown in Figure 3, we found that a thermal steady state can be reached within one minute on the water cooled cold plate.

Below we illustrate the test and its evaluation on an exemplary measurement at 40 A heating current and 2 A sensing current. A related study with more examples was conducted previously [10].

An examination on the validity of using other heating and sensing current values will be given below in Section 3.3.

At these currents we experienced crisp, noise-free thermal transients (shown below in Figure 5).

The calibration (Figure 4) showed that the temperature dependence of the forward voltage is quite linear at the 2 A sensor current, the slope of the plot is $S = -1.645$ mV/K. This mapping scales the temperature axis of Figure 5.

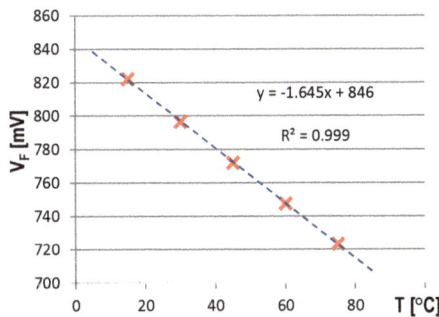

Figure 4. Voltage to temperature mapping at sense current.

In the testing of devices of extreme die size (processors, for example) or devices made of wide band gap semiconductors, the voltage to temperature mapping is nonlinear. Up to date evaluation software products can store the calibration chart and can calculate the absolute temperature value during a transient from the actual voltage value [12]. It is important to emphasize that in further sections of the paper we assume linearity of the thermal parameters of the assembly (thermal conductivity, specific heat) but no linearity of the electric behavior or the voltage-temperature mapping is needed.

The figure reveals that approximately 54 °C temperature elevation occurs at the "dry cold plate" boundary at 40 A/43.5 W heating (curve DRY 40A to 2A). Using thermal grease the temperature change drops to 32 °C (curve WET 40A to 2A; the figure keys correspond to these boundary conditions in subsequent figures).

An elaborated description of this experiment and its evaluation is given in a previous study [11].

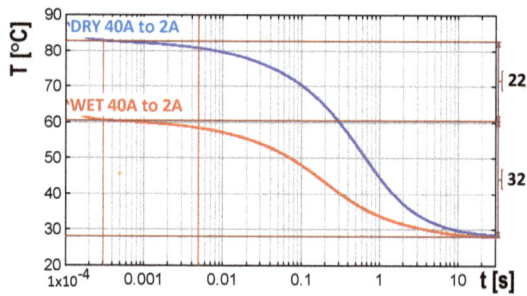

Figure 5. Measured cooling transients at "dry" and "wet" boundary conditions (from [11]).

3. Structure Analysis

3.1. Z_{th} Curves

In Figure 5 we see many details already. Based on former works we can attribute the temperature change in the millisecond range to heat propagation in the die and through the die-attach, in the second range to the cooling mount, in the minute range to heating of the circulated water, etc. This plot, however, characterizes the heat conducting path only at the given powering.

A more general portrayal of the thermal behavior of an assembly is based on the theory of linear time invariant (LTI) systems. While the electronic devices are strongly non-linear in their electric characteristics, the thermal conductivity and specific heat of the device components and of the measurement environment shows only a minor change in the typical temperature range of use. This means that by shifting the base plate temperature we obtain similar recorded curves and by altering the applied power we obtain similar, proportionally magnified records.

Reducing the cooling curves to temperature *change* only and fitting them at their hottest point (Figure 6), we find that the cooling is not influenced by the actual boundary condition until 0.2 s, and the curves coincide perfectly. This can be explained by stating that until 0.2 s the heat propagates inside the package, and it still did not reach the air/grease thermal interface.

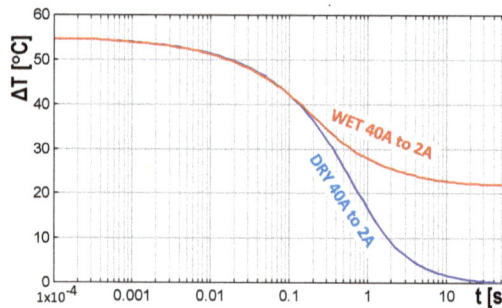

Figure 6. Component on high and low conductance boards, cooling curves fit at hot point (from [11]).

Normalizing the cooling curves with the applied power we obtain the Z_{th} *curves.* (They are sometimes called "thermal impedance", but in a strict interpretation an impedance is defined in the frequency domain.) As a fairly accurate temperature transient for any power step can be produced if we multiply the Z_{th} curves by the actual power, this curve is used frequently for the characterization of the thermal behavior.

This concept of proportionality to power is not fully accurate. At increased power level and higher temperature elevation, the cooling mechanisms (turbulent convection, radiation) become more intensive; the real temperature change is lower than the one extrapolated from the multiplied Z_{th} curve. As such, using Z_{th} for temperature estimation remains on the safe side.

A deeper analysis of nonlinear effects is given in another study [3].

Dividing the curves in Figure 6 by the corresponding negative power step from Table 1 (around −44 W), we get Figure 7.

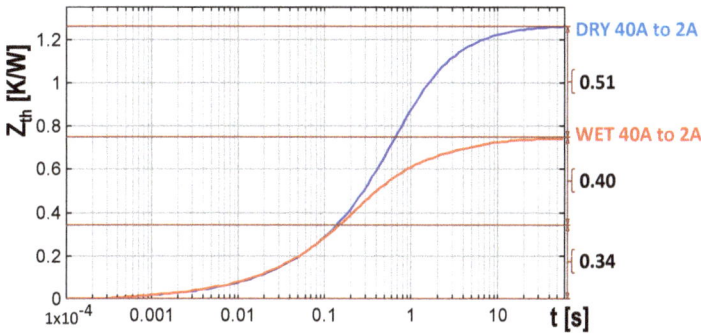

Figure 7. Z_{th} curves, component on cold plate at "dry" and "wet" boundary conditions (from [11]).

We can observe in the plot that until approximately 0.2 seconds and 0.34 K/W, the Z_{th} curves coincide because the heat still proceeds in the internal structures of the device. The gradual divergence of the curves indicates that not all trajectories of the heat flow arrive at the same time to the case surface. A way to give a more definite determination of the separation point will be highlighted in the next section.

The arrangement shows 0.74 K/W total thermal resistance with the "wet" and 1.25 K/W with the "dry" thermal interfaces.

3.2. Structure Functions

A much more informative representation of the assembly can be derived from the transients, constructing an equivalent RC circuit of thermal resistances and capacitances, producing the same Z_{th} as the assembly, at the same excitation.

Actual thermal systems correspond to a complex three-dimensional RC net. In the case of a single heat source, it can be treated as being excited by $P(t)$ power changing in time at a node in the net and terminated by constant temperature at the boundaries. LTI theory claims that such a system behaves in the same way as a reduced set of thermal resistances and capacitances arranged in one of the configurations shown in Figure 8. (In an analogous electric RC network the P power applied on the input equals to an I current and the temperature change in the net corresponds to V voltage on the nodes.)

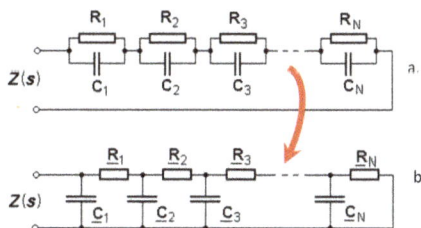

Figure 8. Foster (**a**) and Cauer (**b**)-type representations of a 3D thermal RC net (from [11]).

If constant P_{on} power is switched on at zero time (step function), a single series RC element in Figure 8a produces an exponential growth after switching on the power (current), adding $T(t) = P_{on} R \cdot (1 - e^{-t/\tau})$ temperature (voltage) term to the chain response. The R_i *magnitude* denotes the thermal resistance of the *i*th fragment; $\tau_i = R_i \cdot C_i$ is a *time constant*, where C_i is the thermal capacitance. Accordingly, at the input (in this case at the junction) we get a *sum* of exponential functions,

$$T(t) = P_{on} \cdot \sum_{i=1}^{n} R_i \cdot \left(1 - e^{-t/\tau_i}\right) \tag{1}$$

The Network Identification by Deconvolution (NID) method [1,6] provides a systematic methodology to produce several hundred RC stages, resulting in a highly accurate approximation of the original Z_{th} curve. The time constants distilled from the curves of Figure 7 are shown graphically as an $R(\tau)$ time constant spectrum in Figure 9.

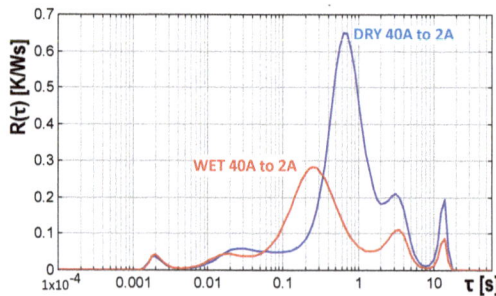

Figure 9. Time constant spectrum calculated from the Z_{th} curve of Figure 7.

The information in time constant spectra is hard to interpret. A more usable representation of the system can be gained converting the RC stages of Figure 8a into an equivalent RC ladder of Figure 8b, as demonstrated in previous studies [1,6,13]. Instead of providing hundreds of R and C values in tabular format we can construct a graphic representation, the *structure function*.

In this chart (Figure 10) we sum up the thermal resistances in the ladder, starting from the heat source (junction) along the x-axis and the thermal capacitances along the y-axis. (The Σ sign in the axis keys refers to the cumulative nature of the function, and to the fact that ladder elements are added up along the axes.)

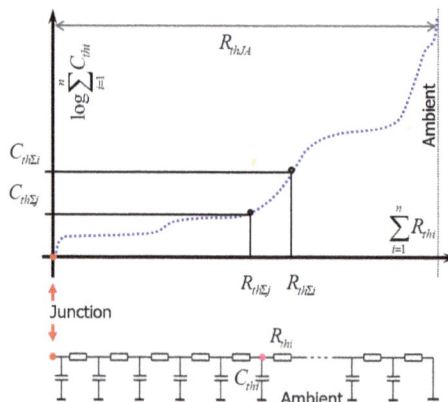

Figure 10. Structure function: the graphic representation of the thermal RC equivalent of the system (from [11]).

Thermal capacitance is proportional to the mass and volume of a material layer through its specific heat and density. *Low gradient sections* in the chart mean that a small amount of material having low capacitance causes large change in the thermal resistance. These regions have *low thermal conductivity* or *small cross-sectional area*. *Steep sections* correspond to material regions of *high thermal conductivity* or *large cross-sectional area*, as even a large bulk of material corresponding to high thermal capacitance is of low thermal resistance only. Sudden breaks of the slope belong to material or geometry changes. Thus, thermal resistance and capacitance values, geometrical dimensions, heat transfer coefficients, and material parameters can be directly read on structure functions.

3.3. Examination of the Magnitudes for Valid Heating and Sensing Currents

The widespread misconceptions in the thermal management community regarding the required current levels during transient testing can be summarized in the following two points:

- The heating current has to be high enough to reach at least 30 °C to 50 °C temperature elevation, otherwise the measurement will be of limited accuracy;
- The sensing current has to be kept low in order to avoid the "self-heating" of the die in the cooling phase.

The first postulation is simply projected from the poor specification of old thermal transient testers. Current test equipment has 10 µV voltage and 0.01 centigrade temperature resolution, which enables it to record a temperature swipe of 5 or 10 centigrade at low noise and full accuracy.

Regarding the second point, in the previous measurements we captured *the temperature of a single point only*, on a single data acquisition channel of the test equipment.

This technique is compliant with the present thermal transient testing standards [6] and is advantageous in many ways. First of all, all channel offset and gain errors of the equipment are cancelled-out during the calibration process of Section 2.3, so the only source of error can be the inaccuracy in the thermostat temperature. On the other hand, as the change of the R_{th} thermal resistance is negligible in the span of the measurement, a higher sensor current just adds a constant (small) temperature shift to the actual transient, not influencing the calculations based on the *change* of the temperature.

The source of the misconception regarding the allowed sensing current is an incorrect interpretation of the standards [4,5]. If the temperature difference at the junction is derived from the temperature of two separate points *in space*, e.g., from the voltage of the hot junction and some data from a different temperature sensor at a distant point, and moreover if these are measured on different tester channels, then:

- We add up offset and gain errors of the tester channels; and
- The temperature surplus caused by higher sensing current really adds to the measurement error.

Sudden change of the R_{th} thermal resistance can occur in special cases, e.g., at phase change in the thermal interface materials. This may undermine the accuracy of the one-point measurement, but then the methods based on temperature measurements at distant points are equally invalid.

As an example, we can calculate the extent of the measurement error in the test presented in Section 2. The equipment recorded $V_{F1} = 1.1$ V forward voltage on the "hot" device at $I_{HEAT} = 40$ A heating current, with $I_{SENSE} = 2$ A added after a longer stabilization period. This resulted in $P_1 = (I_{HEAT} + I_{SENSE}) \cdot V_{F1} = 46.2$ W power on the device just before switching off the heating.

After switching off the heating, the device forward voltage at $I_{SENSE} = 2$ A started at $V_{F2} = 0.75$ V. Then, in 60 seconds this forward voltage grew by approximately 50 mV, to $V_{F\infty} = 0.8$ V. Just after switching off there was $P_2 = I_{SENSE} \cdot V_{F2} = 1.5$ W power on the device. The power step is $P_1 - P_2 = 44.7$ W.

This calculation differs slightly from the actual measured power step in Table 1 due to rounding.

The only real source of inaccuracy that has an effect other than a constant shift in the junction temperature is the slow change of power during cooling, at the end of which the power grows to

$P_\infty = I_{SENSE} \cdot V_{F\infty} = 1.6$ W, and the error in powering during the full cooling period is $P_\infty - P_2 = 0.1$W; compared to the power step this is 0.224%.

Table 2 summarizes the device temperature and powering error based on measured and calculated voltages at fixed heating and different sensing currents. The voltages and temperatures of row **a** corresponding to 2 A sensing current are taken from the actual measurement. Row **b** and **c** contain V_F (I_F) forward voltage data calculated from the exponential characteristics of a pn junction at low current (Shockley equation), which claims that an 1:10 shrink of a (low) forward current causes 60 mV voltage shrinkage at room temperature. Temperatures in rows **b** and **c** are calculated from the R_{th} thermal resistance of the assembly and the S voltage-to-temperature sensitivity of the device, as identified in Section 2 above.

Table 2. Calculated voltage, temperature, and power on the solid state relay device at several heating and sensing currents.

	R_{th} = 0.74 K/W			S = 1.65 mV/°C				
	I_{HEAT}	I_{SENSE}	V_{F1}	V_{F2}	$V_{F\infty}$	P_1	P_2	P_∞
	A	A	V	V	W	W	W	W
a	40	2	1.1	0.75	0.8	46.2	1.5	1.6
b	40	0.2	1.1	0.69	0.74	44.22	0.138	0.148
c	40	0.02	1.1	0.63	0.68	44.022	0.0126	0.0136

	$T_1 = P_1 \cdot R_{th}$	$T_2 = P_2 \cdot R_{th}$	$P_1 - P_2$	$T_1 - T_2$	$P_2 - P_\infty$	Error: $(P_\infty - P_2)/(P_1 - P_2)$
	°C	°C	W	°C	W	%
a	34.19	1.11	44.7	33.08	–0.1	0.224
b	32.72	0.102	44.082	32.62	–0.01	0.023
c	32.58	0.009	44.009	32.57	–0.001	0.002

Figure 11 below demonstrates the absolute error of the calibration in the 10 °C to 90 °C temperature range, at 2 A, 1 A, and 0.2 A sense currents. The maximum error over 80 °C temperature change is 0.2 °C.

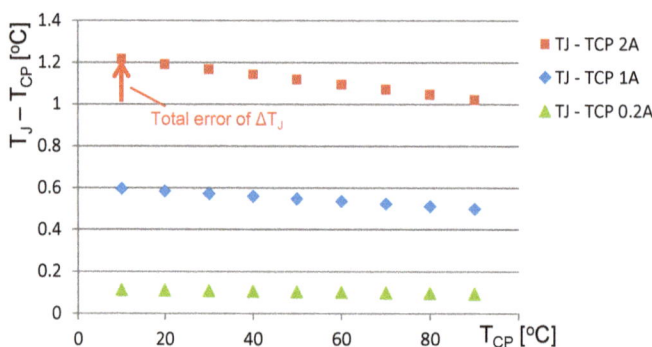

Figure 11. Difference of the actual T_J junction temperature of the device presented in Section 2.1 and the T_{CP} cold plate temperature during calibration.

In the case when an exact T_J junction temperature has to be achieved at high heating current, a corrective step in the cold plate temperature regulation can be performed to reach precisely the target temperature. This occurs at the combined thermal and radiometric and photometric measurements of high-power LED devices, where the optical characterization takes place in the thermal steady state at accurate junction temperature before initiating the cooling. The R_{th} thermal resistance of the assembly can be determined for this correction by transient measurements before starting the actual test [14].

4. Case Studies

Below we present a series of examples where changes in the whole heat conducting path can be identified with the help of structure functions in a different manner.

4.1. Structure Functions of the Solid State Relay Module

Converting the Z_{th} curves of Figure 7 to structure functions we gain Figure 12. In this figure and in other figures below, the steep sections (high change of thermal capacitance along small thermal resistance) belong to silicon or metal layers in a laminate structure, and flat sections (low change of thermal capacitance along large thermal resistance) are thermal interface materials (TIM). Each dot on the curve represents a thermal resistance–thermal capacitance pair.

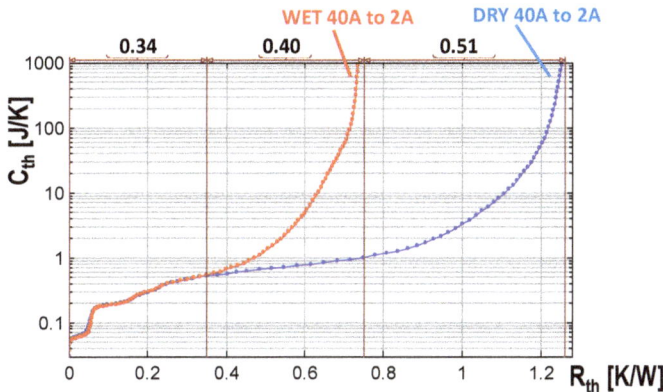

Figure 12. Structure functions: component on a cold plate at "dry" and "wet" boundary conditions (from [11]).

Up to 0.34 K/W the steps in the function indicate the sandwich-like *internal* structural details, such as die, solder, and ceramics (DBC). The physical dimensions, volumes, and distances can be read in the chart if some material parameters are known, or in other cases, the thermal conductivity and specific heat can be determined if the geometry is known. After the junction to case separation point, we see the heat spreading in the grease and the cold plate.

Thermal capacitance values above C_{th} = 1000 J/K are not shown in the figure. Already 100 J/K thermal capacitance would correspond to 4000 mm^3 volume of the aluminum cold plate used in the measurement arrangement, 1000 J/K would correspond to 40000 mm^3 of aluminum. This volume of aluminum is not present in the proximity of the device, so this C_{th} range belongs rather to the capacitance of the water flow in the plate. In this way the structure function portion depicting higher capacitance is not informative regarding the device under test.

The air gap on dry surfaces adds 0.51 K/W to the junction to ambient total thermal resistance.

It has to be mentioned that the structure function analysis is not a "black box" technique. There are three ways to assign actual assembly components to sections in the structure function. These are:

1. The manufacturer of the device may know all internal geometries and material parameters. In such a way a "synthetic" structure function can be built up, for example, superposing slices of material with given thermal resistance and capacitance in a spreadsheet tool, and comparing the measured structure functions to it;

2. An approximate model can be built up in a finite element or a finite difference simulation tool, such as [15]. Thermal transients can be simulated in the tool and structure functions can be composed of those. Geometry and material parameters can be tuned until the simulated and measured structure functions match;

3. Measured structure functions can be compared to an already identified "golden device". This technique is advantageous in production control.

In the case of simulation, the environment farther from the device does not need detailed modeling, as it is often replaced by a surface with a constant heat transfer coefficient (HTC) applied.

Air or liquid cooling is nearly proportional to the area where it is applied.

In the actual measurement the cooling occurs on the copper coated ceramic base plate of the module. The physical samples and also the module geometry in the data sheet indicate approximately 20 mm × 40 mm (0.0008 m^2) as the effective contact area.

From the partial thermal resistance values after the divergence point in Figure 12 we can estimate the quality of the heat sink. We can calculate HTC = 1/(0.4 K/W · 0.0008 m^2) = 3125 W/m^{2}K for the "wet" surface and HTC = 1/[(0.4 K/W + 0.51 K/W)· 0.0008 m^2]= 1373 W/m^{2}K on the "dry" surface. These values are near to the ones published in the literature.

4.2. Measurement of a Large Digital Signal Processor

It is a really challenging task to measure the thermal performance of large processor, memory, or gate array circuits. Their size can be of several cm^2, the die is thin, and the package is optimized for low thermal resistance.

In this example we present the typical obstacles experienced in the measurement of processor devices. These are typically powered and measured in a thermal test on their substrate diode, which covers the whole silicon surface and can be reached by reversing the supply and ground of the circuit.

These devices have separate supplies for their core, internal storage, and peripheries. The different functional parts in the chip and the isolation among them form complex pnpn structures. These parts share different current bias levels in different proportions and tend to inject charge into each other in a thyristor-like manner.

In the actual test a large, experimental digital signal processor (DSP) circuit was measured. The die of the circuit was 20 mm × 20 mm size and was encapsulated in a ball grid array (BGA) package with soldered die-attach, ensuring low thermal resistance.

To obtain a valid thermal signal we had to carry out experiments in a wide I_{SENSE} sensing current range starting from milliamperes up to several amperes. To arbitrate the quality of the result we normalized the measured transients with the applied power, but without correcting the initial electric transient of the signal. In this way we produced "quasi Z_{th} "curves, having no thermal nature in their initial portion (Figure 13).

We found that even in the sensing current range of 0.5 A to 2 A, a time variant redistribution of charge occurs until several milliseconds. At 4 A sensor current we experienced a clean thermal signal from 200 µs onwards.

This measurement also supported the general observation that the longest time interval where a transient of true thermally induced nature can be taken can be achieved with a high sensing current and low heating current. This approach minimizes the electric effects at the current level change, and does not diminish the accuracy of the measurement, as proved in Section 3.3.

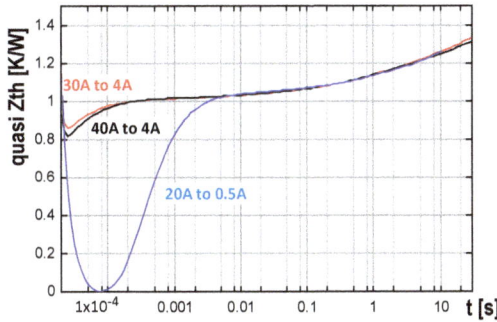

Figure 13. The "quasi Z_{th}" curves of the measurement on the experimental Digital Signal Processor device and normalized transients without correction of the initial electric portion, at different I_{HEAT} and I_{SENSE} levels.

4.3. Die-Attach Analysis

In this study we compare three MOSFET samples (S1, S2, and S3) from the same manufacturing batch. Measuring their structure functions on a cold plate we gained Figure 14.

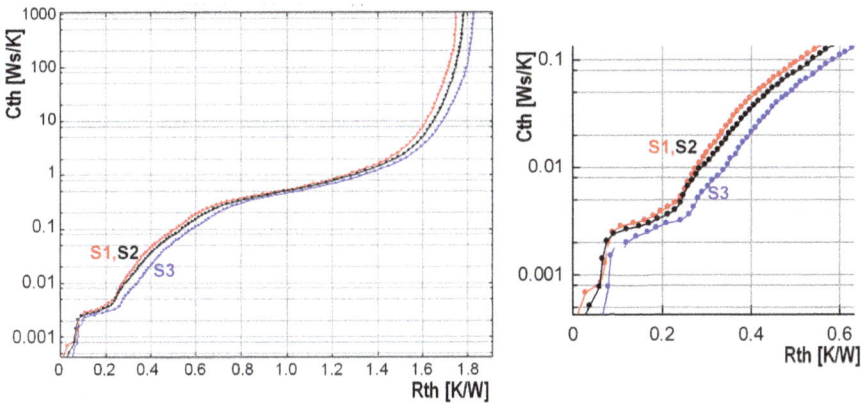

Figure 14. Die-attach comparison of three MOSFET samples (S1, S2, S3).

Sectioning the samples after the measurement revealed that the volume of the silicon die is approximately 2 mm³, having an effective thermal capacitance of approximately 3.2 mJ/K, calculated from the specific heat of the material. We find this C_{th} value around 0.1 K/W in Figure 14. The structure function is very flat until 0.25 K/W, i.e., in this section there is a material of low thermal conductance, in our case a soldered die-attach with less than 20% of the silicon's thermal conductivity.

The end of the die-attach region is clearly indicated by a sharp increase of the structure function at around 0.2 K/W. Differences between the samples develop in the die-attach region and all curves run parallel afterwards.

We verified by scanning acoustic microscopy (SAM) pictures as shown in Figure 15 that S3 (darker in Figure 15) has a thicker solder layer than the other two. The white spots correspond to solder voids, but they have minor importance regarding the thermal resistance.

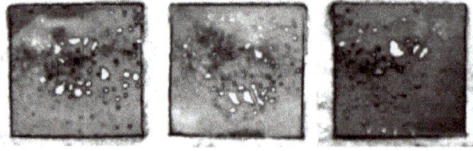

Figure 15. Scanning acoustic microscopy images of the die-attach solder of S1 (left), S2 (center), and S3 (right).

As the die-attach is the most critical part of any electronics assembly from a reliability point of view, thermal transient testing and structure function analysis are used most frequently for die-attach testing.

4.4. Design of the Package or Module Base Plate

In this study we show a measurement example with identical dice encapsulated in packages of different base plate thickness.

The standard LM337 voltage stabilizer exists in TO-220 packages with single gauge (d = 0.51 mm) and double gauge (d = 1.3 mm) tab thickness construction (Figure 16). It is hard to predict whether a thinner heat spreader performs better or worse if the device is mounted on a heat sink.

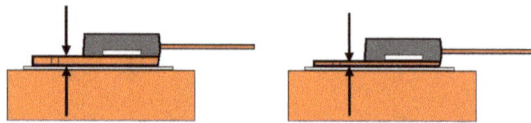

Figure 16. TO-220 package on cold plate: double gauge and single gauge.

Figure 17 already shows that the single gauge solution (LM337 single curve) is of slightly worse thermal performance; however, the reasons remain unclear.

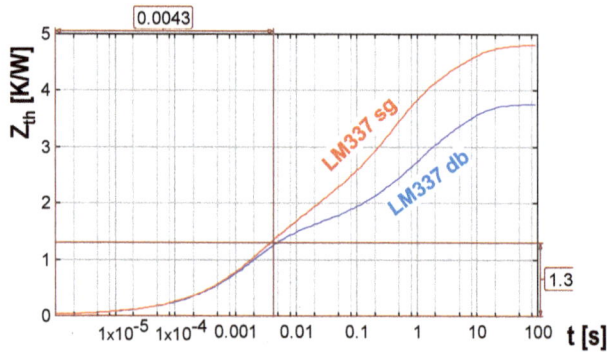

Figure 17. Z_{th} curves of stabilizer IC in TO-220 package: single gauge and double gauge.

Figure 18 demonstrates that up to 1.3 K/W the heat propagates internally in the die and die-attach. The divergence point of the Z_{th} curves indicates that at 4.3 ms the trajectories reach the inner surface of the copper tab. We can observe the quick elevation of the thermal capacitance in the double gauge tab, corresponding to a broad heat spreading cone in which the heat crosses the die to tab interface, resulting in a total of 3.8 K/W junction to ambient thermal resistance, as opposed to 4.8 K/W for the single gauge.

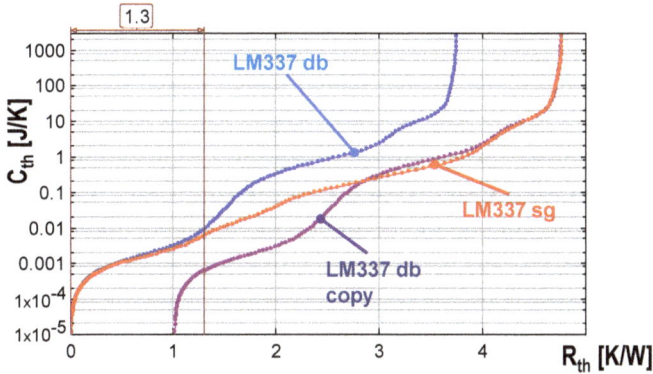

Figure 18. Structure functions of stabilizer IC in TO-220 package: single gauge and double gauge.

Making a copy of the double gauge structure function and shifting it to the ambient end in Figure 18, we can observe that between 4 K/W and 4.8 K/W the curves coincide, as in both cases this final portion of the structure functions already belongs to the measurement environment. The partial resistance of 0.8 K/W belonging to the cold plate can be determined.

4.5. Discovering Run-in and Curing Effects in the TIM Layer at Power Cycling

Let us now turn back to our original assembly of Section 2. The standard junction to case thermal resistance measurement (Sections 2 and 3) already marked out the expected junction to ambient thermal resistance variation with "good" and "poor" thermal interface quality.

Applying a soft thermal paste on the module base we experience a significant improvement in the thermal behavior compared to the "dry" case. With repeated 44 W power pulses of 60 s length, we see a continuous improvement (run-in) of the TIM layer. The curves from TIM.01 to TIM.30 (here the last two numbers give the cycle numbers) show the thinning of the paste due to temperature and pressure.

After 20 cycles the layer stabilizes and we reach the "wetted by thermal grease" boundary (Figure 19). Similar cases are discussed in detail in previous studies [16,17].

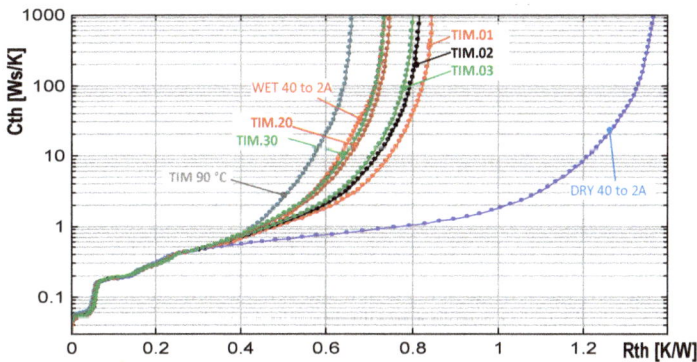

Figure 19. Run-in and curing effect of a soft thermal paste.

A further improvement could be reached by curing the soft thermal paste at a higher temperature. The module was fixed for 24 hours on a plate kept at 90 °C, with its screws pulled at the 0.9 Nm torque proposed in its data sheet. The paste becomes runny and thin at this temperature and pressure, and fills the rough microstructures of both surfaces. The improvement can be well seen in Figure 19 (TIM 90 °C curve).

At this point, we have to define the expected power range that has to be applied during the test. Having a sufficiently sensitive thermal tester (as in the reference [12]), we can see the structural elements in the assembly already at a power level that results in around 4 °C temperature elevation on the die. However, for studying changes within the TIM, especially phase change effects, a real application power level is needed. For accelerated reliability testing, a higher-than-nominal power has to be applied [16,18].

This result emphasizes the importance of proper initial handling of TIM layers in critical applications. A well-designed priming power sequence or previous curing of the assembly at higher temperature can prevent early failure of a device due to overheating when first switched on.

4.6. Testing of Power Modules at Different Heating Currents

We are accustomed to the experience that in the case of discrete power devices, the Z_{th} curves fit perfectly at all power levels, because the internal wiring is optimized for low voltage drop, resulting in low power loss on the wires. An equally good low power loss characterizes the BGA packaged devices, similar to the one presented in Section 4.2.

At power switching modules the internal wiring is more intricate and some compromises cannot be avoided. For this reason we typically see a shrinking and growing "accordion" effect in the Z_{th} curves (Figure 20) and also in structure functions.

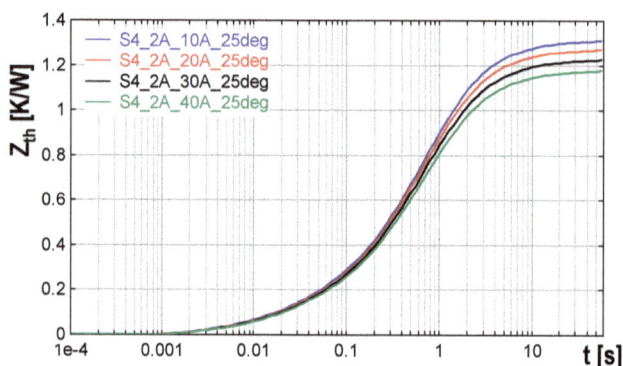

Figure 20. Z_{th} curves calculated from powers of Table 1 at cold plate temperature of 25 °C.

In Table 1 we summarized the steady state power on the solid state relay of Section 2 when heated by different currents. Figure 20 demonstrates the Z_{th} curves recorded at 10 A to 40A heating current and at 25 °C cold plate temperature.

During the cooling we record the correct die temperature. When composing Z_{th} curves or structure functions, we divide by the power which is measured across the whole module, including the portion dissipated in the internal wiring. Supposing that we can neglect this power component at 10 A current, Figure 20 indicates that at 40 A, already 13% of the heating occurs away from the die.

With an appropriate correction factor we can fit the Z_{th} curves again. This factor helps estimation of the actual load on the internal wiring. A scheme of the internal wiring is hinted in Figure 27, for example.

In future test standards it has to be contemplated how appropriate correction factors can be introduced.

4.7. Heat Transfer in Modules between Dice

We have measured a number of power switching modules and studied the heat transfer from one die to its neighbors. Now, we present measurement results on a typical half bridge module, SKM600GB126D, built of two insulated gate bipolar transistors (IGBT).

This module was also mounted on a water-cooled cold plate (Figure 21a), with both "dry" and "wet" surface qualities. As there are two IGBTs and two reverse clamping diodes in the module (Figure 21b), a number of self-heating and transfer-heating configurations can be measured. Now, we limit the test to examining the cross-coupling of the IGBTs.

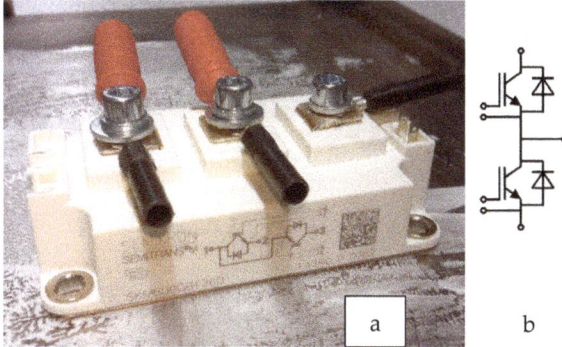

Figure 21. IGBT module on a cold plate on a wet surface (**a**); circuit scheme (**b**).

Applying a high enough V_{GE} voltage on the gate, the IGBT exhibits diode-like "saturation" characteristics between its collector and emitter. A single device can be well measured in saturation mode, as suggested in Figure 22. Self- and transfer-heating can be measured simultaneously, as suggested in Figure 23, maintaining a sensor current on both devices T1 and T2, but heating only T1.

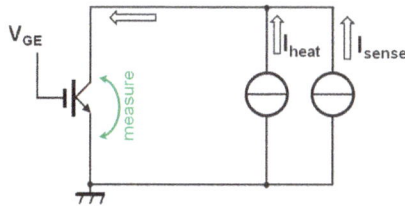

Figure 22. Measurement of an IGBT in saturation mode.

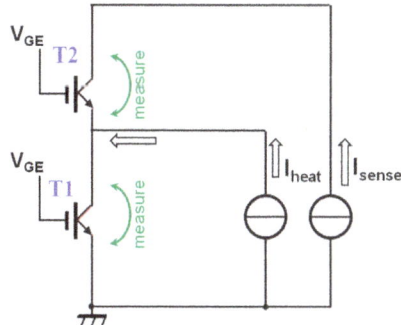

Figure 23. Measurement of an IGBT module with the lower T1 device driven and with self- and transfer-heating measurement.

The obtained Z_{th} curves shown in Figure 24, namely, Z_{11} representing the self-heating and Z_{12} corresponding to the heat transfer from T1 to T2, reveal that the runtime between the two IGBT dice is slightly more than one second, here Z_{12} starts to elevate.

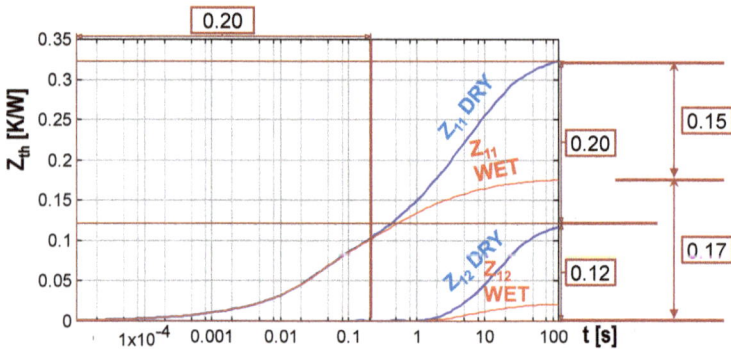

Figure 24. Self- and transfer-heating Z_{th} in the IGBT module ("dry" curves are thick, blue lines, "wet" curves are thin, red lines).

After 100 s the temperature stabilizes, Z_{11} reaches the R_{11} junction to ambient "self" thermal resistance, and Z_{12} reaches the R_{12} "transfer" thermal resistance.

The blue (thicker) Z_{th} curves belonging to the dry cold plate boundary indicate a junction to ambient thermal resistance of $R_{11} = 0.32$ K/W for the self-heating of T1. The heat transfer from T1 to T2 is represented as $R_{12} = 0.12$ K/W. Measuring the full array we find that the dice are of the same size, $Z_{11} \approx Z_{22}$, and the heat transfer is approximately symmetric, $Z_{21} \approx Z_{12}$. However, as the IGBTs are not of the same size as the reverse diodes, their thermal resistance is different.

Figure 25 repeats the previous chart but now with the "wet" curves highlighted in red, thicker curves. For the "wet" case we obtain $R_{11} = 0.17$ K/W thermal resistance for T1, and a transfer of $R_{12} = 0.02$ K/W. This means that the good thermal contact of the base "decouples" the dice and we gain not only serious reduction of the thermal resistance but also much lower thermal stress resulting from the heating of other devices.

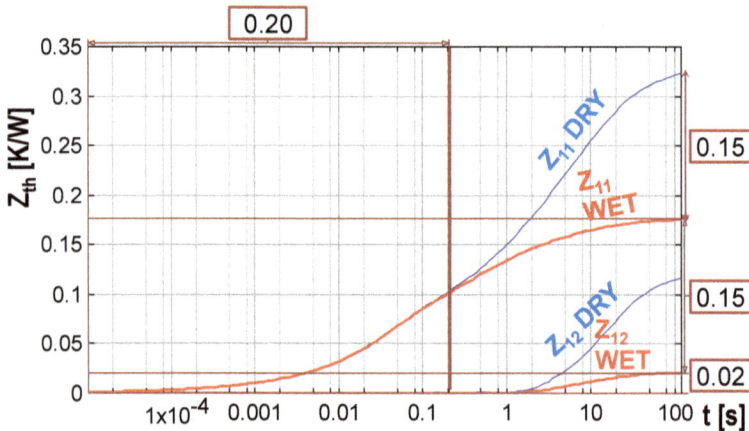

Figure 25. Self- and transfer-heating Z_{th} in the IGBT module ("dry" curves are thin, blue lines, "wet" curves are thick, red lines).

Figure 26 demonstrates the structure functions of T1. We can identify a junction to case thermal resistance of 0.1 K/W. The junction to ambient resistance improves from 0.32 K/W to 0.17 K/W, an improvement of 0.15 K/W, or 46%.

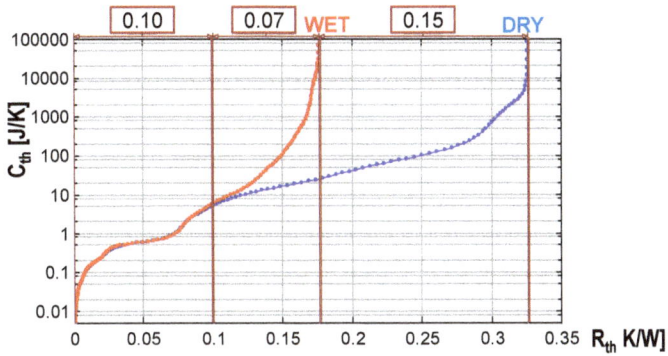

Figure 26. Structure functions of the lower T1 IGBT on wet and dry surfaces.

4.8. Testability of an Automotive Engine Control Unit Module

The circuit scheme of an automotive Engine Control Unit (ECU) module is shown in Figure 27. When attempting to test it, we noticed that there is some access to most switching devices in the module. Unfortunately, this is rather an exception than a rule for power modules.

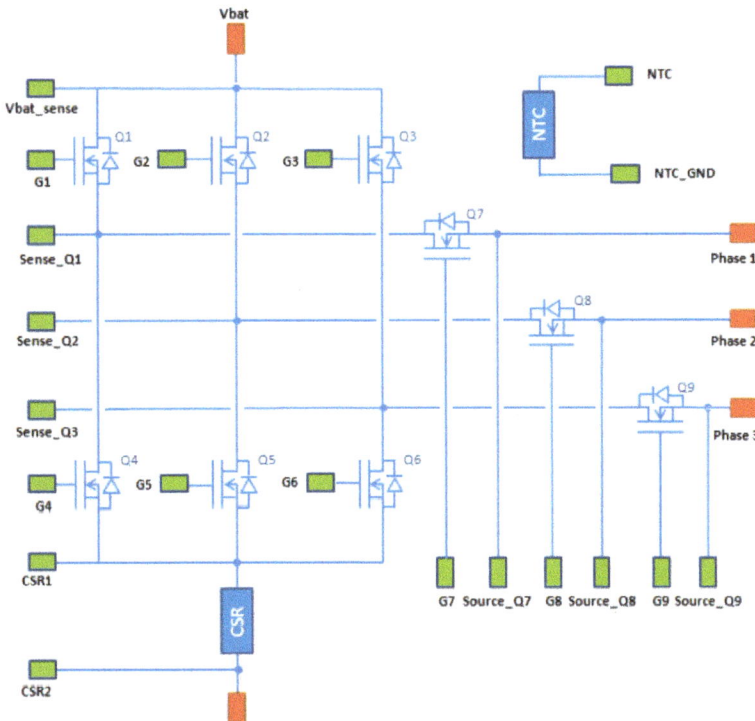

Figure 27. Circuit scheme of an automotive Engine Control Unit module.

For structural analysis we need a relatively low power. For example, for the MOSFET devices in the upper row, namely Q1, Q2, and Q3, we have access through *Vbat* and their emitter sense pins. The current is limited to a few amperes on the sense wiring, but this is more than enough to obtain crisp and reasonable structural results, as seen in Section 3.3.

For reliability tests we need a much higher current. Accordingly, we can apply power only on transistor groups, e.g., Q1 + Q7, through *Vbat* and *Phase1*. This powering corresponds more or less to the normal operation mode of the module.

Figure 28 shows the recorded cooling transients on devices Q2, Q3, and Q7 at 4 A heating and 0.1 A sensing current. The resulting power step is approximately 3.3 W, which already produces a temperature change of 5 °C to 6 °C in a crisp and noise free manner.

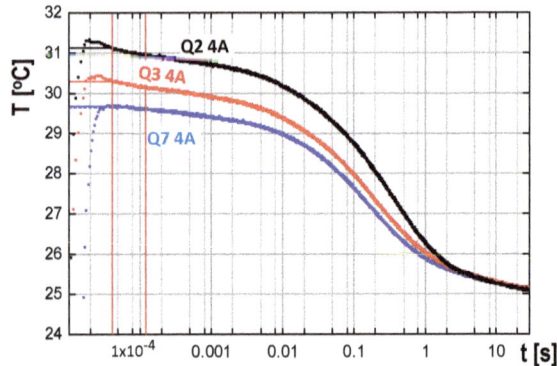

Figure 28. Cooling transients on devices Q2, Q3, and Q7 at 4 A heating and 0.1 A sensing current.

The electric transients are to be corrected at around 30 µs, after which we observe a clean temperature-induced signal.

Figure 29 demonstrates the differences in the heat conducting path when the heat flow starts from Q2 or Q3, both in the upper row, or from Q7, sending the current towards *Phase1*. All transistors were driven by 4 A. A faster increase of the structure function after the divergence point indicates that Q3 has better TIM coverage under the DBC plate than Q2, which has the same size. Q7 is a significantly larger transistor, as the early section of the structure function suggests.

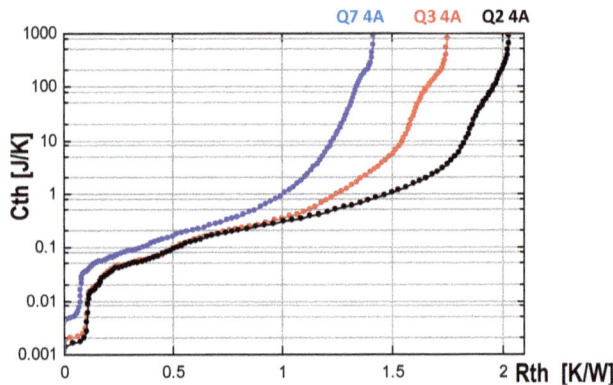

Figure 29. Structure functions: paths from Q2, Q3, and Q7 to the ambient temperature.

5. Discussion

Thermal transient measurements offer a tool for identifying structural details in the heat conducting path belonging to a power assembly.

For the identification of structural details of an assembly, linearity of the thermal parameters (thermal conductivity, specific heat) has to be assumed, but no linearity of the electric behavior or the voltage-temperature mapping is needed.

The actual "thermally induced" nature of a transient can be verified by comparing a set of measurements at different heating and sensing currents and normalizing them with the power change during the test.

Although the well-established framework of existing thermal measurement standards is developed to extract a few characteristic parameters only, such as the junction to case thermal resistance of packages, still it helps in planning the structural analysis process.

Thermal transient testing standards prescribe recording the temperature of a single point on a single data acquisition channel of the test equipment. This technique is advantageous in many ways. First, all channel offset and gain errors of the equipment cancel out at the calibration process, so the only source of error is the deviation from the targeted thermostat temperature during calibration. On the other hand, as the change of the thermal resistance is negligible in the span of the measurement, a higher sensor current adds a constant (small) temperature shift to the actual transient, not influencing the calculations based on the *change* of the temperature.

The longest time interval where a transient of true thermally induced nature can be taken is with a high sensing current and low heating current, as such minimizing electric effects at the current change.

Although the powering and sensing occurs for a single device, the methodology can reveal the composition and integrity of remote structural elements.

The structure function method analyzed in the paper is not a technique that can be used without any preliminary knowledge of the structure. Relevant interpretation can be achieved in different ways:

- Knowing all internal geometries and material parameters a "synthetic" structure function can be built up, superposing slices of material with given thermal resistance and capacitance and comparing the measured structure functions to it;
- Thermal transients can be simulated on an approximate solid model and structure functions can be composed from those. Geometry and material parameters can be tuned until the simulated and measured structure functions match;
- Measured structure functions can be compared to an already identified "golden device". This technique is rather advantageous in production control.

In complex power modules the heating current(s) may flow through more devices, and more dice can be used for sensing, applying a sensor current on them.

In cases of modules, an estimation can be given on the proportion of the internal power dissipation on the wiring based on the "accordion"-like shrinking of normalized cooling curves.

Transient testing can be used to analyze the heat transfer between several dice in a module and their thermal coupling, depending on the external cooling conditions.

Author Contributions: G.F. and Z.S. carried out the transient tests and established the necessary parameters for the structure function calculations from the measured transients. G.F. formulated the bulk of the paper and designed the figures. M.R. provided the concept of the paper, elaborated the mathematical background and confirmed the validity of the results.

Funding: The related research performed at BME, Budapest University of Technology and Economics, Hungary, and reported in this paper has been supported by the Hungarian National Research, Development and Innovation Fund (TUDFO/51757/2019-ITM, Thematic Excellence Program).

Acknowledgments: The authors want to express their special gratitude to Vladimir Szekely, for inventing the theoretical background of the structure function based structural analysis.

Conflicts of Interest: The authors declare no conflict of interest.

Energies **2019**, *12*, 2696

References

1. Szekely, V. Identification of RC networks by deconvolution: Chances and Limits. *IEEE Trans. Circuits Syst. I Fundam. Theory Appl.* **1998**, *45*, 244–258. [CrossRef]
2. Schweitzer, D.; Pape, H.; Chen, L. Transient Measurement of the Junction-To-Case Thermal Resistance Using Structure Functions: Chances and Limits. In Proceedings of the 24th SEMITHERM, San Jose, CA, USA, 16–20 March 2008. [CrossRef]
3. Rencz, M.; Szekely, V. Non-linearity issues in the dynamic compact model generation. In Proceedings of the 19th SEMITHERM, San Jose, CA, USA, 11–13 March 2003; pp. 263–270.
4. MIL-STD-750F. Department of Defense (USA) Test Method Standard: *Test Methods for Semiconductor Devices*. Available online: http://everyspec.com/MIL-STD/MIL-STD-0700-0799/MIL-STD-750F_39654/ (accessed on 8 July 2019).
5. JEDEC Standard JESD51. Methodology for the Thermal Measurement of Component Packages (Single Semiconductor Devices). December 1995. Available online: www.jedec.org/sites/default/files/docs/Jesd51.pdf (accessed on 8 July 2019).
6. JEDEC JESD 51-14 Standard: Transient Dual Interface Test Method for the Measurement of the Thermal Resistance JTC. Available online: www.jedec.org/sites/default/files/docs/JESD51-14_1.pdf (accessed on 8 July 2019).
7. Vass-Varnai, A.; Gao, S.; Sarkany, Z.; Kim, J.; Choi, S.; Farkas, G.; Poppe, A.; Rencz, M. Issues in junction-to-case thermal characterization of power packages with large surface area. In Proceedings of the 26th SEMITHERM, Santa Clara, CA, USA, 21–25 February 2010. [CrossRef]
8. Farkas, G.; Sarkany, Z. Analysis of advanced materials based on measured thermal transients of insulated gate devices in broad temperature range. In Proceedings of the 21st THERMINIC, Paris, France, 30 September–2 October 2015. [CrossRef]
9. Farkas, G. Thermal transient characterization of semiconductor devices with programmed powering. In Proceedings of the 29th SEMITHERM, San Jose, CA, USA, 17–21 March 2013; pp. 248–255.
10. Farkas, G.; Hara, T.; Rencz, M. Thermal transient testing. In *Wide Bandgap Power Semiconductor Packaging*; Suganuma, K., Ed.; Woodhead Publishing: Cambridge, UK, 2018; Chapter 6. [CrossRef]
11. Farkas, G.; Sarkany, Z.; Szel, A.; Rencz, M. Non-destructive analysis of the heat conducting path in power electronics and solid state lighting. In Proceedings of the 13th HEFAT, Portoroz, Slovenia, 17–19 July 2017; pp. 297–302.
12. Available online:. Available online: http://www.mentor.com/products/mechanical/products/t3ster (accessed on 8 July 2019).
13. Rencz, M.; Szekely, V.; Poppe, A.; Farkas, G.; Courtois, B. New methods and supporting tools for the thermal transient testing of packages. In Proceedings of the International Conference on Advances in Packaging (APACK'01). Singapore, 5–7 December 2001; pp. 407–411.
14. Bein, M.C.; Hegedüs, J.; Hantos, G.; Gaál, L.; Farkas, G.; Rencz, M.; Poppe, A. Comparison of two alternative junction temperature setting methods aimed for thermal and optical testing of high power LEDs. In Proceedings of the 23rd THERMINIC, Amsterdam, The Netherlands, 27–29 September 2017. [CrossRef]
15. Available online: https://www.mentor.com/products/mechanical/flotherm (accessed on 8 July 2019).
16. Vass-Varnai, A.; Sarkany, Z.; Rencz, M. Reliability testing of TIM materials with thermal transient measurements. In Proceedings of the 11th Electronics Packaging Technology Conference (EPTC), Singapore, 9–11 December 2009; pp. 823–827. [CrossRef]
17. Sarkany, Z.; Vass-Varnai, A.; Rencz, M. Analysis of concurrent failure mechanisms in IGBT structures during active power cycling tests. In Proceedings of the 16th Electronics Packaging Technology Conference (EPTC 2014), Singapore, 3–5 December 2014; pp. 650–654. [CrossRef]
18. Available online: https://www.mentor.com/products/mechanical/micred/power-tester-1500a/ (accessed on 8 July 2019).

energies

MDPI

Article

Reliability Investigation of a Carbon Nanotube Array Thermal Interface Material [†]

Andreas Nylander [1], Josef Hansson [1], Majid Kabiri Samani [1], Christian Chandra Darmawan [2], Ana Borta Boyon [3], Laurent Divay [3], Lilei Ye [2], Yifeng Fu [1], Afshin Ziaei [3] and Johan Liu [1,*]

[1] Electronics Materials and Systems Laboratory, Department of Microtechnology and Nanoscience (MC2), Chalmers University of Technology, Kemivägen 9, 412 58 Göteborg, Sweden; andreas.nylander@chalmers.se (A.N.); josef.hansson@chalmers.se (J.H.); kabiri@chalmers.se (M.K.S.); Yifeng.fu@chalmers.se (Y.F.)

[2] SHT Smart High Tech AB, Kemivägen 6, 412 58 Göteborg, Sweden; christianchandrad@outlook.com (C.C.D.); lilei.ye@sht-tek.com (L.Y.)

[3] Thales Research and Technology - France, Campus Polytechnique 1, avenue Augustin Fresnel, 91767 Palaiseau cedex, France; ana.borta-boyon@thalesgroup.com (A.B.B.); laurent.divay@thalesgroup.com (L.D.); afshin.ziaei@thalesgroup.com (A.Z.)

* Correspondence: johan.liu@chalmers.se

[†] This manuscript is based on the conference paper "Thermal Reliability Study of Polymer Bonded Carbon Nanotube Array Thermal Interface Materials" in the Proceedings of the 24th International Workshop on Thermal Investigations of ICs and Systems (Therminic 2018).

Received: 10 April 2019; Accepted: 27 May 2019; Published: 31 May 2019

Abstract: As feature density increases within microelectronics, so does the dissipated power density, which puts an increased demand on thermal management. Thermal interface materials (TIMs) are used at the interface between contacting surfaces to reduce the thermal resistance, and is a critical component within many electronics systems. Arrays of carbon nanotubes (CNTs) have gained significant interest for application as TIMs, due to the high thermal conductivity, no internal thermal contact resistances and an excellent conformability. While studies show excellent thermal performance, there has to date been no investigation into the reliability of CNT array TIMs. In this study, CNT array TIMs bonded with polymer to close a Si-Cu interface were subjected to thermal cycling. Thermal interface resistance measurements showed a large degradation of the thermal performance of the interface within the first 100 cycles. More detailed thermal investigation of the interface components showed that the connection between CNTs and catalyst substrate degrades during thermal cycling even in the absence of thermal expansion mismatch, and the nature of this degradation was further analyzed using X-ray photoelectron spectroscopy. This study indicates that the reliability will be an important consideration for further development and commercialization of CNT array TIMs.

Keywords: thermal management; carbon nanotubes; thermal interface material; reliability; thermal aging

1. Introduction

Efficient heat removal is an increasingly important consideration within modern microelectronics. The current trend of 3D packaging and GaN-based high power devices will only exacerbate the demands on thermal management [1]. A significant source of thermal resistance within electronics packages is at the interface between mating surfaces. Thermal interface materials (TIMs) are used to reduce this resistance, by filling out voids and increasing the real contact area between the surfaces [2]. TIMs should preferably have high thermal conductivity, perfect contact with both surfaces, no thermal contact resistance, and should also be able to absorb stresses in the interface, e.g., due to coefficient of thermal expansion (CTE) mismatch. Today, commercially used TIM can be roughly divided into polymer-based TIMs filled with thermally conductive particles, and solder-based TIMs. However, polymer-based TIMs are limited in thermal conductivity, and despite large research effort into novel fillers, the gains remain limited [3]. On the other hand, solder-based TIMs have excellent thermal performance, but have other limitations such as reliability issues at low bond lines.

As an alternative to existing TIM concepts, novel TIMs based on low-dimensional carbon allotropes such as graphene and carbon nanotubes (CNTs) have gained interest recently owing to their excellent thermal conductivity, of up to 6000 and 3000 W/mK for graphene [4] and CNTs, respectively [5]. While 2D graphene allows for excellent in-plane thermal conductivity suitable for heat spreading, arrays of vertically aligned CNTs instead allows for highly anisotropic through-plane thermal transport due to their 1D geometry. In addition, CNTs are flexible [6], and a CNT array can conform to the mating TIM surfaces, and each CNT can span the entire interface in a TIM application. This eliminates internal thermal contact resistances, making CNT arrays a highly attractive option for TIMs. The effective thermal conductivity of CNT arrays is proportional to the CNT packing density, which is typically less than 10% [7]. Previous studies have experimentally found CNT arrays with an effective thermal conductivity of up to 270 W/mK [7], although more typical values are an order of magnitude lower [8]. Despite this, CNT array TIMs have been shown to achieve performance on par with the best solder-based TIM [3].

CNT arrays are synthesized using chemical vapor deposition (CVD), usually on Si substrates. To utilize them in a TIM application, the most basic approach is the dry contact method, in which the as-grown CNT array is pressed against an opposing surface with a constant force applied in order to force the CNT array to conform to the surface [9,10]. By varying CNT growth parameters and applied pressure, the thermal performance can be optimized and tailored towards specific applications. However, uneven CNT lengths prevent every CNT to achieve contact with the opposing substrate even at high pressures. Since the inter-CNT heat transfer is very low, only CNTs in contact with both surfaces contribute to the total heat transfer, severely limiting the performance.

To increase the fraction of CNTs in contact, a bonding agent that partially penetrates the array can be used. Solder is one such possible agent, where the CNT array is pressed against a reflowed metal layer, partially embedding the CNT tips into the metal phase. This results in a TIM with reduced thermal interface resistance and which does not require applied pressure after reflow. However, the thermal contact resistance between CNTs and metal has been shown to be relatively high [11], thus, to reduce the thermal interface resistance further, organic functionalization can be used [12,13]. Organic functionalization of CNT arrays using polymers acts similarly to metal bonding, but can utilize covalent bonds to contact the CNT tips. As the thermal transport in CNTs is largely phononic in nature [11], organic molecules has a much smaller boundary resistance at the tips, and should therefore be suitable for high-performing CNT array TIMs.

The vertical alignment of the CNT array allows for individual CNTs to strain independently in the *x-y* plane. Together with the flexible nature of CNTs [14], a CNT array could mechanically decouple the two joined surfaces in a TIM application. This decoupling could potentially act as a buffer that could absorb stress originating from mismatches of the coefficient of thermal expansion (CTE) between different materials within the electronics system and thereby increase the performance and reliability. However, to date, there has, to the authors' knowledge, been no studies on the reliability of CNT TIMs, which could confirm this potential.

In this study, we subjected fabricated CNT array TIMs to thermal cycling and subsequently measured the thermal performance to investigate the degradation of the TIM. This paper is an extended and revised version of a conference report that was presented at Therminic 2018 [15] and follows the work of Ni et al. [16] and Daon et al. [17] on the HLK5 polymer used to bond CNT arrays for TIM applications. This is the first reported study on the thermal reliability of a CNT array TIM. As CNT technology has matured, including large scale fabrication of CNT arrays [18], ensuring the reliability of fabricated TIM remains an important step towards market realization.

2. Materials and Methods

The TIM setup in this study consistrf of a CVD grown CNT array on a Si substrate, bonded through a polymer layer to a Cu superstrate acting as a heat sink. The assembled structure can be seen in Figure 1.

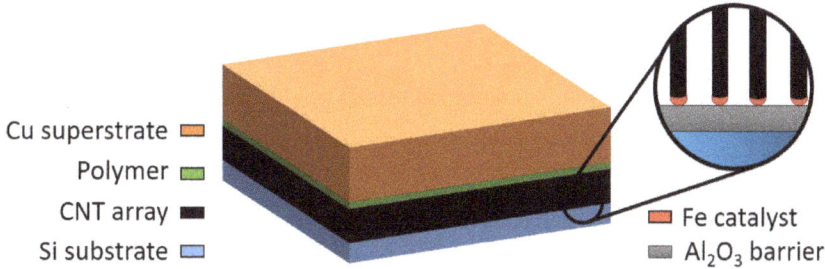

Cu superstrate
Polymer
CNT array
Si substrate

Fe catalyst
Al_2O_3 barrier

Figure 1. Illustration of the thermal interface fabricated for this project and the materials configuration.

The CNT arrays were synthesized on square Si substrates, 280 µm thick and 10 mm on the side. The catalyst structure consisted of a 10 nm thick Al_2O_3 diffusion barrier layer and a 1 nm thick Fe catalyst layer deposited using e-beam evaporation. The CNT synthesis was performed in a commercial Aixtron Black Magic II cold-wall CVD reactor. The substrates were placed on a graphite heater inside a vacuum chamber, and annealed at 500 °C under a flow of 837 sccm of H_2 gas for 3 min. During the growth phase, 200 sccm of C_2H_2 was also introduced into the chamber while the temperature was ramped up to 700 °C. This step was held for 20 s, which yielded a CNT array of 15 µm. The growth was terminated by shutting off the heater and flushing the chamber with N_2. This CVD process resulted in vertically aligned CNTs, with the iron catalyst particles at the roots of each CNT, as shown in the inset of Figure 1. The orientation and uniformity of the CNT array was verified by scanning electron microscope (SEM), and the resulting images can be seen in Figure 2. A more detailed description of the CVD process can be found in Reference [19].

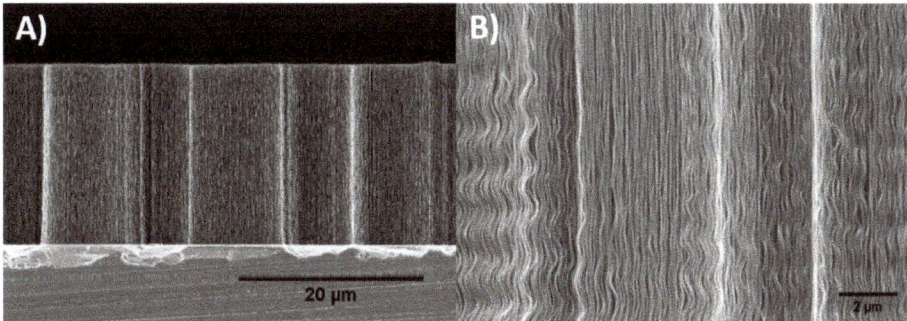

Figure 2. (**A**) Side view SEM image of a CNT array grown on a silicon substrate using the CVD process; and (**B**) high magnification view of the CNT array, which resolves the individual strands.

The grown CNT arrays were subsequently bonded to 10 mm by 10 mm, 2 mm thick Cu pieces using a previously described bonding process [17]. The polymer functionalization was spin coated onto the Cu pieces, and then the CNT array was pressed into the polymer layer. The same process time, temperature and pressure were used for all samples.

The reliability of the CNT array TIMs was investigated by subjecting the assembled interface to thermal cycling. The thermal cycling test condition "B" from the JEDEC standard [20] was chosen, corresponding to a cycling between −55 °C and 125 °C, with a ramp time of 10 min and holding time at each extreme for 20 min. The cycle time for each cycle was 60 min, and the investigation ran for 500 cycles, with thermal measurements every 100 cycles.

The total thermal interface resistance of the CNT array TIM was measured before and during thermal cycling using the laser flash method. A laser pulse induced a heat impulse on one side of the assembled sandwich structure and the transient temperature response on the other side was monitored using an IR detector. By fitting material parameters of the Cu and Si pieces, the remaining thermal interface resistance could be extracted, using a two layer plus contact resistance model [21]. The contact resistance in the model corresponded to the total thermal interface resistance R_{TIM}, as shown in Figure 3. R_{TIM} in turn included contribution from the effective bulk thermal conductivity of the CNT array κ_{CNT}, the thermal boundary resistances R_{C1} and R_{C2} at both the CNT/Si and CNT/Cu interfaces, and the bond line thickness (BLT) of the TIM.

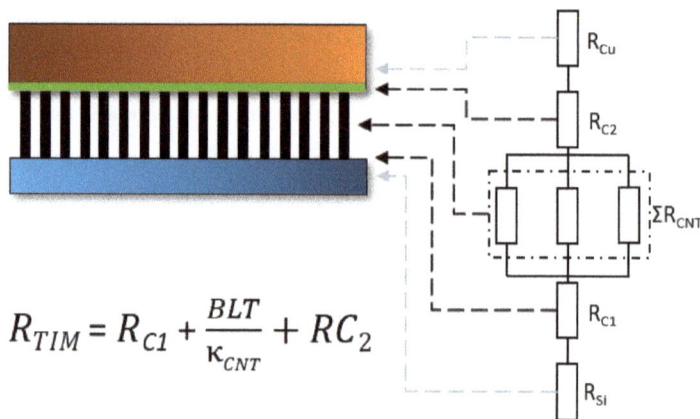

$$R_{TIM} = R_{C1} + \frac{BLT}{\kappa_{CNT}} + RC_2$$

Figure 3. Simple schematic of the origin of the different thermal resistance contributions in the CNT array TIM. The black contributions were included in the thermal interface resistance results that were obtained from the laser flash experiment [15].

To investigate the individual contributions of the thermal resistance components in Figure 3, we used the PPR method on as-grown CNT array samples. In the PPR method that is illustrated in Figure 4A, a picosecond Nd:YAG pump beam laser pulse was used to heat the temperature of the sample, while a He-Ne probe beam laser monitored the transient temperature dependent reflectivity response of the sample surface. The temperature response could be modeled as a one-dimensional heat conduction problem, and by fitting the model to the experimental temperature response the thermal conductivity of up to three different layers, together with the thermal boundary resistance between them, could be extracted. For the method to work, the top layer needed to have a temperature dependent reflectivity, thus 300 nm TiN was sputtered onto the CNT array, as shown in Figure 4B. The PPR measurement setup is described in more detail in References [22–24]. The PPR measurement could determine the thermal resistance components from the bulk CNT array and the Si/CNT thermal boundary resistance R_{C1}, which combined with R_{TIM} from the laser flash measurements allowed for the extraction of the CNT/Cu boundary resistance R_{C2} as well.

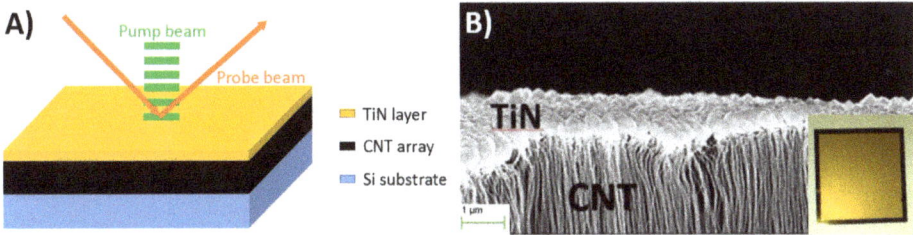

Figure 4. (**A**) Schematic illustration of the three-layer PPR measurement; and (**B**) SEM image of the CNT covered Si samples after sputtering of TiN [15].

To investigate the chemical evolution of the Si/CNT interface during thermal cycling, we used a PHI 5000C X-ray photoelectron spectroscopy (XPS) system to analyze samples before and after thermal cycling. By analyzing the elemental composition at the delaminated interface of the CNT array TIM, information about the interface degradation could be obtained. We analyzed both the CNT roots on the Cu superstrate from the delaminated samples, as well as the silicon substrate before and after thermal cycling. The acquired spectra was shifted to coincide with the 284.8 eV C1s peak and adjusted after a Shirley type background prior to peak fitting.

3. Results and Discussion

The thermal performance of the fabricated CNT array TIM interfaces was investigated using the laser flash and PPR methods. Different sets of samples were prepared for each method, with CNT array TIMs illustrated in Figure 1 intended for laser flash measurements and TiN coated CNT arrays in Figure 4B for PPR measurements. Both sets of samples were measured before and after thermal cycling to investigate how CNT array TIMs age during usage.

The laser flash measurements were first conducted on CNT array TIMs before cycling and then at every consecutive 100th cycle up to the completed 500 cycles. Thermal interface resistance results calculated after measurement for samples before and during cycling are presented in Figure 5A,B, respectively. Before thermal cycling, the CNT array TIMs had an average thermal interface resistance of of 1.7 mm^{2}K/W $\pm$ 0.28 at 20 °C.

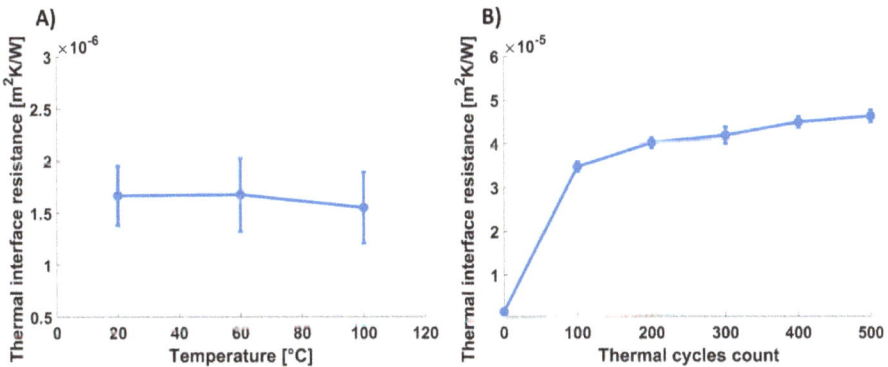

Figure 5. (**A**) Results from the initial laser flash measurement of the CNT array TIM; and (**B**) thermal interface resistance of the CNT array TIMs during thermal aging.

The thermal aging behavior of the CNT array TIM is shown in Figure 5B. Already at 100 thermal cycles, the thermal interface resistance increased 20 times from 1.7 mm^{2}K/W to 34.7 mm^{2}K/W.

Subsequent thermal cycling further increased the thermal resistance from 100 to 500 cycles, although at a slower rate. After 500 cycles, the R_{TIM} increased to 46 mm^{2}K/W.

It should be noted that some of the measured samples exhibited an even sharper increase in terms of thermal interface resistance during the cycling. However, these samples delaminated by themselves before the thermal cycling could be completed and were for that reason disregarded. All of the samples were delaminated by shearing, which showed that all samples, before and after thermal cycling, detached on the Si/CNT side of the interface leaving all the CNTs on the Cu superstrate side, as shown in Figure 6.

Figure 6. Shearing of the CNT array TIMs always resulted in a delamination where the CNTs remained on the Cu superstrate side leaving the Si substrate with a clean surface.

The increased thermal interface resistance shown in Figure 5B was determined to originate from either the CNT array or from the thermal contact on either the Si substrate side or on the Cu superstrate side. The CNTs themselves were considered to be thermally stable at the temperature range of the cycling and the effective thermal conductivity κ_{CNT} of the CNT array should therefore remain constant [25]. Furthermore, the polymer functionalization did not show any signs of deterioration during this study as the Si/CNT interface seemed to fail before any noticeable effects on the CNT/polymer adhesion could be distinguished. This leaves the probable failure of the CNT array TIMs to a degradation of the contact between the CNT and the Si substrate that would result in a similar increase of the thermal contact resistance R_{C1}. This could originate either from a CTE mismatch between the Cu superstrate and the Si substrate that would result in an uprooting of the CNTs or a chemical degradation of the contact point where the CNT was attached to the Si substrate.

To further investigate the degradation of the CNT/catalyst attachment, TiN covered CNT array samples, as presented in Figure 4B, were evaluated using PPR before and after thermal aging. These samples were free standing arrays on a silicon substrate, thus no CTE mismatch would degrade the structure during cycling, which leaves only the Si/CNT contact to affect the thermal performance. The samples were cycled 30 times, as the main degradation of the CNT array TIM was found to occur during the first 100 cycles. The attained temperature response curves were fitted to a numerical model of the sample stack to obtain the thermal conductivity κ_{CNT} of the CNT array as well as the thermal boundary resistance R_{C1} between the CNT and Si chip.

Figure 7 shows the measured temperature response from the PPR measurements. Before thermal cycling, an effective thermal conductivity (κ_{CNT}) of 71 W/mK and a thermal boundary resistance (R_{C1}) of 0.96 mm^{2}K/W were found. Together with the R_{TIM} value obtained from the laser flash measurement presented in Figure 5A, R_{C2} could be calculated using the thermal interface relation stated in Figure 3. This relation gave a R_{C2} of 0.49 mm^{2}K/W for the CNT array TIMs. By comparison, this means that the polymer functionalization was about twice as good as the Si/CNT boundary on the other side of the interface in terms of thermal boundary resistance.

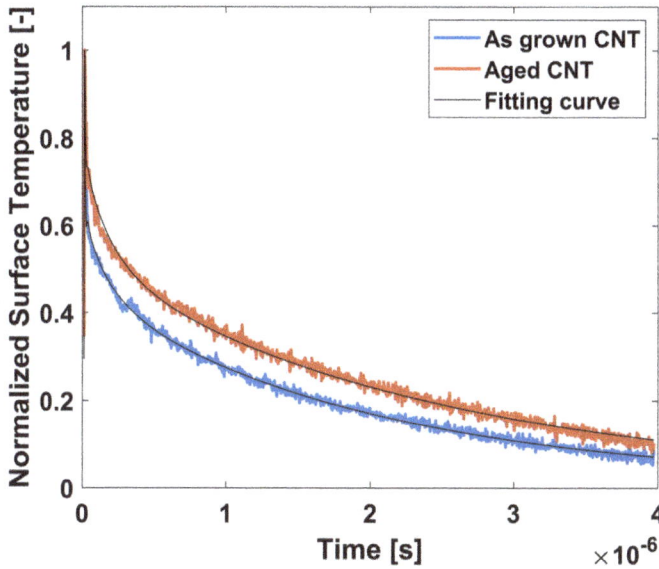

Figure 7. Time–temp response curve from PPR measurements on samples before and after thermal aging together with the numerical fitting curves.

After subjecting the TiN covered CNT arrays to 30 cycles of thermal cycling, R_{C1} increased with 137%, which could be found from the slower temperature decrease of the aged CNTs in Figure 7. While this is a significant degradation, it is not enough to explain the increase of the total thermal interface resistance, as shown in Figure 5B. The remaining increase could be attributed to uprooting of the CNT array due to a combination of CTE mismatch induced stress and chemical degradation of the interface.

To further investigate the degradation of the Si/CNT interface, the elemental composition of the delaminated interfaces was investigated using XPS analysis. According to the illustration in Figure 1, the Si/CNT interface was composed of the Si substrate, the Al_2O_3 barrier layer, Fe catalyst nanoparticles as well as the CNT. The XPS analysis showed no Al signal on the CNT roots on the Cu superstrate after delamination, which indicates that the failure occurred in the Fe interface between the CNTs and the Al_2O_3 layer. The Fe peak from the XPS measurements is shown in Figure 8 for both the CNT roots (Figure 8A) and Si chip (Figure 8B). The Fe traces found shown in Figure 8A could be identified from the $Fe2p_{3/2}$ peak at 709.5 eV as well as $Fe2p_{1/2}$ peak at 723.5 eV, both of which related to iron oxide species [26]. Comparing the average signal found on samples before and after shearing, it was found that traces in the cycled samples were 70% lower than those in the samples that were sheared without thermal cycling. According to these results, the CNT were uprooted together with the Fe catalyst nanoparticle during shearing. After thermal cycling, the bond between the CNT and the Fe catalyst nanoparticle had weakened, thereby leaving the Fe particles on the Si substrate.

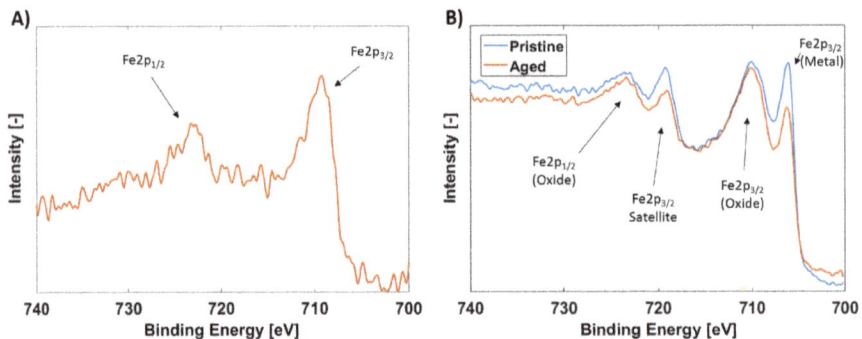

Figure 8. XPS analysis results of the Fe peak: (**A**) from the roots of the CNTs remaining on the Cu superstrate after delamination; and (**B**) from the Si substrate after delamination before and after thermal aging.

The silicon substrates from the TiN covered samples before and after aging were also analyzed using XPS, as shown in Figure 8B. The CNT arrays were removed from the silicon substrate using a tape to reveal the Fe catalyst layer underneath. In both cases, peaks could be identified related to the metallic $Fe2p_{3/2}$ at 706.6 eV, oxide $Fe2p_{3/2}$ at 710 eV, satellite $Fe2p_{3/2}$ peak at 719 eV as well as the oxide $Fe2p_{1/2}$ peak at 723.5 eV [27]. After aging, the metallic $Fe2p_{3/2}$ had decreased in intensity by 15% in comparison to the oxide $Fe2p_{3/2}$ and $Fe2p_{1/2}$ peaks, which indicates that the catalyst layer on the Si substrate had gone through a noticeable oxidation already after 30 thermal cycles. However, additional studies will be required to conclude if this oxidation process directly responsible for the delamination issues found in the CNT array TIM.

Regardless of the cause for delamination, it was obvious that the bonded surface at the CNT tips was stronger than the Si/CNT substrate. This indicates that a process to transfer the array from the growth substrate may result in a more reliable interface. Such an approach, where both the CNT tips and roots would be bonded to a surface, has been demonstrated in several publications [28,29]. This approach would also have the benefit of bypassing the challenge of making the CNT CVD synthesis CMOS compatible, since the CVD process normally requires significantly higher temperatures than the CMOS processes allow [30].

4. Conclusions

In this study, we investigated the thermal cycling behavior of a CNT array TIM sandwiched between a Si growth substrate and a polymer bonded Cu superstrate. The results obtained show that the CNT array TIM is susceptible to degradation after thermal aging. While the TIMs fabricated in this study are on par with similar state-of-the-art TIMs presented previously, laser flash measurements show an increase in thermal interface resistance of 20 times within the first 100 cycles, and a delamination at the Si/CNT interface. In addition to delamination, PPR measurements reveal that the polymer bonded interface has a factor of two lower thermal boundary resistance than the growth interface, which indicates that the Si/CNT interface needs further improvement in order for CNT array TIMs to be viable.

A combination of PPR and XPS measurements further revealed a degradation of the Si/CNT interface. PPR measurements revealed a 137% increase in thermal boundary resistance after 30 cycles. XPS analysis of the uprooted CNTs showed a decrease in iron on the CNT roots after thermal cycling, indicating a weakening of the CNT-Fe bond during thermal cycling. Further analysis on the Si substrate showed oxidation of the Fe catalyst, which may contribute to the weakening of the bond.

While this study revealed a weakening of the Si/CNT bond during thermal cycling, even without CTE mismatch induced stress, CTE mismatch was most likely still a significant contributor to the interface failure, and a more thorough investigation on this effect was required. Nonetheless, even systems without significant CTE mismatch may experience degradation, and transfer and bonding towards both mating TIM surfaces may be required to create a reliable interface.

Author Contributions: Conceptualization, A.N., L.D. and J.L.; Funding acquisition, A.Z. and J.L.; Investigation, A.N. and M.K.S.; Methodology, A.N.; Project administration, A.Z. and J.L.; Resources, M.K.S., C.C.D., A.B.B. and L.D.; Supervision, L.Y., Y.F. and J.L.; Visualization, A.N.; Writing—original draft, A.N.; and Writing—review and editing, J.H.

Funding: This research was funded by the Swedish Foundation for Strategic Research (SSF) under contract (Nos. SE13-0061 and GMT14-0045), the Production Area of Advance at Chalmers University of Technology, the National Science Foundation of Sweden (Contract No: 621-2007-4660), Formas (Contract No: FR-2017/0009), the Swedish National Board for Innovation (Vinnova) under the program of SIOGRAFEN as well as the EU Horizon 2020 program "Smartherm".

Acknowledgments: We would like to thank our project partners through the Horizon 2020 program 'Smartherm' from Thales Research & Technology, Smart High Tech AB, Berliner Nanotest und Design GmbH as well as Thales Microtechnology for fruitful discussions that have resulted in the content published in this paper.

Conflicts of Interest: The authors declare no conflict of interest.

Abbreviations

The following abbreviations are used in this manuscript:

BLT	Bond line thickness
CMOS	Complementary metal oxide semiconductor
CNT	Carbon nanotube
CTE	Coefficient of thermal expansion
CVD	Chemical vapor deposition
PPR	Pulsed photothermal reflectance
SEM	Scanning electron microscope
TIM	Thermal interface material
XPS	X-ray photoelectron spectroscopy

References

1. Moore, A.L.; Shi, L. Emerging challenges and materials for thermal management of electronics. *Mater. Today* **2014**, *17*, 163–174. [CrossRef]
2. Razeeb, K.M.; Dalton, E.; Cross, G.L.W.; Robinson, A.J. Present and future thermal interface materials for electronic devices. *Int. Mater. Rev.* **2018**, *63*, 1–21. [CrossRef]
3. Hansson, J.; Nilsson, T.M.; Ye, L.; Liu, J. Novel nanostructured thermal interface materials: A review. *Int. Mater. Rev.* **2018**, *63*, 22–45. [CrossRef]
4. Balandin, A.A.; Ghosh, S.; Bao, W.; Calizo, I.; Teweldebrhan, D.; Miao, F.; Lau, C.N. Superior thermal conductivity of single-layer graphene. *Nano Lett.* **2008**, *8*, 902–907. [CrossRef]
5. Pop, E.; Mann, D.; Wang, Q.; Goodson, K.; Dai, H. Thermal conductance of an individual single-wall carbon nanotube above room temperature. *Nano Lett.* **2006**, *6*, 96–100. [CrossRef]
6. Yu, M.F.; Lourie, O.; Dyer, M.J.; Moloni, K.; Kelly, T.F.; Ruoff, R.S. Strength and breaking mechanism of multiwalled carbon nanotubes under tensile load. *Science* **2000**, *287*, 637–640. [CrossRef] [PubMed]
7. Tong, T.; Zhao, Y.; Delzeit, L.; Kashani, A.; Meyyappan, M.; Majumdar, A. Dense vertically aligned multiwalled carbon nanotube arrays as thermal interface materials. *IEEE Trans. Compon. Packag. Technol.* **2007**, *30*, 92–100. [CrossRef]
8. Marconnet, A.M.; Panzer, M.A.; Goodson, K.E. Thermal conduction phenomena in carbon nanotubes and related nanostructured materials. *Rev. Mod. Phys.* **2013**, *85*, 1295–1326. [CrossRef]
9. Zhang, K.; Chai, Y.; Yuen, M.M.F.; Xiao, D.; Chan, P. Carbon nanotube thermal interface material for high-brightness light-emitting-diode cooling. *Nanotechnology* **2008**, *19*, 215706. [CrossRef]
10. Cola, B.A.; Xu, X.; Fisher, T.S.; Capano, M.A.; Amama, P.B. Carbon nanotube array thermal interfaces for high-temperature silicon carbide devices. *Nanoscale Microscale Thermophys. Eng.* **2008**, *12*, 228–237. [CrossRef]
11. Li, Q.; Liu, C.; Fan, S. Thermal boundary resistances of carbon nanotubes in contact with metals and polymers. *Nano Lett.* **2009**, *9*, 3805–3809. [CrossRef]

12. Le Khanh, H.; Divay, L.; Le Barny, P.; Leveugle, E.; Chastaing, E.; Demoustier, S.; Ziaei, A.; Volz, S.; Bai, J. Thermal interfaces based on vertically aligned carbon nanotubes: An analysis of the different contributions to the overall thermal resistance. In Proceedings of the IEEE 2010 16th International Workshop on Thermal Investigations of ICs and Systems (THERMINIC), Barcelona, Spain, 6–8 October 2010; pp. 1–4.
13. Kaur, S.; Raravikar, N.; Helms, B.A.; Prasher, R.; Ogletree, D.F. Enhanced thermal transport at covalently functionalized carbon nanotube array interfaces. *Nat. Commun.* **2014**, *5*, 3082. [CrossRef]
14. Teo, E.H.; Yung, W.K.; Chua, D.H.; Tay, B. A carbon nanomattress: A new nanosystem with intrinsic, tunable, damping properties. *Adv. Mater.* **2007**, *19*, 2941–2945. [CrossRef]
15. Nylander, A.; Darmawan, C.C.; Boyon, A.B.; Divay, L.; Samani, M.K.; Ras, M.A.; Fortel, J.; Fu, Y.; Ye, L.; Ziaei, A.; et al. Thermal reliability study of polymer bonded carbon nanotube array thermal interface materials. In Proceddings of the IEEE 2018 24rd International Workshop on Thermal Investigations of ICs and Systems (THERMINIC), Stockholm, Sweden, 26–28 September 2018; pp. 1–5.
16. Ni, Y.; Le Khanh, H.; Chalopin, Y.; Bai, J.; Lebarny, P.; Divay, L.; Volz, S. Highly efficient thermal glue for carbon nanotubes based on azide polymers. *Appl. Phys. Lett.* **2012**, *100*, 193118. [CrossRef]
17. Daon, J.; Sun, S.; Jiang, D.; Leveugle, E.; Galindo, C.; Jus, S.; Ziaei, A.; Ye, L.; Fu, Y.; Liu, J. Chemically enhanced carbon nanotubes based thermal interface materials. In Proceddings of the IEEE 2015 21st International Workshop on Thermal Investigations of ICs and Systems (THERMINIC), Paris, France, 30 September–2 October 2015; pp. 1–4.
18. De Villoria, R.G.; Figueredo, S.; Hart, A.; Steiner Iii, S.; Slocum, A.; Wardle, B. High-yield growth of vertically aligned carbon nanotubes on a continuously moving substrate. *Nanotechnology* **2009**, *20*, 405611. [CrossRef]
19. Wang, T.; Jeppson, K.; Liu, J. Dry densification of carbon nanotube bundles. *Carbon* **2010**, *48*, 3795–3801. [CrossRef]
20. Standard, J. *Temperature Cycling*; JESD22-A104D; JEDEC Solid State Technology Association: Arlington, VA, USA, 2009; pp. 158–162.
21. Milošević, N.; Raynaud, M.; Maglić, K. Estimation of thermal contact resistance between the materials of double-layer sample using the laser flash method. *Inverse Probl. Eng.* **2002**, *10*, 85–103. [CrossRef]
22. Sun, S.; Samani, M.K.; Fu, Y.; Xu, T.; Ye, L.; Satwara, M.; Jeppson, K.; Nilsson, T.; Sun, L.; Liu, J. Improving thermal transport at carbon hybrid interfaces by covalent bonds. *Adv. Mater. Interfaces* **2018**, *5*, 1800318. [CrossRef]
23. Jing, L.; Samani, M.K.; Liu, B.; Li, H.; Tay, R.Y.; Tsang, S.H.; Cometto, O.; Nylander, A.; Liu, J.; Teo, E.H.T.; et al. Thermal conductivity enhancement of coaxial carbon@ boron nitride nanotube arrays. *ACS Appl. Mater. Interfaces* **2017**, *9*, 14555–14560. [CrossRef]
24. Han, H.; Zhang, Y.; Wang, N.; Samani, M.K.; Ni, Y.; Mijbil, Z.Y.; Edwards, M.; Xiong, S.; Sääskilahti, K.; Murugesan, M.; et al. Functionalization mediates heat transport in graphene nanoflakes. *Nat. Commun.* **2016**, *7*, 11281. [CrossRef]
25. Wei, X.; Wang, M.S.; Bando, Y.; Golberg, D. Thermal stability of carbon nanotubes probed by anchored tungsten nanoparticles. *Sci. Technol. Adv. Mater.* **2011**, *12*, 044605. [CrossRef]
26. Lin, T.C.; Seshadri, G.; Kelber, J.A. A consistent method for quantitative XPS peak analysis of thin oxide films on clean polycrystalline iron surfaces. *Appl. Surf. Sci.* **1997**, *119*, 83–92. [CrossRef]
27. Biesinger, M.C.; Payne, B.P.; Grosvenor, A.P.; Lau, L.W.; Gerson, A.R.; Smart, R.S.C. Resolving surface chemical states in XPS analysis of first row transition metals, oxides and hydroxides: Cr, Mn, Fe, Co and Ni. *Appl. Surf. Sci.* **2011**, *257*, 2717–2730. [CrossRef]
28. Yao, Y.; Tey, J.N.; Li, Z.; Wei, J.; Bennett, K.; McNamara, A.; Joshi, Y.; Tan, R.L.S.; Ling, S.N.M.; Wong, C. High-quality vertically aligned carbon nanotubes for applications as thermal interface materials. *IEEE Trans. Compon. Packag. Manuf. Technol.* **2014**, *4*, 232–239. [CrossRef]
29. Cross, R.; Cola, B.A.; Fisher, T.; Xu, X.; Gall, K.; Graham, S. A metallization and bonding approach for high performance carbon nanotube thermal interface materials. *Nanotechnology* **2010**, *21*, 445705. [CrossRef]
30. Kumar, M.; Ando, Y. Chemical vapor deposition of carbon nanotubes: A review on growth mechanism and mass production. *J. Nanosci. Nanotechnol.* **2010**, *10*, 3739–3758. [CrossRef]

energies

MDPI

Review

Thermal Management of High-Power Density Electric Motors for Electrification of Aviation and Beyond

David C. Deisenroth and Michael Ohadi *

Department of Mechanical Engineering, University of Maryland, 8228 Paint Branch Dr., Rm 3131, College Park, MD 20740, USA; ddeisenr@umd.edu
* Correspondence: ohadi@umd.edu; Tel.: +301-405-5263

Received: 16 April 2019; Accepted: 24 August 2019; Published: 20 September 2019

Abstract: Enhanced cooling, coupled with novel designs and packaging of semiconductors, has revolutionized communications, computing, lighting, and electric power conversion. It is time for a similar revolution that will unleash the potential of electrified propulsion technologies to drive improvements in fuel-to-propulsion efficiency, emission reduction, and increased power and torque densities for aviation and beyond. High efficiency and high specific power (kW/kg) electric motors are a key enabler for future electrification of aviation. To improve cooling of emerging synchronous machines, and to realize performance and cost metrics of next-generation electric motors, electromagnetic and thermomechanical co-design can be enabled by innovative design topologies, materials, and manufacturing techniques. This paper focuses on the most recent progress in thermal management of electric motors with particular focus on electric motors of significance to aviation propulsion.

Keywords: electric aircraft; motor cooling; thermal management

1. Introduction

The global fleet of aircraft is projected to more than double to 48,000 in the next 20 years. The increase in air travel is driven by population, economic growth, and the growing global middle class [1]. One of the largest single expenditures of airlines is jet fuel, accounting for around 20% of annual expenditures [2]. Electric motor propulsion may reduce this cost by improving aerodynamic efficiency through wider implementation of distributed propulsion, beyond the typical maximum of four propulsors per plane in production jet-engine aircraft, as shown in Figure 1 [3–6]. Other benefits of electric motor propulsion include reduced noise and vibration.

Figure 1. Example of distributed propulsion aircraft architecture.

For widespread electric propulsion to be feasible, the motors must be highly power-dense. High gravimetric power density increases aircraft efficiency by decreasing the takeoff weight, while increased volumetric power density decreases frontal area and drag [7]. Currently, though, commercial high-power density motors can continuously operate at only about 5 kW/kg. The thermal limitation of the motor is demonstrated by peak power output of more than 10 kW/kg, but the duration is limited to about 20 s before the temperature rise becomes unmanageable [8–10]. High conversion efficiency is also desirable, because, for example, a motor producing 60 kW of rotor power and operating at 95% efficiency produces 3 kW of heat. In aircraft, heat generation is onboard energy (storage mass) that is not used for propulsion, but higher capacity cooling systems are generally larger and heavier. Therefore, to increase the feasibility of electric aircraft via the integration of high-performance cooling systems, continuous motor power density must be increased while maintaining efficiency. Developments in high-power density motors may also increase the feasibility of electric personal transporters, multi-passenger vehicles, and trains, as well as compact electricity generators.

1.1. Electric Propulsion Motor Architectures

The intrinsic compactness of permanent magnet motors—generated without bulky electromagnets within the rotor—coupled with more efficient operation make this motor architecture advantageous for aircraft propulsion. Most permanent magnet motors have an electromagnetic stator that synchronously drives the permanent magnet rotor, which then transmits the power to the propulsion blades. Figure 2a shows a constructed radial flux permanent magnet motor that could drive an aircraft propeller. Figure 2b shows a computer aided design (CAD) radial cross-section of a similar, simplified, motor for illustration of the primary heat generation sites in such a motor.

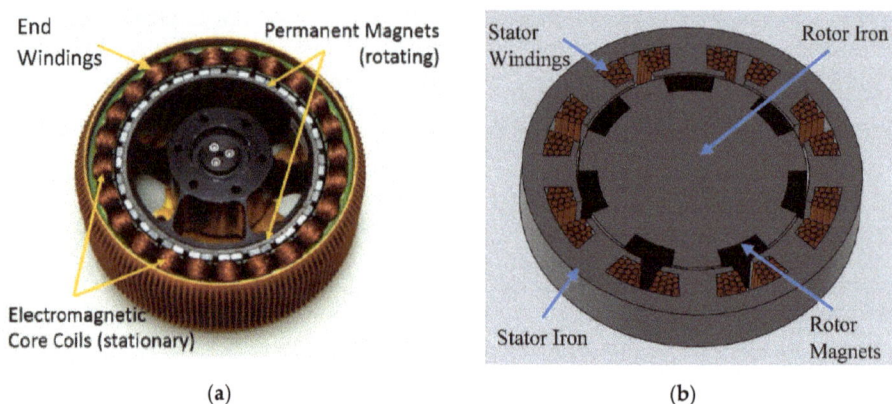

(a)   (b)

Figure 2. Radial flux permanent magnet electric motors: (**a**) digital image, and (**b**) computer aided design section view [9].

Figure 3a shows an exploded view of an axial flux permanent magnet motor, while Figure 3b shows an axial cross section of the same motor. Axial flux motor architectures produce magnetic flux along the motor axis, necessitating that the rotor and stator be the same diameter and adjacent along the axis of the motor. Axial flux motors driving a single font propeller have been demonstrated in high performance electric aircraft racing [11].

As shown in Figure 4, four common types of motor architectures have been demonstrated for electric propulsion. Figure 4a is perhaps the most "traditional" motor architecture, in which the stator circumferentially surrounds the rotor, which drives a shaft. The magnetic flux occurs radially from the stator windings to the permanent magnets. Figure 4b shows a somewhat similar configuration to Figure 4a, except the rotor is hollow, allowing propulsion blades to be contained inside the rotor. This is said to be an "inrunner" rotor design. Figure 4c illustrates an "outrunner" rotor design, in which the stator is contained within the rotor, which may have blades on its exterior. "Double runner" rotors combine the inrunner and outrunner configurations, and may result in increased efficiency at the cost of increased manufacturing complexity [12]. Figure 4d shows a similar cross-section of an axial flux permanent magnet motor, which most often is available in a shaft-drive configuration. Propulsion configurations include fixed pitch propellers, variable pitch propellers, contra-rotating (counter-rotating) fans or propellers, ducted fans, and rim driven fans [13]. Each of these configurations may be compatible with distributed propulsion.

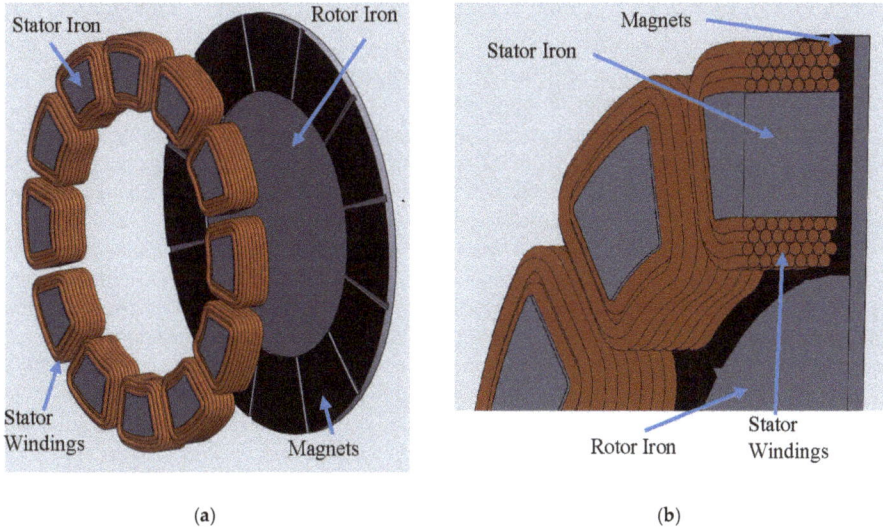

(a)   (b)

Figure 3. CAD illustration of an axial flux permanent magnet electric motor: (**a**) exploded view of rotor and stator, and (**b**) axial cross-section of the axial flux permanent magnet motor.

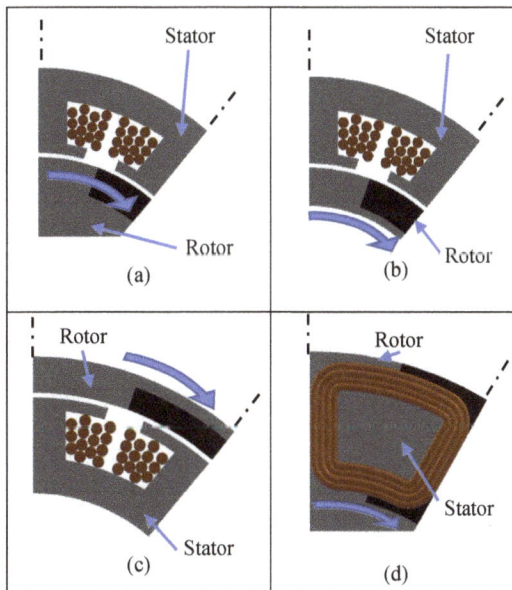

Figure 4. Electric propulsion motor architectures: (**a**) Radial flux shaft drive; (**b**) radial flux inrunner rim drive; (**c**) radial flux outrunner drive; (**d**) axial flux shaft drive.

1.2. Heat Generation and Loss Mitigation

Like any electrical device, permanent magnet motors generate heat. For example, a motor producing 60 kW of rotor power and operating at 95% efficiency produces 3 kW of heat. Without adequate cooling, heat generation causes temperature rise, which has many detrimental effects. First, excess temperature may cause catastrophic failure by the loss of integrity of the polymer electrical insulation and bonding materials. Second, temperature cycles cause mechanical stresses due to thermal expansion, resulting in fatigue, which is exacerbated by the magnitude of the thermal cycles. In electric motors, the coefficient of thermal expansion of the permanent magnet materials is about double that of silicon steel, which can result in mechanical failure of magnets with excess temperature cycling [14]. Loss of permanent-magnet functionality may also occur via demagnetization at high temperatures [14].

Temperature rise is also associated with decreased efficiency of electric motors. The electrical resistivity of copper increases with temperature, increasing the thermal load generated by Joule heating as temperature rises. Reduced efficiency may also occur due to decreased permanent magnet performance (reduced remanence and coercivity) at elevated temperatures [14].

A heat generation (loss) mitigation strategy that has long been integrated into motors is use of laminated rotor and stator iron. The electrical insulation between the high magnetic permeance silicon steel sheets increases the electrical flow path resistivity within the material, and therefore decreases eddy currents and iron losses. The same strategy may be applied to permanent magnets to reduce magnet losses [14]. It should be noted, though, that electrical insulation is often also thermal insulation, which reduces cooling performance for the remaining heat load. The bonding material that holds permanent magnets in place is also a thermal resistance, but in most cases, heat generation in the permanent magnets is less than either the copper or iron [14].

In modern motors, the largest contributor to heat generation is the copper stator windings. For example, in the optimized air-cooled motors for the NASA SCEPTOR program, the copper (resistive) losses contributed 43% of the total thermal load, while the iron losses (primarily due to Joule heating by eddy currents) contributed 37%, and the permanent-magnet rotor load contributed the balance of 20% [12]. Depending on the motor configuration and operating conditions, copper losses may be the dominant source of heat generation, with values upward of 64% of the total [15].

The primary approach to maximizing efficiency is to model and optimize the motor parameters. The NASA SCEPTOR motors were modeled with a two-dimensional finite element analysis [12]. The variables in the optimization include rotor configuration (inrunner, outrunner, or double runner), number of stator slots and rotor poles, ratio of number stator slots to rotor poles, motor mass, operating speed, torque, voltage, lamination dimensions, magnet type, stator fill factor, yoke material, stack length, rotor and stator inner and outer diameters, magnet thickness, airgap distance, slot height, tooth width, and tooth tip width [12]. The importance of optimization for aircraft electric propulsion was well illustrated by the Pareto fronts, which clearly indicate that there is a tradeoff between motor mass and efficiency, as shown in Figure 5.

After optimization, the selected design was an outrunner with a stationary interior housing, which contained the stator and cooling fins. The outrunner was chosen for the best tradeoff between mass, efficiency, and manufacturability. The propulsion blades were mounted on the exterior of the outrunner, which was contained within a nacelle. In the outrunner design, centrifugal force aids in retention of the permanent magnets within the rotor, requiring only weak bonding material, which may improve reliability and reduce thermal resistance to magnet heat dissipation. It was found that the air-cooling system contributed 1.6% to cruise drag. Furthermore, the limitations of air cooling are highlighted by the 60 kW motor with a power density of only 2.7 kW/kg [12]. The next sections will describe liquid cooling techniques that increase the levels of heat dissipation beyond those practical with air cooling.

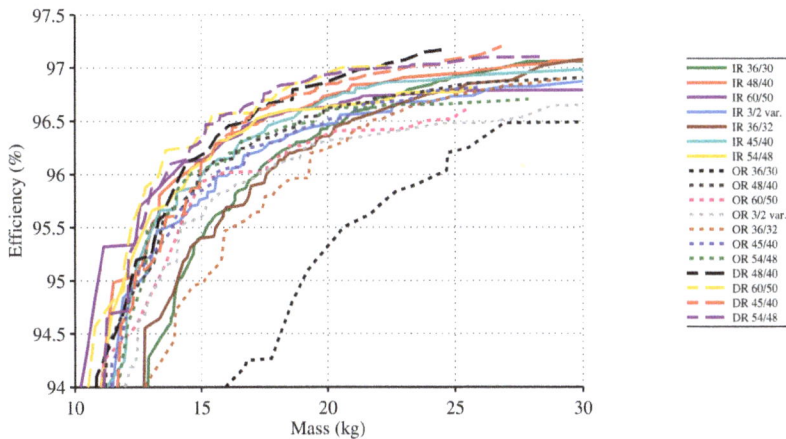

Figure 5. Edges of Pareto optimal fronts for design of the NASA SCEPTOR electric propulsion system. In the legend, IR indicates inrunner, OR indicates outrunner, and DR indicates double runner. The ratios in the legend are the number of stator slots to the number of rotor poles (stator slots/rotor poles).

1.3. Heat Rejection

The focus of this review is on the approaches that have been developed to extract heat from high-power density motors. A problem that remains, though, is heat rejection from the plane to the environment, which is not trivial. Current aircraft reject heat for a variety of purposes through a variety of mechanisms.

Fuel-oil heat exchangers are commonplace on commercial aircraft. Oil cooling is essential for engine oil to retain appropriate viscosity for lubrication, and fuel is an excellent coolant. Therefore, fuel cooling may be useful for hybrid aircraft, although it must be assured that proper fuel flow is maintained for motor cooling even when the turbofans and electric motors are not used simultaneously [16]. Another approach to oil cooling is an air-oil heat exchanger. These heat exchangers have been demonstrated in the EM-11 ORKA aircraft, in which multiple kilowatts of heat are dissipated from the oil-air heat exchanger [17]. For electric aircraft using ducted fans, a similar bleed air scheme could also potentially be used to reject heat from the motor coolant.

Until the roll-out of the Boeing 787, most commercial jetliners used engine compressor bleed air for cabin cooling [18]. The 787 uses a dedicated electric compressor for the air cycle machine environmental control system. A similar concept may be used to reject heat from the motor coolant in electric aircraft. Regardless of the approach, heat rejection from the electric motors is an essential area of research to bolster the technologies that enable more-electric aircraft. The next sections of this review will focus on heat extraction from motors and other power dense electric devices.

2. Established Cooling Approaches

2.1. Oil Bath Cooling

Oil bath cooling increases the heat dissipation ability of the cooling system beyond air cooling. With this configuration, oil is contained within the motor housing and is exposed to the stator windings, stator iron, and rotor. Having the stator and rotor partially or fully submerged in oil causes two problems: the electromagnetic efficiency may be decreased by the fluid in the airgap, and viscous losses occur because of the stirring of the oil by the rotor. Due to these conditions, the fluid must also be chemically stable and non-flammable, and have high magnetic permeability and high electrical resistivity [19]. Automatic transmission fluid (ATF) is often chosen for this application. The ATF may either be passively or actively pumped to an external heat exchanger.

In the case of a production traction motor, the motor was partially filled with oil. In this arrangement, oil may be entrained with the rotor and sprayed about within the enclosure, generating an effect similar to jet impingement or non-evaporative spray cooling [19]. Alternatively, the motor enclosure may be entirely submerged [15]. Although oil bath cooling is an improvement from air cooling, commercial motors with power densities of 5 kW/kg require water jacket cooling.

2.2. Water Jacket Cooling

The highest performance commercial motors can power racing aircraft, sustain more than 5 kW/kg power density, and sprint for power densities up to 12.1 kW/kg [10]. The peak power of the 14 kg motor was 170 kW, and the peak torque density was 17.9 N-m/kg. These axial flux permanent magnet motors are cooled by a water jacket. Axial flux architectures are particularly compatible with circumferential water jacket cooling because a large area of the high thermal conductivity copper windings can be put in contact with the water jacket. This contrasts with radial flux motors, which have a low thermal conductivity steel stator body between most of the winding area and the water jacket. Water jacket cooling may be achieved by circumferential flow paths, helical ducts, axial serpentine channels, or circumferential serpentine channels. Although water jacket cooling is a robust design that has achieved quite high performance, there remains room for improvement in motor cooling, which will next be discussed.

2.3. Heat Pipe Cooling

Modern portable devices increasingly use "heat pipes." The flow in heat pipes is typically driven by a capillary wick, which transports liquid from the condenser back to the evaporator. A burgeoning area of study is "pulsating heat pipes," which are not limited by relatively weak capillary forces driving standard heat pipes, while still utilizing passive (pumped by internal fluid forces) phase change cooling. In pulsating heat pipes, the flow is driven by differential bubble expansion and contraction in the evaporator and condenser sections [20]. Water is often the working fluid for these two types of heat pipes. A recent study used micropillar wicks for passive, capillary-pumped evaporation of water, achieving a peak semiconductor chip-scale heat flux of 45 W/cm^2 at 19 °C superheat before becoming limited by the efficacy of capillary pumping [21]. These passive systems may produce HTC's comparable to pool boiling, on the order of 10,000 W/m^{2}K, which are suitable for many modern applications.

The concept of heat pipe cooling has been applied to electric motors for many decades [22]. It has been shown that heat pipes integrated into the external fins of a motor can decrease the external temperature of the motor from 102 °C to 68 °C, reducing the thermal resistance from 0.5 °C/W to 0.28 °C/W with a heat load of 150 W. Studies and patents have been developed for integrating capillary heat pipes into stator windings [23]. Higher power-output and power-density motors require rotor cooling as well [24]. As a result, rotating capillary [25,26] and rotating pulsating heat pipes [27] have been developed and applied to electric motor cooling. There is even interest in implementation of heat pipes in traction motors of modern automobiles [28].

However, as motors become more power dense, the need for higher cooling density than can be provided by heat pipes is increasing. Pumped two-phase cooling is advantageous because two-phase pumped cooling can remove more heat per pumping power than single-phase pumped cooling when configured correctly, although much research remains to be done in predictive modeling of heat transfer and pressure drop in diabatic two-phase flows [29–31]. The primary focus of the research to be presented in the remainder of this publication is on pumped single-phase cooling systems, which—with some modification—may also be compatible with high heat flux pumped two-phase cooling.

3. Experimental Thermal Improvements

The following discussion will describe experimental heat transfer improvements that have been applied in research environments to push the envelope of electric motor performance. As discussed previously, the majority of heat production occurs in the copper windings of the stator, and the second-largest heat production source is iron losses in the stator and rotor. The following sections will describe improvements to the thermal performance of the stator iron, as well as varying approaches to winding cooling, including inter-winding cooling (between windings) and intra-winding cooling (within windings).

3.1. Stator Iron Thermal Enhancement

Pyrhonen et al. [32] demonstrated an intra-iron cooling improvement in an axial flux motor, dual stator, single rotor configuration. In this study, copper rods were embedded in stator iron to significantly improve the heat conductivity from the stator iron losses to the cooling water heat exchanger on the opposite side of stators from the rotor, as illustrated in Figure 6. This thermal enhancement is significant, because the high thermal conductivity copper paths are perpendicular to the thermal resistances generated by electrical insulation between the stator iron laminates. The copper heat conductors created a continuous thermal path from the interior of the stator iron into the cooling water, where they protruded into the flow. In addition to the copper heat conductors, aluminum oxide based Ceramacast 675-N potting material with thermal conductivity of 100 W/m-K was added between the stator end windings and water jacket, further improving the thermal performance of the stator.

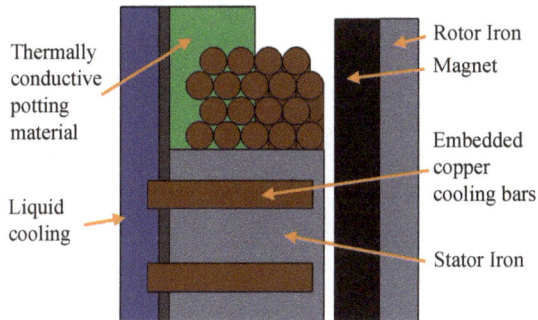

Figure 6. Illustration of intra-iron copper cooling paths and thermally conductive end winding potting material in an axial flux, permanent magnet motor.

The study by Pyrhonen et al. [32] independently showed the degree to which one copper bar per stator tooth, three conductive bars per stator tooth, and the potting material on the end windings improved the thermal performance. A single copper bar in each stator tooth decreased the winding temperatures by about 15 °C, while three copper bars further decreased the winding temperatures by about 10 °C. Under the same conditions, the potting material had an effect of cooling the windings by less than 2 °C. Ultimately, the improved motor could operate continuously at 75 kW power output, while in stock configuration it exceeded the temperature limit of 130 °C after 70 min of operation at the same power output.

Pyrhonen et al. [32] further showed an important codesign consideration for intra-iron improvements. It was found that a detriment of a 3% increase in stator iron losses was caused by decreased iron area that was occupied by the copper bars. Therefore, the addition of copper bars caused a decrease in electromagnetic performance of the stator for an improvement in the overall performance of the motor. This example illustrates the sometimes-conflicting performance interaction of thermal and electromagnetic systems; and provides further evidence of the importance of accurate, combined thermal and electromagnetic modeling of further development of high-performance motors.

3.2. Inter-Winding Cooling

A simple but effective approach to inter-winding cooling was demonstrated by Rhebergen et al. [33]. The approach brings heat exchange channels into direct contact with the stator windings by placing the channels in the interior of the winding in the gap between stator teeth at the air gap. This position is illustrated in Figure 7, with the IW position. In the design, the channels ran in a serpentine path, cooling several windings in one fluid pass. The cooling channels were modeled as a thermally conductive polymer in order to have a relative permeability near unity. Contact resistance was eliminated by bonding the cooling channels to the windings with the thermally enhanced polymer with thermal conductivity of 2 W/m-K. Simulations predicted that with water supplied at room temperature, the peak temperatures of the stator and windings could be reduced by nearly 100 °C with a 5 kW heat load, when compared to the standard natural convection case, to approximately 133 °C. The inter-winding cooled motor, therefore, could operate at much higher power than it could with its original configuration while maintaining the integrity of its electrical insulators.

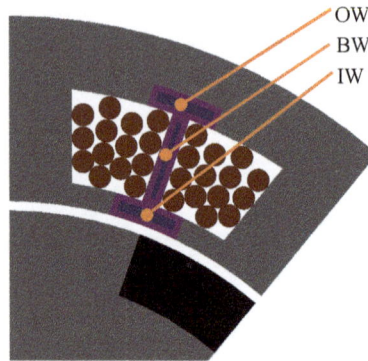

Figure 7. Illustration of three heat exchanger positions in radial flux motor inter-winding positions demonstrated in the literature: outside of windings (OW), between windings (BW), and interior of windings (IW).

Sixel et al. [34] additively manufactured a polymer heat exchanger that facilitated flow channels between the coil windings (BW in Figure 7) and on the outside of the coil windings (OW in Figure 7). The flow channels were designed to be aluminum filled polycarbonate with thermal conductivity of 1 W/m-K by fused deposition modeling [34]. The water-cooling channels occupied the OW and BW positions shown in Figure 7. It was concluded that that a 44% decrease in the winding temperature rise can be achieved by this inter-winding heat exchanger when it is used as an enhancement from the standard water jacket cooling.

Semidey and Mayor [35] designed, fabricated, and tested an internally micro-featured heat exchanger that was placed between the windings (BW position in Figure 7). The heat exchanger was fabricated by machining microfeatures into copper plates. Two halves were then assembled to form a channel for water flow. It was found that the base motor (Golden Motor HPM-10KW) could handle a steady-state current density of 8.24 A/mm^2 and a transient current density of 14.7 A/mm^2. After installation of the inter-winding heat exchanger between the stator windings, the machine was capable of a steady current density in excess of 24.7 A/mm^2 and transient current density of more than 40 A/mm^2 with class F insulation, which is limited to an operating temperature of 155 °C. Although it could not be confirmed due to test bench limitations, it may be estimated that the improved motor produced approximately three times more power than in the stock configuration.

3.3. Intra-Winding Cooling

Lindh et al. [36] demonstrated an intra-winding cooling method for an axial flux permanent magnet synchronous machine, as illustrated in Figure 8. Stainless steel coolant conduits were tightly wrapped with stranded Litz wire, and the coaxial tubing and electrical conductor served as the stator windings. A shortcoming of the long coaxial fluid flow path occurred with polyalphaolefin oil as the cooling fluid, in which the pressure drop was nearly 6 bar. Nevertheless, it was found that compared to the stock motor, intra-winding oil cooling decreased the average end winding temperature by 67 °C, while intra-winding water cooling decreased the average end winding temperature by 97 °C. The stock winding current density reached 4 A/mm^2, while the thermally improved motor produced 14 A/mm^2, an improvement that corresponds to more than a factor of three in motor power and torque density.

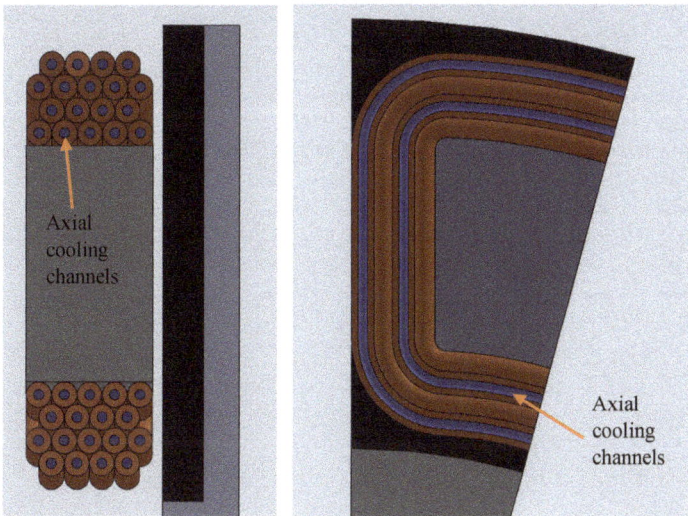

Figure 8. Illustration of intra-winding coaxial cooling paths within stator windings of an axial flux, permanent magnet motor.

Wohlers et al. [37] demonstrated a different approach to intra-winding cooling than Lindh et al. [36] via additive manufacturing of the coil windings with integrated cooling channels. In the design, a single coil occupied about half the width of stator slot, with 11 wraps around each stator tooth. Typically, this large electrical cross section in the winding would cause large current density disparities, or current crowding, leading to reduced efficiency. The geometrical freedom of additive manufacturing, though, allowed Wohlers et al. [37] to locally tune the winding cross section to significantly increase the current density uniformity. Furthermore, the cooling channels were not constrained to a coaxial configuration but were instead routed to have inlets and outlets on either end of the stator, which could significantly reduce the pressure drop per cooling capacity of the system. The windings were laser sintered from AlSi10Mg powder, an alloy that has approximately double the electrical resistance, and therefore, heat generation per current density as copper. It was found that the steady-state current density of the coil geometry could reach 58.5 A/mm^2. Extrapolating from those results, it was determined that with a similar configuration, cast copper could possibly reach current density of 130 A/mm^2, which is about five times the current state of the art in electric motors.

4. Integrated Motor Drives

Motor drives are essential for realizing the benefit of permanent magnet synchronous motors. The motor drive takes power from a source and can use feedback control to convert that power into transient control of the magnetic field produced by the stator windings. Automotive companies have already integrated motor drives (IMD) with transaxle assemblies, eliminating large power and communication cables running between the motor and controller, as well as eliminating the separate enclosure required for the power and logic components of the motor drive. In the production hybrid-electric automobile, the motors were cooled by ATF, while the power electronics were water cooled [38].

The same advantages that make IMD's essential in automotive applications also make IMD's essential in electric aircraft. In continued development of IMD technology, it is projected that in the near future, motor drives may be completely embedded in the motor enclosures, generating a negligible effect on the motor mass and volume [38]. The advantages of IMD's may be further realized by wireless signal control, which would eliminate motor control signal cabling weight and space requirements in aircraft, which is particularly important for distributed propulsion designs with many motors.

Perhaps the most important technological innovation that enables integrated motor drives is wide bandgap electronics, such as SiC and GaN, because of their comparatively high-power density over more traditional Si devices. The benefits of wide bandgap devices also include higher switching speed and higher temperature operation [38]. Although higher temperature operation somewhat eases the need for highly effective thermal management, the factor of five increase in power density of motor drives enabled by wide SiC devices often necessitates the need for liquid cooling while Si devices may more readily be air cooled [38].

In addition to decreasing the size and weight required by the drive electronics, wide bandgap electronics (GaN) can be used to develop motors with different manufacturing approaches from commonly used methods. As reported by Jahns et al. [38], Wang [39] developed motor drive modules that were integrated into segmented sections of the stator, bringing the vision of fully integrated drive electronics closer to reality. From the increasing importance of IMD's in electric motors, and their clear advantages for aerospace applications, the challenge of thermal management of motors becomes coupled with that of high heat flux power and logic electronics. Jahn et al. [38] concluded that in order for the vision of fully integrated, highly compact motor drives to become realizable, multi-disciplinary, multi-physics approaches are needed.

5. Lessons from High-Performance Electronics

As power electronics increasingly need pumped fluid cooling and drive electronics are increasingly integrated into motors, the symbiosis between high heat flux electronics and high-power density motors is increasing. Further, with wide bandgap semiconductors, the need for pumped single- and two-phase cooling for logic and power devices is increasing [40,41]. As well, at the same time, logic and power devices are increasingly integrated with electric machines [38]. So, although the thermal management challenges of motors and semiconductors are intercoupled, industry and research in thermal developments in motors lags that of semiconductors.

The motivation for thermal management in semiconductors is comparable with that of motors. At temperatures approach their upper operation limit, semiconductor devices typically exhibit decreased functionality and efficiency, producing less desirable power output per input, and then total loss of functionality [42,43]. Mechanical fatigue considerations in semiconductor packages are similar to that for motors, in which thermal cycling fatigue occurs due to bonded ceramic, metallic, and polymer materials with vastly differing coefficients of thermal expansion [42].

Historically, the cooling capability and performance of semiconductor thermal management systems has increased [44]. Early semiconductor devices utilized crude cooling systems, primarily relying on spreading across a somewhat increased surface area to dissipate to the ambient. As devices have innovated, so have the thermal management systems, which have evolved from rudimentary heat sinks to optimized natural convection heat sinks, forced air coolers, liquid coolers, and heat pipes. Although these cooling approaches are sufficient for keeping typical devices below about 130 °C with heat fluxes less than 100 W/cm^2, the DARPA ICECool program sought to push the envelope of possible heat fluxes from semiconductor devices beyond 1 kW/cm^2 while keeping pumping power low [45]. This extreme cooling performance goal was realized via three fundamental heat transfer considerations: materials, topology, and fluids.

Desirable material properties for semiconductor thermal management include high thermal conductivity, heat capacity, and mechanical resistance to fatigue. Copper (k = 400 W/m-K, ρ = 9 g/cm^3) is often the material of choice for thermal management of high heat flux electronics, since while aluminum is more than three times lighter, it is also about half as thermally conductive. For extremely high near-junction thermal transport, diamond (k = 2000 W/m-K) has shown increasing utility and manufacturability [46–48]. Altman et al. [46] showed that microchannels can be formed in diamond with an epitaxial GaN semiconductor on the active side of the GaN on diamond device.

Advantageous heat transfer topologies for semiconductor thermal management include high surface area to volume ratios, surface features that promote mixing for heat transfer enhancement, and short flow paths for minimal pressure drop. High surface area to volume features can be realized by fabricated microscale features, including microchannels, in the heat transfer surface via deep reactive ion etching [49,50], microdeformation [51,52], and additive manufacturing [53,54]. Additive manufacturing has also repeatedly shown excellent utility in forming complex fluid manifolding paths for high heat flux electronics [53,55]. Other topologies that can realize extreme heat flux thermal management systems for semiconductors are those that eliminate thermal resistance, such as die attach, heat spreaders, and thermal interface materials. This "embedded cooling" approach brings the fluid as close as possible to the heat generation, increasing the manageable heat fluxes and cooling effectiveness [45].

Desirable fluid properties for highly effective thermal management include high thermal conductivity, high heat capacity, and high latent heat of vaporization (in two-phase cooling). Simultaneously, low viscosity is desirable for low pumping power. Through historical development of semiconductor coolers, the cooling fluid has evolved with increasing heat transfer coefficients in thermal management devices. Electronics coolers have progressed from passive air cooling to forced air cooling, then to forced liquid cooling and to passive two-phase cooling [56,57]. This trend is likely to continue, with a growing need for actively pumped two-phase cooling, as suggested by the growing number of studies on the topic. Pumped two-phase cooling is advantageous because it can remove more heat per pumping power than single-phase pumped cooling when configured correctly, and heat transfer coefficients generated by two-phase flows are considerably larger than those of single-phase flows under typical cooling conditions [29,58]. Further design possibilities for very high heat transfer coefficients include solid-liquid phase change materials, spray cooling, liquid metal cooling, and thermoelectric cooling [50,59,60].

5.1. High Heat Flux Two-Phase Coolers in Copper

Each of the following high heat flux two-phase chip coolers are constructed of copper, making them approaches that could be applied to thermal management of power dense motors for aircraft propulsion. Each cooling approach shown in Table 1 utilizes area enhancement to maximize the heat transfer area exposed to diabatic two-phase flow. Each approach also uses manifolding in order to reduce the pressure drop while producing high heat transfer. The chip-scale coolers produced heat fluxes ranging from 200 W/cm^2 to 1230 W/cm^2 with temperature rises above the fluid reference temperature ranging from 22 °C to about 56 °C, pressure drops ranging from 6 kPa to 60 kPa, demonstrating the efficacy of localized pumped two-phase cooling. The differences in the studies include the approach by which surface area and other surface enhancements were achieved, as well as the type of manifolding and the cooling fluid.

Table 1. High heat flux chip-scale two-phase cooling approaches in copper.

Ref.	Cooling Approach	Fluid	Sub-Cooling (°C)	Peak Heat Flux (W/cm²)	Saturation Temperature (°C)	Peak Super-Heat (°C)	Superheat Reference Temperatures	Flow Rate	Pressure Drop (kPa)
[61]	Array of impinging jets on copper pin-fins coated with microporous surface enhancement	HFE-7100	10	206	61[1]	35	$T_{heater} - T_{sat}$	30 mL/s	10.9
[62]	Array of impinging jets on copper pin-fins coated with microporous surface enhancement	R245fa	5	218	45	22	$T_{heater} - T_{in}$	10 g/s	6.4
[63]	Manifold microchannels in copper	HFE-7100	-	300	60	50	$T_{surface} - T_{sat}$	4.2 mL/s	-
[64]	Parallel microchannels in copper with non-contacting, tapered manifold expanding in streamwise direction	Water	10	506	100[1]	26.2	$T_{wall} - T_{sat}$	393 kg/m²s	-
[52]	Manifold microchannels in copper	R245fa	8.5	1230	36	56	$T_{wall} - T_{sat,av}$	1400 kg/m²s	60

[1] At atmospheric pressure.

The approaches used by Rau et al. [61] and Joshi and Dede [62] incorporated macroscale and microscale surface enhancements. The macroscale features were pin-fins, and a microporous coating was added, forming a multiscale surface enhancement. Both studies used similar manifolds, in which an array of 25 orifices produced jet impingement on the multiscale enhanced surface. Better performance was produced in the second study, which used R245fa in contrast to HFE-7100 [62]. In the second study, pressure drop and temperature rise were also reduced while peak heat flux was increased.

The approach reported by Kalani and Kandlikar [63] used a tapered manifold, which increased the cross-sectional flow area moving downstream with a non-contacting manifold above the microchannels. This approach may increase vapor venting and was reported to reduce the pressure drop while fundamentally altering the flow regimes occurring within the channels. A peak heat flux of 506 W/cm^2 was reported with a wall temperature of about 130 °C with water as the working fluid.

Manifold-microchannels (MMC), the area enhancement geometry of interest in this dissertation for high heat flux two-phase cooling, were used by Baummer et al. [64] to achieve a base heat flux of 300 W/cm^2 with HFE-7100 as the working fluid. A later generation of the design achieved 1230 W/cm^2 with the low-pressure refrigerant R245fa as the working fluid [52]. So, it can be observed that MMC's can achieve very high heat fluxes with low superheats and low pressure drops in remote, chip-scale coolers.

5.2. Comparison of Working Fluids

As shown in Table 2, water is an excellent heat transfer fluid among several common fluids used for two-phase electronics cooling. The latent heat of vaporization and thermal conductivity of water are an order of magnitude greater than those of electronics cooling fluids (FC-72 and HFE-7100) and refrigerants often researched for electronics cooling (R236fa, R245fa, and R134a). Furthermore, the specific heat of water is more than three times greater than those fluids. Furthermore, high-purity water is orders of magnitude less expensive than any of the engineered fluids. Therefore, water is in many ways a natural choice for high heat flux cooling.

However, water is not without drawbacks for electronics cooling. Firstly, unless consistently deionized, water is an electrical conductor. Secondly, water is a solvent, and therefore should be carefully isolated from electronic devices. Furthermore, the atmospheric pressure saturation temperature of water is 100 °C, meaning that for two-phase operation, at atmospheric pressure, junction temperatures will be quite high. The saturation temperature can be reduced by operating the system under vacuum, but this approach raises reliability concerns of air ingress over time, a concern which is less relevant in systems operating with positive pressure. Water also has a higher freezing temperature than most engineering fluids, which could result in catastrophic failure in low ambient temperature environments; ethylene glycol reduces freezing temperature but is detrimental to thermal performance.

Engineered fluids can mitigate many of the concerns presented by water. These fluids are quite inert, dielectric, are available in a variety of saturation temperatures, and have a more favorable liquid-vapor density ratio. FC-72 has been considered the industry standard cooling fluid for single- and two-phase electronics cooling because of these favorable properties. HFE-7100, also manufactured by 3M, has a lower global warming potential and 27% greater latent heat, but tends to be more corrosive than FC-72. R134a is an industry standard refrigerant and may soon be replaced by R1234ze in some applications.

The lower pressure refrigerants are reported more often for embedded semiconductor cooling, with R245fa being most common for its low pressure, high latent heat, and high thermal conductivity compared to R236fa. The room temperature saturation pressure of R245fa is about half that of R236fa, 1.5 atm compared to 2.7 atm, respectively. This reduced pressure requirement reduces the structural strength required by the embedded cooling package, which may also be used to reduce package size and weight. Another advantage is that the latent heat of vaporization of R245fa exceeds that of R236fa by 31% (190 kJ/kg compared to 145 kg/kg) while the liquid thermal conductivity is also 17%

higher (0.081 W/m-K compared to 0.069 W/m-K). These higher fluid property values both improve effectiveness of two-phase heat removal.

Automatic transmission fluid is not typically used in two-phase systems, in part due to its (intentionally) prohibitively high boiling temperature. It is, though, a potent single-phase cooling liquid, having specific heat about 30% greater, and thermal conductivity double that of common refrigerants. In addition to heat transfer, ATF is designed for hydrodynamic power transmission and lubrication [65].

The most common refrigerant used in vapor compression cycles is R134a, while CO_2 is an increasingly commonly used fluid for transcritical and supercritical cycles. In contrast to vapor compression, transcritical cooling cycles pressurize the working fluid beyond the critical pressure and temperature of the fluid. This supercritical fluid is as dense as a liquid but expands and compresses much like a gas. For an R134a cycle rejecting heat at a fluid temperature of 70 °C, the highest pressure occurring in the cycle is 21 bar. In contrast, a transcritical CO_2 cycle rejecting heat at a fluid temperature of 70 °C requires pressure to exceed 160 bar. This high pressure introduces some engineering challenges but provides significantly higher cooling system power density.

In particular, the size and weight of the heaviest and often bulkiest component of the refrigeration system, the compressor, can be significantly reduced. It has been shown that the compressor displacement can be reduced in excess of 80% when compared to a conventional R134a vapor compression cycle [66]. In pressure containment devices, including the compressor, thicker walls are required to contain the pressure, but the reduced internal volume decreases the total amount of material needed, resulting in decreased size and weight.

The benefits of transcritical CO_2 as a refrigerant (also called R744) have been under investigation for two decades for automotive applications, and it has been shown that R744 can reduce the size, weight, and power required by the air conditioning system [67]. Mercedes-Benz committed to introducing transcritical CO_2 cycles into their luxury automobiles for increased cooling performance, as well as the very low refrigerant global warming potential [68]. It should be noted that in order to realize the benefits, special precautions must be taken to mitigate the corrosive nature of supercritical CO_2.

Table 2. Comparison of common working fluids for embedded two-phase coolers. All properties at 25 °C unless otherwise noted.

Fluid	Saturation Temperature at Atmospheric Pressure (°C)	Saturation Pressure (kPa)	Specific Heat (kJ/kg-K)	Latent Heat of Vaporization [1] (kJ/kg)	Liquid Thermal Conductivity (W/m-K)	Liquid Density (kg/m^3)	Vapor Density [1] (kg/m^3)
FC-72 [69]	56	-	1.1	88	0.057	1680	13
HFE-7100 [69]	60	-	1.18	112	0.069	1510	9.9
R236fa [70]	-	272	1.24	145	0.069	1360	18
R245fa [70]	-	148	1.32	190	0.081	1340	8.5
R134a [70]	-	666	1.43	178	0.083	1210	32
ATF [65]	-	-	1.95	-	0.16	870	-
sCO2 [2] [71]	-	-	4.00	290	0.078	710 [3]	-
Water [70]	100	-	4.18	2250	0.59	1000	0.6

[1] At normal boiling point; [2] Properties taken at 35 °C and 10 MPa; [3] Supercritical fluid.

6. Summary

Novel, emerging electric propulsion systems for hybrid and electric aviation call for distributed electric propulsion to reduce drag and promote fuel efficiency and emission reductions. High efficiency and high specific power (kW/kg) electric motors are a key enabler for future electrification of aviation. This review focused on the role of thermal management and most recent improvements to the current thermal limitations of electric motors, with emphasis on approaches to increase the electrification of air transport. The study demonstrates that the commercial state-of-the art electric motors are thermally limited and that a substantial increase in power and torque density, as well as efficiency of electric motors can be achieved by mitigating the thermal limitations in electric motors through a co-design approach utilizing electromagnetics and thermomechanical design aspects.

Current methods for commercial motor cooling—such as oil bath or water jacket cooling—impose limitations on power density due to lack of sufficient absorption of heat. This is particularly important for motors that benefit from operation at high voltages, which in turn require thick electrical insulation layers that often further limit heat dissipation. Researchers have shown that there exists significant potential for improved cooling. Specifically, the stator windings generate the majority of heat losses in electric motors, and examples show that from 43% to 64% of total motor heat generation occurs due to these resistive losses. By employing inter-winding and intra-winding thermal management, the power density of electric motors can be increased by a factor of three to five or more. Supercritical fluids that offer the density of a liquid and viscosity of a gas can also contribute to substantial additional cooling of motors with direct impact on enhancing power and torque densities and efficiency.

Recent studies have shown that the heat flux levels from semiconductor devices can be increased by more than ten-fold by placing thermal management as the top priority in the design and fabrication of high-performance devices. For electric motors the potential for such improvement can be even more significant realizing the same principles of thermally potent materials, topologies, and fluids. These approaches can be used to increase the power density of electric motors so that thermal management of electric motors will no longer lag that of semiconductors. An effective thermal management solution will optimally cool the motor itself as well as the integrated drive, the power electronics and the power conditioning system.

Author Contributions: Conceptualization, writing—review and editing, M.O.; investigation, writing—original draft preparation, D.C.D.

Funding: This research received no external funding.

Conflicts of Interest: The authors declare no conflicts of interest.

References

1. Airbus. Nearly 37,400 New Aircraft Valued at US$5.8 Trillion Required over 20 Years. Available online: https://www.airbus.com/newsroom/press-releases/en/2018/07/nearly-37-400-new-aircraft-valued-at-us-5-8-trillion-required-ov.html (accessed on 6 September 2019).
2. Airlines for America. A4A Passenger Airline Cost Index (PACI). Available online: http://airlines.org/dataset/a4a-quarterly-passenger-airline-cost-index-u-s-passenger-airlines/ (accessed on 6 September 2019).
3. Gohardani, A.S.; Doulgeris, G.; Singh, R. Challenges of future aircraft propulsion: A review of distributed propulsion technology and its potential application for the all electric commercial aircraft. *Prog. Aerosp. Sci.* **2011**, *47*, 369–391. [CrossRef]
4. Madavan, N.; Heidmann, J.; Bowman, C.; Kascak, P.; Jankovsky, A.; Jansen, R. A NASA perspective on electric propulsion technologies for commercial aviation. In Proceedings of the Workshop on Technology Roadmap for Large Electric Machines, Urbana-Champaign, IL, USA, 5–6 April 2016; pp. 5–6.
5. Kasem, A.; Gamal, A.; Hany, A.; Gaballa, H.; Ahmed, K.; Romany, M.; Abdelkawy, M.; Abdelrahman, M.M. Design and implementation of an unmanned aerial vehicle with self-propulsive wing. *Adv. Mech. Eng.* **2019**, *11*, 1687814019857299. [CrossRef]
6. Moore, K.; Ning, A. Distributed Electric Propulsion Effects on Traditional Aircraft through Multidisciplinary Optimization. In Proceedings of the 2018 AIAA/ASCE/AHS/ASC Structures, Structural Dynamics, and Materials Conference, Kissimmee, FL, USA, 8–12 January 2018.
7. Bennion, K. *Electric Motor Thermal Management R&D 2016 NREL Annual Merit Review*; National Renewable Energy Laboratory: Golden, CO, USA, 2016.
8. AVID Technology Limited. EVO Axial Flux Electric Motors. Available online: https://avidtp.com/product/evo-motors/ (accessed on 6 September 2019).
9. YASA Limited. YASA 750R E-Motor. Available online: https://www.yasa.com/wp-content/uploads/2018/01/YASA-750-Product-Sheet.pdf:2018 (accessed on 6 September 2019).
10. Moreels, D. Axial-Flux Motors and Generators Shrink Size, Weight. Available online: https://www.powerelectronics.com/automotive/axial-flux-motors-and-generators-shrink-size-weight (accessed on 6 September 2019).

11. Robinson, T. Bright Sparks—The Quest for Electric Speed. Available online: https://www.aerosociety.com/news/bright-sparks-the-quest-for-electric-speed/ (accessed on 6 September 2019).

12. Dubois, A.; van der Geest, M.; Bevirt, J.; Clarke, S.; Christie, R.J.; Borer, N.K. Design of an Electric Propulsion System for SCEPTOR's Outboard Nacelle. In Proceedings of the 16th AIAA Aviation Technology, Integration, and Operations Conference, Washington, DC, USA, 13–17 June 2016.

13. Bolam, R.C.; Vagapov, Y.; Anuchin, A. Review of Electrically Powered Propulsion for Aircraft. In Proceedings of the 2018 53rd International Universities Power Engineering Conference (UPEC), Glasgow, UK, 4–7 September 2018; pp. 1–6.

14. Yang, Y.Y.; Bilgin, B.; Kasprzak, M.; Nalakath, S.; Sadek, H.; Preindl, M.; Cotton, J.; Schofield, N.; Emadi, A. Thermal management of electric machines. *IET Electr. Syst. Transp.* **2017**, *7*, 104–116. [CrossRef]

15. Ponomarev, P.; Polikarpova, M.; Pyrhonen, J. Thermal Modeling of Directly-Oil-Cooled Permanent Magnet Synchronous Machine. In Proceedings of the 2012 XXth International Conference on Electrical Machines (ICEM), Marseille, France, 2–5 September 2012; pp. 1882–1887.

16. Gonzalez, C. The Future of Electric Hybrid Aviation. Available online: https://www.machinedesign.com/batteriespower-supplies/future-electric-hybrid-aviation (accessed on 6 September 2019).

17. Musto, M.; Bianco, N.; Rotondo, G.; Toscano, F.; Pezzella, G. A simplified methodology to simulate a heat exchanger in an aircraft's oil cooler by means of a Porous Media model. *Appl. Therm. Eng.* **2016**, *94*, 836–845. [CrossRef]

18. Sinnett, M. 787 No-Bleed Systems: Saving Fuel and Enhancing Operational Efficiencies. Available online: http://www.boeing.com/commercial/aeromagazine/articles/qtr_4_07/article_02_1.html (accessed on 6 September 2019).

19. Popescu, M.; Staton, D.; Boglietti, A.; Cavagnino, A.; Hawkins, D.; Goss, J. Modern Heat Extraction Systems for Electrical Machines—A Review. In Proceedings of the 2015 IEEE Workshop on Electrical Machines Design, Control and Diagnosis (Wemdcd), Torino, Italy, 26–27 March 2015; pp. 289–296.

20. Charoensawan, P.; Khandekar, S.; Groll, M.; Terdtoon, P. Closed loop pulsating heat pipes: Part A: Parametric experimental investigations. *Appl. Therm. Eng.* **2003**, *23*, 2009–2020. [CrossRef]

21. Adera, S.; Antao, D.; Raj, R.; Wang, E.N. Design of micropillar wicks for thin-film evaporation. *Int. J. Heat Mass Transf.* **2016**, *101*, 280–294. [CrossRef]

22. Corman, J.; Tompkins, R.; Edgar, R.; Mclaughlin, M. Rotating Electrical Machine Having Rotor and Stator Cooled by Means of Heat Pipes. U.S. Patent 3,801,843, 2 April 1974.

23. Hassett, T.; Hodowanec, M. Electric Motor with Heat Pipes. U.S. Patent 7,569,955, 4 August 2009.

24. McCluskey, F.P.; Saadon, Y.; Yao, Z.; Camacho, A. Cooling for Electric Aircraft Motors. In Proceedings of the 2019 18th IEEE Intersociety Conference on Thermal and Thermomechanical Phenomena in Electronic Systems (ITherm), Las Vegas, NV, USA, 28–31 May 2019; pp. 1134–1138.

25. Song, F.; Ewing, D.; Ching, C.Y. Heat transfer in the evaporator section of moderate-speed rotating heat pipes. *Int. J. Heat Mass Transf.* **2008**, *51*, 1542–1550. [CrossRef]

26. Faghri, A.; Gogineni, S.; Thomas, S. Vapor flow analysis of an axially rotating heat pipe. *Int. J. Heat Mass Transf.* **1993**, *36*, 2293–2303. [CrossRef]

27. Dehshali, M.E.; Nazari, M.; Shafii, M. Thermal performance of rotating closed-loop pulsating heat pipes: Experimental investigation and semi-empirical correlation. *Int. J. Therm. Sci.* **2018**, *123*, 14–26. [CrossRef]

28. Fedoseyev, L.; Pearce, E.M. Rotor Assembly with Heat Pipe Cooling System. U.S. Patent 9,331,552, 3 May 2016.

29. Agostini, B.; Thome, J.R.; Fabbri, M.; Michel, B. High heat flux two-phase cooling in silicon multimicrochannels. *IEEE Trans. Compon. Packag. Technol.* **2008**, *31*, 691–701. [CrossRef]

30. Garimella, S.V.; Sobhan, C.B. Transport in microchannels—A critical review. *Annu. Rev. Heat Transf.* **2003**, *13*, 1–50. [CrossRef]

31. Thome, J.R. Boiling in microchannels: A review of experiment and theory. *Int. J. Heat Fluid Flow* **2004**, *25*, 128–139. [CrossRef]

32. Pyrhonen, J.; Lindh, P.; Polikarpova, M.; Kurvinen, E.; Naumanen, V. Heat-transfer improvements in an axial-flux permanent-magnet synchronous machine. *Appl. Therm. Eng.* **2015**, *76*, 245–251. [CrossRef]

33. Rhebergen, C.; Bilgin, B.; Emadi, A.; Rowan, E.; Lo, J. Enhancement of Electric Motor Thermal Management through Axial Cooling Methods: A Materials Approach. In Proceedings of the 2015 IEEE Energy Conversion Congress and Exposition (ECCE), Montreal, QC, Canada, 20–24 September 2015; pp. 5682–5688.

34. Sixel, W.; Liu, M.D.; Nellis, G.; Sarlioglu, B. Cooling of Windings in Electric Machines via 3D Printed Heat Exchanger. In Proceedings of the 2018 IEEE Energy Conversion Congress and Exposition (ECCE), Portland, OR, USA, 23–27 September 2018; pp. 229–235.

35. Semidey, S.A.; Mayor, J.R. Experimentation of an Electric Machine Technology Demonstrator Incorporating Direct Winding Heat Exchangers. *IEEE Trans. Ind. Electron.* **2014**, *61*, 5771–5778. [CrossRef]

36. Lindh, P.; Petrov, I.; Jaatinen-Värri, A.; Grönman, A.; Martinez-Iturralde, M.; Satrústegui, M.; Pyrhönen, J. Direct Liquid Cooling Method Verified With an Axial-Flux Permanent-Magnet Traction Machine Prototype. *IEEE Trans. Ind. Electron.* **2017**, *64*, 6086–6095. [CrossRef]

37. Wohlers, C.; Juris, P.; Kabelac, S.; Ponick, B. Design and direct liquid cooling of tooth-coil windings. *Electr. Eng.* **2018**, *100*, 2299–2308. [CrossRef]

38. Jahns, T.M.; Dai, H. The past, present, and future of power electronics integration technology in motor drives. *CPSS Trans. Power Electron. Appl.* **2017**, *2*, 197–216. [CrossRef]

39. Wang, J. Design of Multilevel Integrated Modular Motor Drive with Gallium Nitride Power Devices. Ph.D. Thesis, The University of Wisconsin-Madison, Madison, WI, USA, 2015.

40. Ditri, J.; Cadotte, R.; Fetterolf, D.; McNulty, M. Impact of microfluidic cooling on high power amplifier RF performance. In Proceedings of the 2016 15th IEEE Intersociety Conference on Thermal and Thermomechanical Phenomena in Electronic Systems (ITherm), Las Vegas, NV, USA, 31 May–3 June 2016; pp. 1501–1504.

41. Schultz, M.; Yang, F.H.; Colgan, E.; Polastre, R.; Dang, B.; Tsang, C.; Gaynes, M.; Parida, P.; Knickerbocker, J.; Chainer, T.; et al. Embedded Two-Phase Cooling Of Large 3d Compatible Chips With Radial Channels. In Proceedings of the International Technical Conference and Exhibition on Packaging and Integration of Electronic and Photonic Microsystems, San Francisco, CA, USA, 6–9 July 2015; Volume 3.

42. McCluskey, F.P.; Grzybowski, R.; Podlesak, T. *High Temperature Electronics*; CRC Press: Boca Raton, FL, USA, 1997.

43. Han, B.; Jang, C.; Bar-Cohen, A.; Song, B. Coupled Thermal and Thermo-Mechanical Design Assessment of High Power Light Emitting Diode. *IEEE Trans. Compon. Packag. Technol.* **2010**, *33*, 688–697. [CrossRef]

44. Bar-Cohen, A.; Robinson, F.L.; Deisenroth, D.C. Challenges and opportunities in Gen3 embedded cooling with high-quality microgap flow. In Proceedings of the 2018 International Conference on Electronics Packaging and iMAPS All Asia Conference (ICEP-IAAC), Mie, Japan, 17–21 April 2018; pp. K-1–K-12.

45. Bar-Cohen, A.; Maurer, J.J.; Felbinger, J.G. DARPA's intra/interchip enhanced cooling (ICECool) program. In Proceedings of the CS MANTECH Conference, New Orleans, LA, USA, 13–16 May 2013.

46. Altman, D.H.; Gupta, A.; Tyhach, M. Development of a Diamond Microfluidics-Based Intra-Chip Cooling Technology for GaN. In Proceedings of the ASME 2015 International Technical Conference and Exhibition on Packaging and Integration of Electronic and Photonic Microsystems collocated with the ASME 2015 13th International Conference on Nanochannels, Microchannels, and Minichannels, San Francisco, CA, USA, 6–9 July 2015.

47. Twitchen, D.J.; Pickles, C.S.J.; Coe, S.E.; Sussmann, R.S.; Hall, C.E. Thermal conductivity measurements on CVD diamond. *Diam. Relat. Mater.* **2001**, *10*, 731–735. [CrossRef]

48. Bar-Cohen, A.; Maurer, J.J.; Sivananthan, A.J.M.A. Near-junction microfluidic cooling for wide bandgap devices. *MRS Adv.* **2016**, *1*, 181–195. [CrossRef]

49. Deisenroth, D.C.; Mandel, R.K.; Greve, H.; Dessiatoun, S.V.; McCluskey, P.; Ohadi, M.M. Direct bonding of a titanium header to an embedded two-phase FEEDS cooling device for high-flux electronics. In Proceedings of the 2016 15th IEEE Intersociety Conference on Thermal and Thermomechanical Phenomena in Electronic Systems (ITherm), Las Vegas, NV, USA, 31 May–3 June 2016; pp. 1072–1077.

50. Bae, D.G.; Mandel, R.K.; Dessiatoun, S.V.; Rajgopal, S.; Roberts, S.P.; Mehregany, M.; Ohadi, M.M.; Orlando, F.L.U.S. Embedded two-phase cooling of high heat flux electronics on silicon carbide (SiC) using thin-film evaporation and an enhanced delivery system (FEEDS) manifold-microchannel cooler. In Proceedings of the 2017 16th IEEE Intersociety Conference on Thermal and Thermomechanical Phenomena in Electronic Systems (ITherm), Orlando, FL, USA, 30 May–2 June 2017; pp. 466–472.

51. Ohadi, M.; Choo, K.; Dessiatoun, S.; Cetegen, E. *Next Generation Microchannel Heat Exchangers*; Springer: New York, USA, 2013.

52. Cetegen, E. Force Fed Microchannel High Heat Flux Cooling Utilizing Microgrooved Surfaces. Ph.D. Thesis, University of Maryland, College Park, MD, USA, 2010.

53. Dede, E.M.; Ishigaki, M.; Joshi, S.N.; Zhou, F. Design for additive manufacturing of wide band-gap power electronics components. In Proceedings of the 2016 International Symposium on 3D Power Electronics Integration and Manufacturing (3D-PEIM), Raleigh, NC, USA, 13–15 June 2016; pp. 1–20.

54. Arie, M.A.; Shooshtari, A.H.; Dessiatoun, S.V.; Al-Hajri, E.; Ohadi, M.M. Numerical modeling and thermal optimization of a single-phase flow manifold-microchannel plate heat exchanger. *Int. J. Heat Mass Transf.* **2015**, *81*, 478–489. [CrossRef]

55. Mandel, R.; Dessiatoun, S.; McCluskey, P.; Ohadi, M. Embedded Two-Phase Cooling of High Flux Electronics via Micro-Enabled Surfaces and Fluid Delivery Systems (FEEDS). In Proceedings of the ASME 2015 International Technical Conference and Exhibition on Packaging and Integration of Electronic and Photonic Microsystems collocated with the ASME 2015 13th International Conference on Nanochannels, Microchannels, and Minichannels, San Francisco, CA, USA, 6–9 July 2015.

56. Soule, C.A. Future trends in heat sink design. *Electron. Cool.* **2001**, *7*, 18–27.

57. Wilson, J. Cooling Solutions in the Past Decade. *Electron. Cool.* **2005**, *11*, 4.

58. Deisenroth, D.C. Two-Phase Flow Regimes and Heat Transfer in a Manifolded-Microgap. Ph.D. Thesis, University of Maryland, College Park, MD, USA, 2018.

59. McCluskey, P.; Saadon, Y.; Yao, Z.; Shah, J.; Kizito, J. Thermal Management Challenges in Turbo-Electric and Hybrid Electric Propulsion. In Proceedings of the 2018 International Energy Conversion Engineering Conference, Cincinnati, OH, USA, 9–11 July 2018; p. 4695.

60. Zhang, R.; Hodes, M.; Lower, N.; Wilcoxon, R. Water-Based Microchannel and Galinstan-Based Minichannel Cooling Beyond 1 kW/cm^2 Heat Flux. *IEEE Trans. Compon. Packag. Manuf. Technol.* **2015**, *5*, 762–770. [CrossRef]

61. Rau, M.J.; Garimella, S.V.; Dede, E.M.; Joshi, S.N. Boiling Heat Transfer from an Array of Round Jets with Hybrid Surface Enhancements. *J. Heat Transf. Trans. Asme* **2015**, *137*, 071501. [CrossRef]

62. Joshi, S.N.; Dede, E.M. Two-phase jet impingement cooling for high heat flux wide band-gap devices using multi-scale porous surfaces. *Appl. Therm. Eng.* **2017**, *110*, 10–17. [CrossRef]

63. Baummer, T.; Cetegen, E.; Ohadi, M.; Dessiatoun, S. Force-fed evaporation and condensation utilizing advanced micro-structured surfaces and micro-channels. *Microelectron. J.* **2008**, *39*, 975–980. [CrossRef]

64. Kalani, A.; Kandlikar, S.G. Flow patterns and heat transfer mechanisms during flow boiling over open microchannels in tapered manifold (OMM). *Int. J. Heat Mass Transf.* **2015**, *89*, 494–504. [CrossRef]

65. Quaiyum, M.A. Experimental investigation of Automatic Transmission Fluid (ATF) in an air cooled minichannel heat exchanger. Master's Thesis, University of Windsor, Ottawa, Canada, 2012.

66. Kim, M.H.; Pettersen, J.; Bullard, C.W. Fundamental process and system design issues in CO2 vapor compression systems. *Prog. Energy Combust. Sci.* **2004**, *30*, 119–174. [CrossRef]

67. California Air Resources Board. CO$_2$ (R744) Systems Technology Update. In Proceedings of the Mobile Air Conditioning Summit, Sacramento, CA, USA, 15–16 March 2005.

68. Daimler. From 2017: First Vehicles with CO$_2$ Air Conditioning. Available online: https://www.daimler.com/sustainability/product/further-environmental-technologies/co2-air-conditioning-system.html (accessed on 4 January 2018).

69. Geisler, K. Dielectric Liquid Cooling of Immersed Components. In *Encyclopedia of Thermal Packaging*, 1st ed.; Bar-Cohen, A., Ed.; World Scientific Publishing Co. Pte. Ltd.: Hackensack, NJ, USA, 2013.

70. Klein, S.; Alvardo, F. *EES-Engineering Equation Solver: User's Manual for Microsoft Windows Operating Systems*; F-Chart Software: Madison, WI, USA, 2015.

71. Linstrom, P.J.; Mallard, W. *NIST Chemistry Webbook*; NIST standard reference database No. 69; NIST: Gaithersburg, MD, USA, 2018.

MDPI

St. Alban-Anlage 66

4052 Basel

Switzerland

Tel. +41 61 683 77 34

Fax +41 61 302 89 18

www.mdpi.com

Energies Editorial Office

E-mail: energies@mdpi.com

www.mdpi.com/journal/energies

www.ingramcontent.com/pod-product-compliance
Lightning Source LLC
Chambersburg PA
CBHW051842210326
41597CB00033B/5744